Basic Electrical Installation Work

8th Edition

Trevor Linsley

Routledge
Taylor & Francis Group
LONDON AND NEW YORK

Eighth edition published 2015
by Routledge
2 Park Square, Milton Park, Abingdon, Oxon OX14 4RN

and by Routledge
711 Third Avenue, New York, NY 10017

Routledge is an imprint of the Taylor & Francis Group, an informa business

Seventh edition published by Routledge 2013

British Library Cataloguing in Publication Data
A catalogue record for this book is available from the British Library

Library of Congress Cataloging in Publication Data
Linsley, Trevor.
Basic electrical installation work / Trevor Linsley. -- City & Guilds edition, Eighth edition.
pages cm
Includes index.
1. Electric apparatus and appliances--Installation--Textbooks. 2. Electronics--Textbooks.
3. Electric wiring, Interior--Textbooks. I. Title.
TK452.L563 2015
621.319'24--dc23
2014045561

IISBN: 978-1-138-84884-9 (pbk)
ISBN: 978-1-315-72596-3 (ebk)

Typeset in Helvetica by
Servis Filmsetting Ltd, Stockport, Cheshire

Printed by Bell and Bain Ltd, Glasgow

MIX
Paper from responsible sources
FSC® C007785

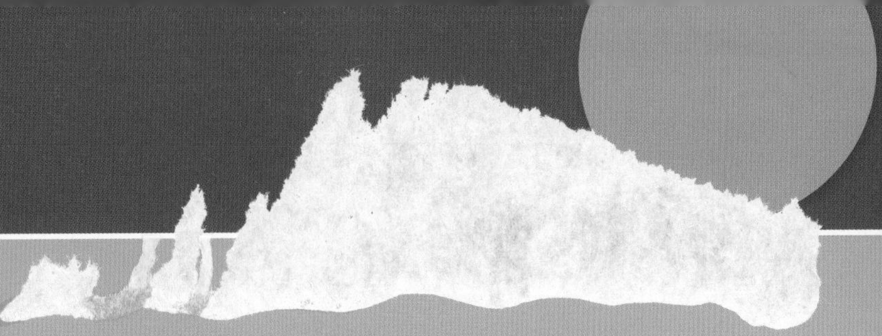

Basic Electrical Installation Work

Everything needed to pass the first part of the City & Guilds 2365 Diploma in Electrical Installations.

Updated in line with the 3rd Amendment of the 17th Edition IET Wiring Regulations, this new edition covers the City & Guilds 2365-02 course. Written in an accessible style with a chapter dedicated to each unit of the syllabus, this book helps you to master each topic before moving on to the next. End-of-chapter revision questions enable learners to check their understanding and consolidate key concepts learned in each chapter. With a companion website containing videos, animations, worksheets and lesson plans, this resource will be invaluable to both students and lecturers alike.

The 8th edition contains:

- Full-colour diagrams and photographs to explain difficult concepts
- Clear definitions of technical terms to make the book a quick and easy reference
- Extensive online material to help both students and lecturers.

The companion website material is available at www.routledge.com/cw/linsley

Trevor Linsley was formerly a senior lecturer at Blackpool and the Fylde College and Head of the NVQ Assessment Centre. Over the last 20 years his books have helped thousands of students to pass their electrical qualifications.

To Joyce, Samantha and Victoria

Contents

Preface

The 8th Edition of *Basic Electrical Installation Work* has been completely rewritten in five chapters to closely match the five units of the City and Guilds Level 2 Diploma in Electrical Installations (Building and Structures) (2365–02).

This book of electrical installation theory and practice will be of value to the electrical trainee working towards:

- The City and Guilds 2365–02 Level 2 Diploma in Electrical Installations (Buildings, and Structures).
- The City and Guilds 2365 Level 3 Diploma in Electrical Installations (Building and Structures) (2365–03).
- The EAL Level 3 Diploma in Electrotechnical Services.
- The SCOTVEC and BTEC Electrical Utilisation Units at Levels I and II.
- Those taking Engineering and Modern Apprenticeship courses.

Basic Electrical Installation Work provides a sound basic knowledge of electrical theory and practice which other trades in the construction industry will find of value, particularly those involved in multi-skilling activities.

The book incorporates the requirements of the latest regulations, particularly:

- 17th Edition IET Wiring Regulations (Incorporating Amendment 3: 2015);
- British Standards BS 7671: 2008 (Incorporating Amendment 3: 2015);
- Part P of the Building Regulations, Electrical Safety in Dwellings 2006;
- Hazardous Waste Regulations 2005;
- Work at Height Regulations 2005.

Trevor Linsley
2015

Acknowledgements

I would like to acknowledge the assistance given by the following manufacturers and professional organizations in the preparation of this book:

- The Institution of Engineering and Technology for permission to reproduce regulations and tables from the 17th Edition IET Regulations;
- The British Standards Institution for permission to reproduce material from BS 7671: 2008;
- Crabtree Electrical Industries for technical information and data;
- Legrand photographs. Visit legrand.co.uk for more information;
- RS Components Limited for technical information and photographs;
- Stocksigns Limited for technical information and photographs;
- Wylex Electrical Components for technical information and photographs;
- Martindale-electric.co.uk for technical information and photographs;
- Tenby electrical products for photographs.

I would like to thank the many college lecturers who responded to the questionnaire from Taylor & Francis the publishers regarding the proposed new edition of this book. Their recommendations have been taken into account in producing this improved 8th Edition.

I would also like to thank the editorial and production staff at Taylor & Francis the publishers for their enthusiasm and support, particularly my Editor, Gavin Fidler. They were able to publish this 8th Edition within the very short time-scale created by the publication of the new 3rd Amendment to the 17th Edition of the IET Regulations.

A special thank you must also go to John Gallagher, an electrical installation lecturer at Blackpool and The Fylde College, for writing the new section on preparing for assessment, and for providing all of the top tips used in the current edition.

Finally, I would like to thank Joyce, Samantha and Victoria for their support and encouragement.

CHAPTER 1

C&G Unit 201/501

Health and safety in building services engineering

Unit 201/501 of the City and Guilds 2365-02 syllabus

Learning outcomes – when you have completed this chapter you should:

- know about health and safety legislation;
- know how to handle hazardous situations;
- know the electrical safety requirements when working in the building services industry;
- know the safety requirements for working with gases and heat-producing equipment;
- know the safety requirements when using access equipment;
- know the safety requirements for working safely in confined spaces and excavations;
- be able to apply safe working practices to manual handling and using access equipment.

This chapter has free associated content, including animations and instructional videos, to support your learning.

When you see the logo, visit the companion website for more on this topic www.routledge.com/cw/linsley

Safety regulations and laws

At the beginning of the nineteenth century children formed a large part of the working population of Great Britain. They started work early in their lives and they worked long hours for unscrupulous employers or masters.

The Health and Morals of Apprentices Act of 1802 was introduced by Robert Peel in an attempt at reducing apprentice working hours to 12 hours per day and improving the conditions of their employment. The Factories Act of 1833 restricted the working week for children aged 13–18 years to 69 hours in any working week.

With the introduction of the Factories Act of 1833, the first four full-time Factory Inspectors were appointed. They were allowed to employ a small number of assistants and were given the responsibility of inspecting factories throughout England, Scotland, Ireland and Wales. This small, overworked band of men were the forerunners of the modern HSE Inspectorate, enforcing the safety laws passed by Parliament. As the years progressed, new Acts of Parliament increased the powers of the Inspectorate and the growing strength of the trade unions meant that employers were increasingly being pressed to improve health, safety and welfare at work.

The most important recent piece of health and safety law was passed by Parliament in 1974 called the Health and Safety at Work Act. This Act gave added powers to the Inspectorate and is the basis of all modern statutory health and safety laws. This law not only increased the employer's liability for safety measures, but also put the responsibility for safety on employees too.

Health, safety and welfare legislation has increased the awareness of everyone to the risks involved in the workplace. All statutes within the Acts of Parliament must be obeyed and, therefore, we all need an understanding of the laws as they apply to the electrical industry.

Statutory laws

Acts of Parliament are made up of Statutes. **Statutory Regulations** have been passed by Parliament and have, therefore, become laws. Non-compliance with the laws of this land may lead to prosecution by the Courts and possible imprisonment for offenders.

Definition

Statutory Regulations have been passed by Parliament and have, therefore, become laws.

We shall now look at some of the Statutory Regulations as they apply to the electrical industry.

The Health and Safety at Work Act 1974

Many governments have passed laws aimed at improving safety at work, but the most important recent legislation has been the Health and Safety at Work Act 1974. The purpose of the Act is to provide the legal framework for stimulating and encouraging high standards of health and safety at work; the Act puts the responsibility for safety at work on both workers and managers.

The employer has a duty to care for the health and safety of employees (Section 2 of the Act). To do this he or she must ensure that:

- the working conditions and standard of hygiene are appropriate;
- the plant, tools and equipment are properly maintained;
- the necessary safety equipment – such as personal protective equipment (PPE), dust and fume extractors and machine guards – is available and properly used;
- the workers are trained to use equipment and plant safely.

Employees have a duty to care for their own health and safety and that of others who may be affected by their actions (Section 7 of the Act). To do this they must:

- take reasonable care to avoid injury to themselves or others as a result of their work activity;
- cooperate with their employer, helping him or her to comply with the requirements of the Act;
- not interfere with or misuse anything provided to protect their health and safety.

Failure to comply with the Health and Safety at Work Act is a criminal offence and any infringement of the law can result in heavy fines, a prison sentence or both.

Figure 1.1 Both workers and managers are responsible for health and safety on site.

Enforcement of health and safety regulations

Laws and rules must be enforced if they are to be effective. The system of control under the Health and Safety at Work Act comes from the Health and Safety Executive (HSE) which is charged with enforcing the law. The HSE is divided into a number of specialist inspectorates or sections which operate from local offices throughout the United Kingdom. From the local offices the inspectors visit individual places of work.

The HSE inspectors have been given wide-ranging powers to assist them in the enforcement of the law. They can:

1. Enter premises unannounced and carry out investigations, take measurements or photographs.
2. Take statements from individuals.
3. Check the records and documents required by legislation.
4. Give information and advice to an employee or employer about safety in the workplace.
5. Demand the dismantling or destruction of any equipment, material or substance likely to cause immediate serious injury.
6. Issue an improvement notice which will require an employer to put right, within a specified period of time, a minor infringement of the legislation.
7. Issue a prohibition notice which will require an employer to stop immediately any activity likely to result in serious injury, and which will be enforced until the situation is corrected.

8 Prosecute all persons who fail to comply with their safety duties, including employers, employees, designers, manufacturers, suppliers and the self-employed.

Safety documentation

Under the Health and Safety at Work Act, the employer is responsible for ensuring that adequate instruction and information is given to employees to make them safety conscious. Part 1, Section 3 of the Act instructs all employers to prepare a written health and safety policy statement and to bring this to the notice of all employees. Figure 1.2 shows a typical Health and Safety Policy Statement of the type which will be available within your company. Your employer must let you know who your safety representatives are, and the new Health and Safety poster shown in Fig. 1.3 has a blank section into which the names and contact information of your specific representatives can be added. This is a large laminated poster, 595 × 415 mm, suitable for wall or notice-board display.

All workplaces employing five or more people had to display the type of poster shown in Fig. 1.3 after 30 June 2000.

To promote adequate health and safety measures the employer must consult with the employees' safety representatives. In companies which employ more than 20 people this is normally undertaken by forming a safety committee which is made up of a safety officer and employee representatives, usually nominated by a trade union. The safety officer is usually employed full-time in that role. Small companies might employ a safety supervisor who will have other duties within the company, or alternatively they could join a 'safety group'. The safety group then shares the cost of employing a safety adviser or safety officer, who visits each company in rotation. An employee who identifies a dangerous situation should initially report to his site safety representative. The safety representative should then bring the dangerous situation to the notice of the safety committee for action which will remove the danger. This may mean changing company policy or procedures or making modifications to equipment. All actions of the safety committee should be documented and recorded as evidence that the company takes its health and safety policy seriously.

The Electricity Safety, Quality and Continuity Regulations 2002 (formerly Electricity Supply Regulations 1989)

The Electricity Safety, Quality and Continuity Regulations 2002 are issued by the Department of Trade and Industry. They are statutory regulations which are enforceable by the laws of the land. They are designed to ensure a proper and safe supply of electrical energy up to the consumer's terminals.

These regulations impose requirements upon the regional electricity companies regarding the installation and use of electric lines and equipment. The regulations are administered by the Engineering Inspectorate of the Electricity Division of the Department of Energy and will not normally concern the electrical contractor, except that it is these regulations which lay down the earthing requirement of the electrical supply at the meter position.

FLASH-BANG ELECTRICAL

Statement of Health and Safety at Work Policy in accordance with the Health and Safety at Work Act 1974

Company objective

The promotion of health and safety measures is a mutual objective for the Company and for its employees at all levels. It is the intention that all the Company's affairs will be conducted in a manner which will not cause risk to the health and safety of its members, employees or the general public. For this purpose it is the Company policy that the responsibility for health and safety at work will be divided between all the employees and the Company in the manner outlined below.

Company's responsibilities

The Company will, as a responsible employer, make every endeavour to meet its legal obligations under the Health and Safety at Work Act to ensure the health and safety of its employees and the general public. Particular attention will be paid to the provision of the following:

1 Plant equipment and systems of work that are safe.
2 Safe arrangements for the use, handling, storage and transport of articles, materials and substances.
3 Sufficient information, instruction, training and supervision to enable all employees to contribute positively to their own safety and health at work and to avoid hazards.
4 A safe place of work, and safe access to it.
5 A healthy working environment.
6 Adequate welfare services.

Note: Reference should be made to the appropriate safety etc. manuals.

Employees' responsibilities

Each employee is responsible for ensuring that the work which he/she undertakes is conducted in a manner which is safe to himself or herself, other members of the general public, and for obeying the advice and instructions on safety and health matters issued by his/her superior. If any employee considers that a hazard to health and safety exists it is his/her responsibility to report the matter to his/her supervisor or through his/her Union Representative or such other person as may be subsequently defined.

Management and supervisors' responsibilities

Management and supervisors at all levels are expected to set an example in safe behaviour and maintain a constant and continuing interest in employee safety, in particular by:

1 acquiring the knowledge of health and safety regulations and codes of practice necessary to ensure the safety of employees in the workplace,
2 acquainting employees with these regulations on codes of practice and giving guidance on safety matters,
3 ensuring that employees act on instructions and advice given.

General Managers are ultimately responsible to the Company for the rectification or reporting of any safety hazard which is brought to their attention.

Joint consultations

Joint consultation on health and safety matters is important. The Company will agree with its staff, or their representatives, adequate arrangements for joint consultation on measures for promoting safety and health at work, and make and maintain satisfactory arrangements for the participation of their employees in the development and supervision of such measures. Trade Union representatives will initially be regarded as undertaking the role of Safety Representatives envisaged in the Health and Safety at Work Act. These representatives share a responsibility with management to ensure the health and safety of their members and are responsible for drawing the attention of management to any shortcomings in the Company's health and safety arrangements. The Company will in so far as is reasonably practicable provide representatives with facilities and training in order that they may carry out this task.

Review

A review, addition or modification of this statement may be made at any time and may be supplemented as appropriate by further statements relating to the work of particular departments and in accordance with any new regulations or codes of practice.
This policy statement will be brought to the attention of all employees.

Figure 1.2 Typical Health and Safety Policy Statement.

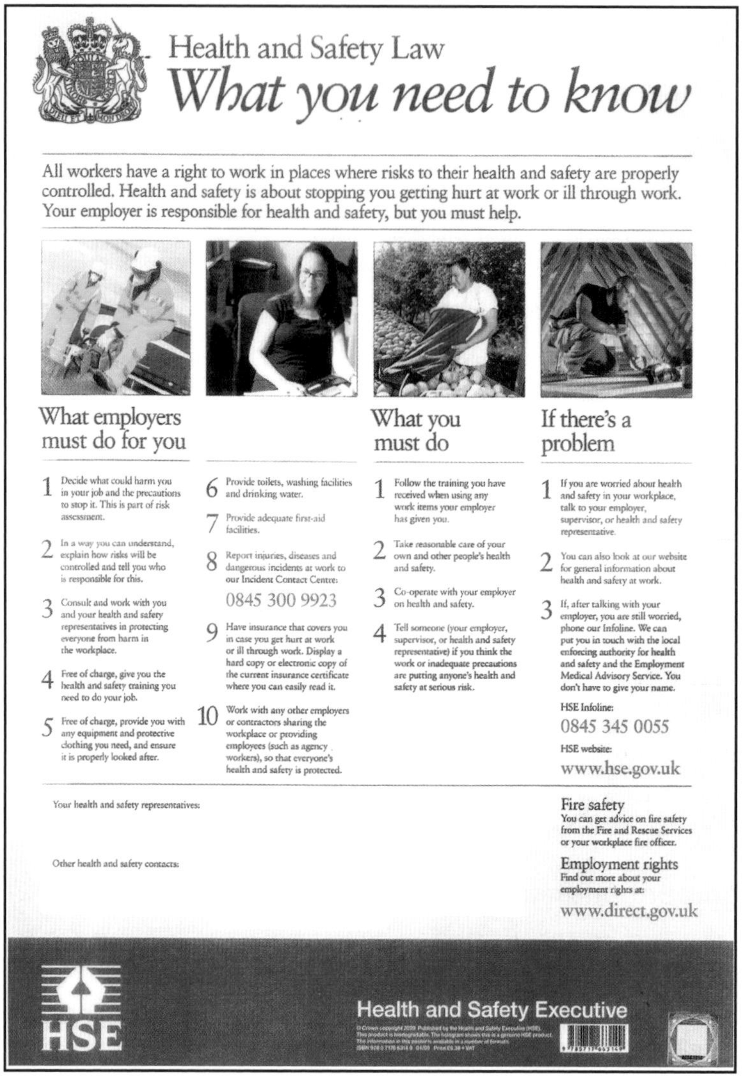

Figure 1.3 Health and Safety law poster. *Source*: HSE © Crown copyright material is reproduced with the permission of the Controller of HMSO and Her Majesty's Stationery Office, Norwich.

The regional electricity companies must declare the supply voltage and maintain its value between prescribed limits or tolerances.

The government agreed on 1 January 1995 that the electricity supplies in the United Kingdom would be harmonized with those of the rest of Europe. Thus the voltages used previously in low-voltage supply systems of 415 V and 240 V have become 400 V for three-phase supplies and 230 V for single-phase supplies. The permitted tolerances to the nominal voltage have also been changed from ±6% to +10% and −6%. This gives a voltage range of 216–253 V for a nominal voltage of 230 V and 376–440 V for a nominal supply voltage of 400 V.

The next proposed change is for the tolerance levels to be adjusted to ±10% of the declared nominal voltage (IET Regulation, Appendix 2:14).

The frequency is maintained at an average value of 50 Hz over 24 hours so that electric clocks remain accurate.

Regulation 29 gives the area boards the power to refuse to connect a supply to an installation which in their opinion is not constructed, installed and protected to an appropriately high standard. This regulation would only be enforced if the installation did not meet the requirements of the IET Regulations for Electrical Installations.

The Electricity at Work Regulations 1989 (EWR)

This legislation came into force in 1990 and replaced earlier regulations such as the Electricity (Factories Act) Special Regulations 1944. The regulations are made under the Health and Safety at Work Act 1974, and enforced by the Health and Safety Executive. The purpose of the regulations is to 'require precautions to be taken against the risk of death or personal injury from electricity in work activities'.

Section 4 of the EWR tells us that 'all systems must be constructed so as to prevent danger … and be properly maintained. … Every work activity shall be carried out in a manner which does not give rise to danger. … In the case of work of an electrical nature, it is preferable that the conductors be made dead before work commences.'

The EWR do not tell us specifically how to carry out our work activities and ensure compliance, but if proceedings were brought against an individual for breaking the EWR, the only acceptable defence would be 'to prove that all reasonable steps were taken and all diligence exercised to avoid the offence' (Regulation 29).

An electrical contractor could reasonably be expected to have 'exercised all diligence' if the installation was wired according to the IET Wiring Regulations (see below). However, electrical contractors must become more 'legally aware' following the conviction of an electrician for manslaughter at Maidstone Crown Court in 1989. The court accepted that an electrician had caused the death of another man as a result of his shoddy work in wiring up a central heating system. He received a nine-month suspended prison sentence. This case has set an important legal precedent, and in future any tradesman or professional who causes death through negligence or poor workmanship risks prosecution and possible imprisonment.

Duty of care

The Health and Safety at Work Act and the Electricity at Work Regulations make numerous references to employer and employees having a '**duty of care**' for the health and safety of others in the work environment. In this context the Electricity at Work Regulations refer to a person as a '**duty holder**'. This phrase recognizes the level of responsibility which electricians are expected to take on as a part of their job in order to control electrical safety in the work environment.

Everyone has a duty of care, but not everyone is a duty holder. The regulations recognize the amount of control that an individual might exercise over the whole electrical installation. The person who exercises 'control over the whole systems, equipment and conductors' and is the electrical company's representative on-site, is the *duty holder*. He might be a supervisor or manager, but he will have a duty of care on behalf of his employer for the electrical, health, safety and environmental issues on that site.

Duties referred to in the regulations may have the qualifying terms '**reasonably practicable**' or '**absolute**'. If the requirement of the regulation is absolute, then that regulation must be met regardless of cost or any other consideration. If the regulation is to be met 'so far as is reasonably practicable', then risks, cost, time, trouble and difficulty can be considered.

Definitions

'*Duty holder*' – this phrase recognizes the level of responsibility which electricians are expected to take on as a part of their job in order to control electrical safety in the work environment.

Everyone has a *duty of care*, but not everyone is a *duty holder*.

The person who exercises 'control over the whole systems, equipment and conductors' and is the electrical company's representative on-site is a *duty holder*.

'*Reasonably practicable*' or '*absolute*' – if the requirement of the regulation is absolute, then that regulation must be met regardless of cost or any other consideration. If the regulation is to be met 'so far as is reasonably practicable', then risks, cost, time, trouble and difficulty can be considered.

Safety first

Duty holder

- This person has the responsibility to control electrical safety in the work environment.

Often there is a cost-effective way to reduce a particular risk and prevent an accident from occurring. For example, placing a fireguard in front of the fire at home when there are young children in the family is a reasonably practicable way of reducing the risk of a child being burned.

If a regulation is not qualified with 'so far as is reasonably practicable', then it must be assumed that the regulation is absolute. In the context of the Electricity at Work Regulations, where the risk is very often death by electrocution, the level of duty to prevent danger more often approaches that of an absolute duty of care.

The Management of Health and Safety at Work Regulations 1999

The Health and Safety at Work Act 1974 places responsibilities on employers to have robust health and safety systems and procedures in the workplace. Directors and managers of any company who employ more than five employees can be held personally responsible for failures to control health and safety.

The Management of Health and Safety at Work Regulations 1999 tell us that employers must systematically examine the workplace, the work activity and the management of safety in the establishment through a process of 'risk assessments'. A record of all significant risk assessment findings must be kept in a safe place and be available to an HSE Inspector if required. Information based on these findings must be communicated to relevant staff and, if changes in work behaviour patterns are recommended in the interests of safety, they must be put in place. The process of risk assessment is considered in detail later in this chapter.

Risks, which may require a formal assessment in the electrical industry, might be:

- working at heights;
- using electrical power tools;
- falling objects;
- working in confined places;
- electrocution and personal injury;
- working with 'live' equipment;
- using hire equipment;
- manual handling – pushing – pulling – lifting;
- site conditions – falling objects – dust – weather – water – accidents and injuries.

And any other risks which are particular to a specific type of workplace or work activity.

The Control of Substances Hazardous to Health Regulations 2002 (COSHH)

The original COSHH Regulations were published in 1988 and came into force in October 1989. They were re-enacted in 1994 with modifications and improvements, and the latest modifications and additions came into force in 2002.

The COSHH Regulations control people's exposure to hazardous substances in the workplace. Regulation 6 requires employers to assess the risks to health from working with hazardous substances, to train employees in techniques which will reduce the risk and provide personal protective equipment (PPE) so that employees will not endanger themselves or others through exposure to hazardous substances. Employees should also know what cleaning, storage and disposal procedures are required and what emergency procedures to follow. The necessary information must be available to anyone using hazardous substances as well as to visiting HSE Inspectors.

Hazardous substances include:

1. any substance which gives off fumes causing headaches or respiratory irritation;
2. man-made fibres which might cause skin or eye irritation (e.g. loft insulation);
3. acids causing skin burns and breathing irritation (e.g. car batteries, which contain dilute sulphuric acid);
4. solvents causing skin and respiratory irritation (strong solvents are used to cement together PVC conduit fittings and tubes);
5. fumes and gases causing asphyxiation (burning PVC gives off toxic fumes);
6. cement and wood dust causing breathing problems and eye irritation;
7. exposure to asbestos – although the supply and use of the most hazardous asbestos material is now prohibited, huge amounts were installed between 1950 and 1980 in the construction industry and much of it is still in place today. In their latest amendments, the COSHH Regulations focus on giving advice and guidance to builders and contractors on the safe use and control of asbestos products. These can be found in Guidance Notes EH 71 or visit www.hse.uk/hiddenkiller.

Where PPE is provided by an employer, employees have a duty to use it to safeguard themselves.

Provision and Use of Work Equipment Regulations 1998

These regulations tidy up a number of existing requirements already in place under other regulations such as the Health and Safety at Work Act 1974, the Factories Act 1961 and the Offices, Shops and Railway Premises Act 1963.

The Provision and Use of Work Equipment Regulations 1998 place a general duty on employers to ensure minimum requirements of plant and equipment. If an employer has purchased good-quality plant and equipment which is well maintained, there is little else to do. Some older equipment may require modifications to bring it into line with modern standards of dust extraction, fume extraction or noise, but no assessments are required by the regulations other than those generally required by the Management Regulations 1999 discussed previously.

The Construction (Health, Safety and Welfare) Regulations 1996

An electrical contractor is a part of the construction team, usually as a subcontractor, and therefore the regulations particularly aimed at the construction industry also influence the daily work procedures and environment of an electrician. The most important recent piece of legislation is the Construction Regulations.

Figure 1.4 All workers on site must wear head protection.

The temporary nature of construction sites makes them one of the most dangerous places to work. These regulations are made under the Health and Safety at Work Act 1974 and are designed specifically to promote safety at work in the construction industry. Construction work is defined as any building or civil engineering work, including construction, assembly, alterations, conversions, repairs, upkeep, maintenance or dismantling of a structure.

The general provision sets out minimum standards to promote a good level of safety on-site. Schedules specify the requirements for guardrails, working platforms, ladders, emergency procedures, lighting and welfare facilities. Welfare facilities set out minimum provisions for site accommodation: washing facilities, sanitary conveniences and protective clothing. There is now a duty for all those working on construction sites to wear head protection, and this includes electricians working on-site as subcontractors.

Personal Protective Equipment (PPE) at Work Regulations 1998

Definition

PPE is defined as all equipment designed to be worn, or held, to protect against a risk to health and safety.

PPE is defined as all equipment designed to be worn, or held, to protect against a risk to health and safety. This includes most types of protective clothing, and equipment such as eye, foot and head protection, safety harnesses, life-jackets and high-visibility clothing.

Under the Health and Safety at Work Act, employers must provide free of charge any PPE and employees must make full and proper use of it. Safety signs such as those shown at Fig. 1.5 are useful reminders of the type of PPE to be used in a particular area. The vulnerable parts of the body which may need protection are the head, eyes, ears, lungs, torso, hands and feet; in addition, protection from falls may need to be considered. Objects falling from a height present the major hazard against which head protection is provided. Other hazards include striking

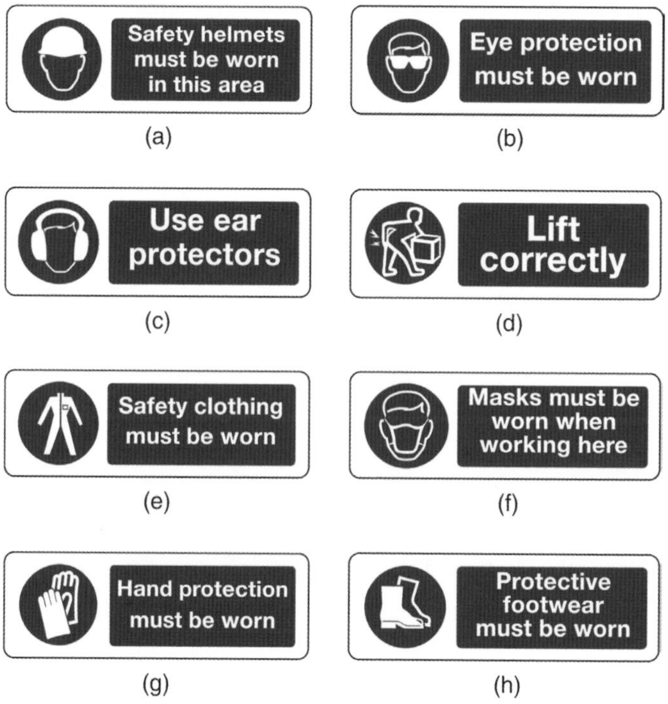

Figure 1.5 Safety signs showing type of PPE to be worn.

the head against projections and hair becoming entangled in machinery. Typical methods of protection include helmets, light-duty scalp protectors called 'bump caps' and hairnets.

The eyes are very vulnerable to liquid splashes, flying particles and light emissions such as ultraviolet light, electric arcs and lasers. Types of eye protectors include safety spectacles, safety goggles and face shields. Screen-based workstations are being used increasingly in industrial and commercial locations by all types of personnel. Working with VDUs (visual display units) can cause eye strain and fatigue.

Noise is accepted as a problem in most industries and surprisingly there has been very little control legislation. The Health and Safety Executive have published a 'Code of Practice' and 'Guidance Notes' HSG 56 for reducing the exposure of employed persons to noise. A continuous exposure limit of below 85 dB for an eight-hour working day is recommended by the Code.

Noise may be defined as any disagreeable or undesirable sound or sounds, generally of a random nature, which do not have clearly defined frequencies. The usual basis for measuring noise or sound level is the decibel scale. Whether noise of a particular level is harmful or not also depends on the length of exposure to it. This is the basis of the widely accepted limit of 85 dB of continuous exposure to noise for eight hours per day.

A peak sound pressure of above 200 pascals or about 120 dB is considered unacceptable and 130 dB is the threshold of pain for humans. If a person has to shout to be understood at 2 metres, the background noise is about 85 dB. If the distance is only 1 metre, the noise level is about 90 dB. Continuous noise at work causes deafness, makes people irritable, affects concentration, causes fatigue and accident proneness, and may mask sounds which need to be heard in order to work efficiently and safely.

It may be possible to engineer out some of the noise, for example, by placing a generator in a separate sound-proofed building. Alternatively, it may be possible to provide job rotation, to rearrange work locations or provide acoustic refuges.

Where individuals must be subjected to some noise at work, it may be reduced by ear protectors. These may be disposable ear plugs, reusable ear plugs or ear muffs. The chosen ear protector must be suited to the user and suitable for the type of noise, and individual personnel should be trained in its correct use.

Breathing reasonably clean air is the right of every individual, particularly at work. Some industrial processes produce dust which may present a potentially serious hazard. The lung disease asbestosis is caused by the inhalation of asbestos dust or particles and the coal dust disease pneumoconiosis, suffered by many coal-miners, has made people aware of the dangers of breathing in contaminated air.

Some people may prove to be allergic to quite innocent products such as flour dust in the food industry or wood dust in the construction industry. The main effect of inhaling dust is a measurable impairment of lung function. This can be avoided by wearing an appropriate mask, respirator or breathing apparatus as recommended by the company's health and safety policy and indicated by local safety signs.

A worker's body may need protection against heat or cold, bad weather, chemical or metal splash, impact or penetration and contaminated dust. Alternatively, there may be a risk of the worker's own clothes causing

Safety first

PPE

Always wear or use the PPE (personal protective equipment) provided by your employer for your safety.

Safety first

Safety signs

Always follow the instructions given in the safety signs where you are working – it will help to keep you safe.

contamination of the product, as in the food industry. Appropriate clothing will be recommended in the company's health and safety policy. Ordinary working clothes and clothing provided for food hygiene purposes are not included in the Personal Protective Equipment at Work Regulations.

Hands and feet may need protection from abrasion, temperature extremes, cuts and punctures, impact or skin infection. Gloves or gauntlets provide protection from most industrial processes, but should not be worn when operating machinery because they may become entangled in it. Care in selecting the appropriate protective device is required; for example, barrier creams provide only a limited protection against infection.

Try this

PPE

- Make a list of any PPE which you have used at work.
- What was this PPE protecting you from?

Boots or shoes with in-built toe-caps can give protection against impact or falling objects and, when fitted with a mild steel sole plate, can also provide protection from sharp objects penetrating through the sole. Special slip-resistant soles can also be provided for employees working in wet areas.

Whatever the hazard to health and safety at work, the employer must be able to demonstrate that he or she has carried out a risk analysis, made recommendations which will reduce that risk and communicated these recommendations to the workforce. Where there is a need for PPE to protect against personal injury and to create a safe working environment, the employer must provide that equipment and any necessary training which might be required and the employee must make full and proper use of such equipment and training.

Non-statutory regulations

Definition

Statutory laws and regulations are written in a legal framework.

Definition

Non-statutory regulations and codes of practice interpret the statutory regulations, telling us how we can comply with the law.

Statutory laws and regulations are written in a legal framework; some don't actually tell us how to comply with the laws at an everyday level.

Non-statutory regulations and codes of practice interpret the statutory regulations, telling us how we can comply with the law.

They have been written for every specific section of industry, commerce and situation, to enable everyone to comply with or obey the written laws.

When the Electricity at Work Regulations (EWR) tell us to 'ensure that all systems are constructed so as to prevent danger' they do not tell us how to actually do this in a specific situation. However, the IET Regulations tell us precisely how to carry out our electrical work safely in order to meet the statutory requirements of the EWR. In Part 1 of the IET Regulations, at 114, it states: 'the Regulations are non-statutory. They may, however, be used in a court of law in evidence to claim compliance with a statutory requirement.' If your electrical installation work meets the requirements of the IET Regulations, you will also meet the requirements of EWR.

Over the years, non-statutory regulations and codes of practice have built upon previous good practice and responded to changes by bringing out new editions of the various regulations and codes of practice to meet the changing needs of industry and commerce.

We will now look at one non-statutory regulation, what is sometimes called 'the electrician's bible', the most important set of regulations for anyone working in the electrical industry, the BS 7671: 2008 Requirements for Electrical Installations, IET Wiring Regulations 17th Edition.

The IET Wiring Regulations 17th edition requirements for electrical installations to BS 7671: 2008

The Institution of Engineering and Technology Requirements for Electrical Installations (the IET Regulations) are non-statutory regulations. They relate principally to the design, selection, erection, inspection and testing of electrical installations, whether permanent or temporary, in and about buildings generally and to agricultural and horticultural premises, construction sites and caravans and their sites. Paragraph 7 of the introduction to the EWR says: 'the IET Wiring Regulations is a code of practice which is widely recognized and accepted in the United Kingdom and compliance with them is likely to achieve compliance with all relevant aspects of the Electricity at Work Regulations.' The IET Wiring Regulations are the national standard in the United Kingdom and apply to installations operating at a voltage up to 1000 V a.c. They do not apply to electrical installations in mines and quarries, where special regulations apply because of the adverse conditions experienced there.

The current edition of the IET Wiring Regulations is the 3rd Amendment to the 17th edition 2015. The main reason for incorporating the IET Wiring Regulations into British Standard BS 7671: 2008 was to create harmonization with European Standards.

The IET Regulations take account of the technical intent of the CENELEC European Standards, which in turn are based on the IEC International Standards.

The purpose in harmonizing British and European Standards is to help develop a single European market economy so that there are no trade barriers to electrical goods and services across the European Economic Area.

To assist electricians in their understanding of the regulations a number of guidance notes have been published. The guidance notes which I will frequently make reference to in this book are those contained in the *On Site Guide.* Eight other guidance notes booklets are also currently available. These are:

- *Selection and Erection*;
- *Isolation and Switching*;
- *Inspection and Testing*;
- *Protection against Fire*;
- *Protection against Electric Shock*;
- *Protection against Overcurrent*;
- *Special Locations*;
- *Earthing and Bonding*.

These guidance notes are intended to be read in conjunction with the regulations.

The IET Wiring Regulations are the electrician's bible and provide the authoritative framework of information for anyone working in the electrical industry.

Figure 1.6 This kind of thing can happen if you're not up to date with IET wiring regulations.

Health and safety responsibilities

We have now looked at statutory and non-statutory regulations which influence working conditions in the electrical industry today. So, who has *responsibility* for these workplace Health and Safety Regulations?

In 1970, a Royal Commission was set up to look at the health and safety of employees at work. The findings concluded that the main cause of accidents at work was apathy on the part of *both* employers and employees.

The Health and Safety at Work Act 1974 was passed as a result of recommendations made by the Royal Commission and, therefore, the Act puts legal responsibility for safety at work on *both* the employer and employee.

In general terms, the employer must put adequate health and safety systems in place at work and the employee must use all safety systems and procedures responsibly.

In specific terms the employer must:

- provide a Health and Safety Policy Statement if there are five or more employees such as that shown in Fig. 1.2;
- display a current employers liability insurance certificate as required by the Employers Liability (Compulsory Insurance) Act 1969;
- report certain injuries, diseases and dangerous occurrences to the enforcing authority (HSE area office – see Appendix B for address);
- provide adequate first aid facilities (see Tables 1.1 and 1.2);
- provide PPE;

Key fact

IET Regulations

- They are the UK National Standard for all electrical work.
- They are the 'electrician's bible'.
- Comply with the IET Regulations and you also comply with Statutory Regulations (IET Regulation 114).

Table 1.1 Suggested numbers of first aid personnel

Category of risk	Numbers employed at any location	Suggested number of first aid personnel
Lower risk e.g. shops and offices, libraries	Fewer than 50 50–100 More than 100	At least one appointed person At least one first aider One additional first aider for every 100 employed
Medium risk e.g. light engineering and assembly work, food processing, warehousing	Fewer than 20 20–100 More than 100	At least one appointed person At least one first aider for every 50 employed (or part thereof) One additional first aider for every 100 employed
Higher risk e.g. most construction, slaughterhouses, chemical manufacture, extensive work with dangerous machinery or sharp instruments	Fewer than five 5–50 More than 50	At least one appointed person At least one first aider One additional first aider for every 50 employed

Table 1.2 Contents of first aid boxes

Item	Number of employees				
	1–5	**6–10**	**11–50**	**51–100**	**101–150**
Guidance card on general first aid	1	1	1	1	1
Individually wrapped sterile adhesive dressings	10	20	40	40	40
Sterile eye pads, with attachment (Standard Dressing No. 16 BPC)	1	2	4	6	8
Triangular bandages	1	2	4	6	8
Sterile covering for serious wounds (where applicable)	1	2	4	6	8
Safety-pins	6	6	12	12	12
Medium-sized sterile unmedicated dressings (Standard Dressings No. 9 and No. 14 and the Ambulance Dressing No. 1)	3	6	8	10	12
Large sterile unmedicated dressings (Standard Dressings No. 9 and No. 14 and the Ambulance Dressing No. 1)	1	2	4	6	10
Extra-large sterile unmedicated dressings (Ambulance Dressing No. 3)	1	2	4	6	8
Where tap water is not available, sterile water or sterile normal saline in disposable containers (each holding a minimum of 300ml) must be kept near the first aid box. The following minimum quantities should be kept:					
Number of employees	1–10	11–50	51–100	101–150	
Quantity of sterile water	1 × 300ml	3 × 300ml	6 × 300ml	6 × 300ml	

- provide information, training and supervision to ensure staffs' health and safety;
- provide adequate welfare facilities;
- put in place adequate precautions against fire, provide a means of escape and means of fighting fire;
- ensure plant and machinery are safe and that safe systems of operation are in place;
- ensure articles and substances are moved, stored and used safely;
- make the workplace safe and without risk to health by keeping dust, fumes and noise under control.

In specific terms the employee must:

- take reasonable care of his/her own health and safety and that of others who may be affected by what they do;
- cooperate with his/her employer on health and safety issues by not interfering with or misusing anything provided for health, safety and welfare in the working environment;
- report any health and safety problem in the workplace to, in the first place, a supervisor, manager or employer.

Categories of safety signs

The rules and regulations of the working environment are communicated to employees by written instructions, signs and symbols. All signs in the

Figure 1.7 Text-only safety signs do not comply.

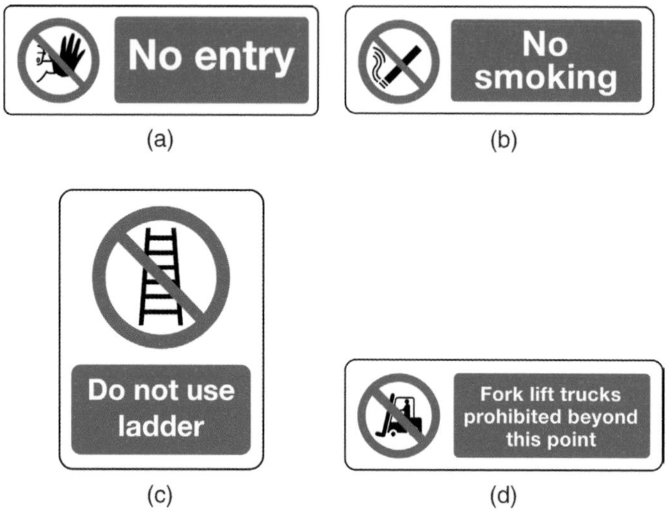

Figure 1.8 Prohibition signs. These are *must not* do signs.

working environment are intended to inform. They should give warning of possible dangers and must be obeyed. At first there were many different safety signs, but British Standard BS 5499 Part 1 and the Health and Safety (Signs and Signals) Regulations 1996 have introduced a standard system which gives health and safety information with the minimum use of words. The purpose of the regulations is to establish an internationally understood system of safety signs and colours which draw attention to equipment and situations that do, or could, affect health and safety. Text-only safety signs became illegal from 24 December 1998. From that date, all safety signs have had to contain a pictogram or symbol such as those shown in Fig. 1.7. Signs fall into four categories: prohibited activities; warnings; mandatory instructions, and safe conditions.

Prohibition signs

These are *must not do* signs. These are circular white signs with a red border and red cross-bar, and are given in Fig. 1.8. They indicate an activity *which must not* be done.

Warning signs

These give safety information. These are triangular yellow signs with a black border and symbol, and are given in Fig. 1.9. They *give warning* of a hazard or danger.

Mandatory signs

These are *must do* signs. These are circular blue signs with a white symbol, and are given in Fig. 1.10. They *give instructions* which must be obeyed.

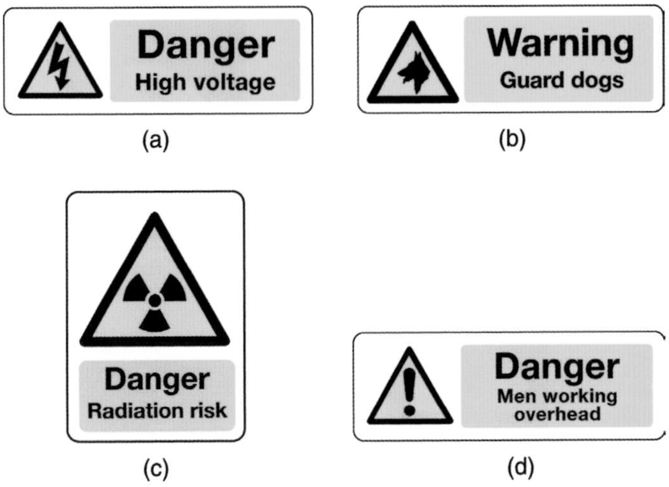

Figure 1.9 Warning signs. These give safety information.

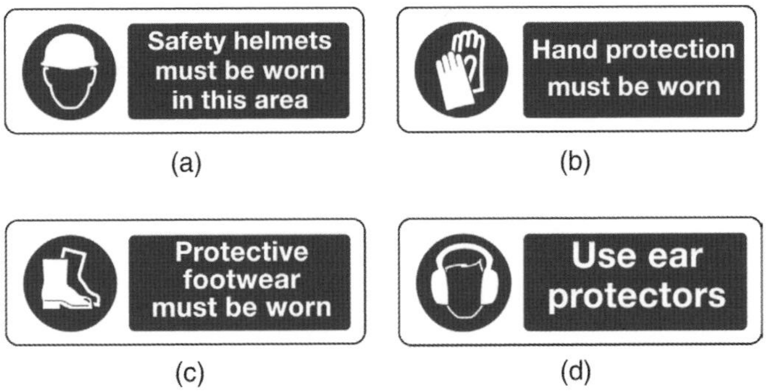

Figure 1.10 Mandatory signs. These are *must do* signs.

Advisory or safe condition signs

These give safety information. These are square or rectangular green signs with a white symbol, and are given in Fig. 1.11. They *give information* about safety provision.

Accidents at work

Despite new legislation, improved information, education and training, accidents at work do still happen. An **accident** may be defined as an uncontrolled event causing injury or damage to an individual or property. An accident can nearly always be avoided if correct procedures and methods of working are followed. Any accident which results in an absence from work for more than three days or causes a major injury or death is notifiable to the HSE. There are more than 40,000 accidents reported to the HSE each year which occur as a result of some building-related activity. To avoid having an accident you should:

1. follow all safety procedures (e.g. fit safety signs when isolating supplies and screen off work areas from the general public);
2. not misuse or interfere with equipment provided for health and safety;
3. dress appropriately and use PPE when appropriate;

Definition

An *accident* may be defined as an uncontrolled event causing injury or damage to an individual or property.

Figure 1.11 Advisory or safe condition signs. These also give safety information.

4. behave appropriately and with care;
5. avoid over-enthusiasm and foolishness;
6. stay alert and avoid fatigue;
7. not use alcohol or drugs at work;
8. work within your level of competence;
9. attend safety courses and read safety literature;
10. take a positive decision to act and work safely.

If you observe a hazardous situation at work, first make the hazard safe, using an appropriate method, or screen it off, but only if you can do so without putting yourself or others at risk, then report the situation to your safety representative or supervisor.

Definition

Fire is a chemical reaction which will continue if fuel, oxygen and heat are present.

Visit the companion website for more on this topic.

Fire control

Fire is a chemical reaction which will continue if fuel, oxygen and heat are present. To eliminate a fire *one* of these components must be removed. This is often expressed by means of the fire triangle shown in Fig. 1.12; all three corners of the triangle must be present for a fire to burn.

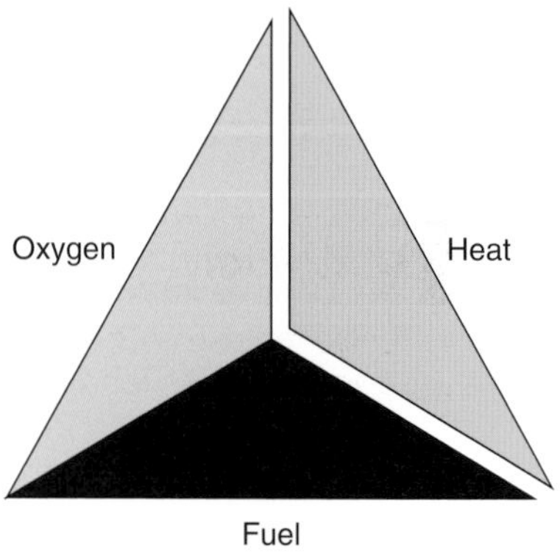

Figure 1.12 The fire triangle.

Fuel

Fuel is found in the construction industry in many forms: petrol and paraffin for portable generators and heaters; bottled gas for heating and soldering. Most solvents are flammable. Rubbish also represents a source of fuel: offcuts of wood, roofing felt, rags, empty solvent cans and discarded packaging will all provide fuel for a fire.

To eliminate fuel as a source of fire, all flammable liquids and gases should be stored correctly, usually in an outside locked store. The working environment should be kept clean by placing rags in a metal bin with a lid. Combustible waste material should be removed from the work site or burned outside under controlled conditions by a competent person.

Oxygen

Oxygen is all around us in the air we breathe, but can be eliminated from a small fire by smothering with a fire blanket, sand or foam. Closing doors and windows but not locking them will limit the amount of oxygen available to a fire in a building and help to prevent it from spreading.

Most substances will burn if they are at a high enough temperature and have a supply of oxygen. The minimum temperature at which a substance will burn is called the 'minimum ignition temperature' and for most materials this is considerably higher than the surrounding temperature. However, a danger does exist from portable heaters, blowtorches and hot-airguns which provide heat and can cause a fire by raising the temperature of materials placed in their path above the minimum ignition temperature. A safe distance must be maintained between heat sources and all flammable materials.

Figure 1.13 Once started, fires can spread rapidly if the conditions are favourable.

Heat

Heat can be removed from a fire by dousing with water, but water must not be used on burning liquids since the water will spread the liquid and the fire. Some fire extinguishers have a cooling action which removes heat from the fire.

Fires in industry damage property and materials, injure people and sometimes cause loss of life. Everyone should make an effort to prevent fires, but those which do break out should be extinguished as quickly as possible.

In the event of fire you should:

- raise the alarm;
- turn off machinery, gas and electricity supplies in the area of the fire;
- close doors and windows but without locking or bolting them;
- remove combustible materials and fuels away from the path of the fire, if the fire is small, and if this can be done safely;
- attack small fires with the correct extinguisher.

Only attack the fire if you can do so without endangering your own safety in any way. Always leave your own exit from the danger zone clear. Those not involved in fighting the fire should walk to a safe area or assembly point.

Fires are divided into four classes or categories:

- Class A are wood, paper and textile fires;
- Class B are liquid fires such as paint, petrol and oil;
- Class C are fires involving gas or spilled liquefied gas;
- Class D are very special types of fire involving burning metal.

Electrical fires do not have a special category because, once started, they can be identified as one of the four above types.

Fire extinguishers are for dealing with small fires, and different types of fire must be attacked with a different type of extinguisher. Using the wrong type of extinguisher could make matters worse. For example, water must not be used on a liquid or electrical fire. The normal procedure when dealing with electrical fires is to cut off the electrical supply and use an extinguisher which is appropriate to whatever is burning. Figure 1.14 shows the correct type of extinguisher to be used on the various categories of fire. The colour coding shown is in accordance with BS EN3: 1996.

Safety first

Fire

If you discover a fire:

- raise the alarm;
- attack small fires with an extinguisher;
- BUT only if you can do so without risk to your own safety.

Definition

Fire extinguishers remove heat from a fire and are a first response for small fires.

Visit the companion website for more on this topic.

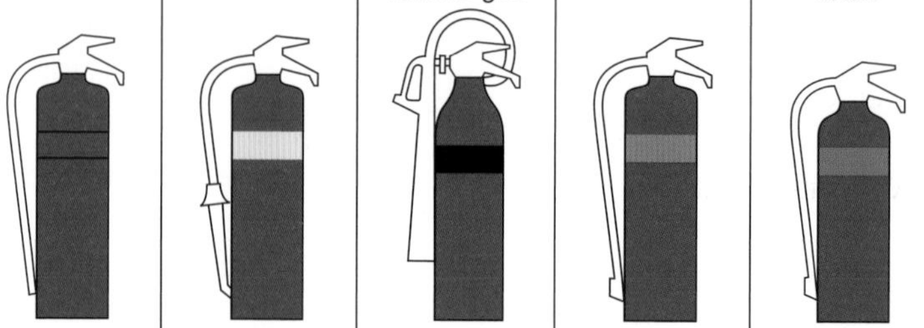

Type of fire extinguisher / Type of fire	(i) Water — Signal red flash on red	(ii) Foam — Pale cream flash on red	(iii) Carbon dioxide gas — Black flash on red	(iv) Dry powder — French blue flash on red	(v) Vaporizing foam — Emerald green flash on red
Class A Paper, wood and fabric	✓ Yes	✓ Yes	✗ No	✓ Yes	✓ Yes
Class B Flammable liquids	✗ No	✓ Yes	✓ Yes	✓ Yes	✓ Yes
Class C Flammable gases	✗ No	✗ No	✓ Yes	✓ Yes	✓ Yes
Electrical fires	✗ No	✗ No	✓ Yes	✓ Yes	✓ Yes
Motor vehicle protection	✗ No	✓ Yes	✗ No	✓ Yes	✓ Yes

Figure 1.14 Fire extinguishers and their applications (colour codes to BS EN3: 1996). The base colour of all fire extinguishers is red, with a different-coloured flash to indicate the type.

Heat-producing equipment

Electrical installers usually make their connections 'cold', terminating cables in joint boxes or using crimping tools for larger cable connections. However, mechanical service trades on site, such as plumbers and heating and ventilation engineers, make many of their connections by 'hot' working.

Hot working is potentially more hazardous than cold working because of the additional risk of fire. Operatives carrying out hot working must follow strict procedures to reduce the risk of fire. These safety procedures must be written down in a method statement and the work procedure strictly followed.

The fuel source of the hot working might be:

- bottled gas – butane in blue containers or propane in red containers are often used with a blowtorch or boiling ring;
- oxy-acetylene gas – this is always used with a torch which mixes the two gases for cutting or welding steel. This is not work which is normally carried out by an electrician.

When hot working is carried out, operatives must be trained and competent to use the equipment. They must also follow all safety procedures to reduce the risk of fire. For example, a fire extinguisher must be placed in the immediate work area and hot working must be completed at least one hour before the operator leaves the site.

Bottles, torches and flexible hoses must be in good condition and tested. If the bottled gas is stored on site it must be placed in a secured outdoor location open to the elements to reduce the risk of a gas buildup from a possibly faulty or leaking on-off valve.

See also Appendix F (B2 .11 .1 and B2 .11.2) of the Electricians Guide to the Building Regulations.

Figure 1.15 Fire exit sign

Evacuation procedures

When the fire alarm sounds you must leave the building immediately by any one of the escape routes indicated. **Exit routes** are usually indicated by a green and white 'running man' symbol. Evacuation should be orderly; do not run but walk purposefully to your designated assembly point.

The purpose of an **assembly point** is to get you away from danger to a place of safety where you will not be in the way of the emergency services. It also allows for people to be accounted for and to make sure that no one is left in the building. You must not re-enter the building until a person in authority gives permission to do so.

An evacuation in a real emergency can be a frightening experience, especially if you do not really know what to do, so take time to familiarize yourself with the fire safety procedures where you are working before an emergency occurs.

Definition

Exit routes are usually indicated by a green and white 'running man' symbol. Evacuation should be orderly; do not run but walk purposefully to your designated assembly point.

Definition

The purpose of an *assembly point* is to get you away from danger to a place of safety where you will not be in the way of the emergency services.

Health and safety risks, precautions and procedures

Earlier in this chapter, we looked at some of the health and safety rules and regulations. In particular, we now know that the Health and Safety at Work Act is the most important piece of recent legislation, because it places responsibilities for safety at work on both employers and employees. This responsibility is enforceable by law. We know what the regulations say about the control of substances that might be hazardous to our health at work, because we briefly looked at the COSHH Regulations 2002 earlier in this chapter. We also know that if there is a risk to health and safety at work our employer must provide personal protective equipment (PPE) free of charge, for us to use so that we are safe at work. The law is in place, we all apply the principles of health and safety at work and we always wear the appropriate PPE, so what are the risks? Well, getting injured at work is not a pleasant subject to think about but each year about 300 people in Great Britain lose their lives at work. In addition, there are about 158,000 non-fatal injuries reported to the Health and Safety Executive (HSE) each year and an estimated 2.2 million people suffer ill health caused by, or made worse by, work. It is a mistake to believe that these things only happen in dangerous occupations such as deep-sea diving, mining and quarrying, fishing industry, tunnelling and fire-fighting, or that it only happens in exceptional circumstances such as would never happen in your workplace. This is not the case. Some basic thinking and acting beforehand could have prevented most of these accident statistics from happening.

The most common categories of risk and causes of accidents at work are:

- slips, trips and falls;
- manual handling, that is, moving objects by hand;
- using equipment, machinery or tools;
- storage of goods and materials which then become unstable;
- fire;
- electricity;
- mechanical handling.

Precautions taken to control risks:

- eliminate the cause;
- substitute a procedure or product with less risk;
- enclose the dangerous situation;
- put guards around the hazard;
- use safe systems of work;
- supervise, train and give information to staff;
- if the hazard cannot be removed or minimized then provide PPE.

Let us now look at the application of some of the procedures that make the workplace a safer place to work, but first I want to explain what I mean when I use the words hazard and risk.

Hazard and risk

Definition

A *hazard* is something with the 'potential' to cause harm, for example, chemicals, electricity or working above ground.

Definition

A *risk* is the 'likelihood' of harm actually being done.

Definition

A *competent person* is anyone who has the necessary technical skills, training and expertise to safely carry out the particular activity.

A **hazard** is something with the 'potential' to cause harm, for example, chemicals, electricity or working above ground.

A **risk** is the 'likelihood' of harm actually being done.

Competent persons are often referred to in Electricity at Work Regulations and the Health and Safety at Work Regulations, but who is 'competent'? For the purposes of the Act, a competent person is anyone who has the necessary technical skills, training and expertise to safely carry out the particular activity. Therefore, a competent person dealing with a hazardous situation reduces the risk. The 3rd Amendment to the 17th Edition of the IET Regulations has chosen not to use the word 'competent person' in the IET Regulations, replacing it with the more specific definitions of skilled (electrically) or instructed (electrically) persons.

Think about your workplace and each stage of what you do, then think about what might go wrong. Some simple activities may be hazardous. Here are some typical activities where accidents might happen.

Typical activity	Potential hazard
Receiving materials	Lifting and carrying
Stacking and storing	Falling materials
Movement of people	Slips, trips and falls
Building maintenance	Working at heights or in confined spaces
Movement of vehicles	Collisions

How high are the risks? Think about what might be the worst result; is it a broken finger or someone suffering permanent lung damage or being killed? How likely is it to happen? How often is that type of work carried out and how close do people get to the hazard? How likely is it that something will go wrong?

How many people might be injured if things go wrong? Might this also include people who do not work for your company?

Employers of more than five people must document the risks at work and the process is known as **hazard risk assessment**.

Definition

Employers of more than five people must document the risks at work and the process is known as *hazard risk assessment*.

Hazard risk assessment: the procedure

The Management of Health and Safety at Work Regulations 1999 tells us that employers must systematically examine the workplace, the work activity and the management of safety in the establishment through a process of risk assessments. A record of all significant risk assessment findings must be kept in a safe place and be made available to an HSE Inspector if required. Information based on the risk assessment findings must be communicated to relevant staff and, if changes in work behaviour patterns are recommended in the interests of safety, they must be put in place.

So risk assessment must form a part of any employer's robust policy of health and safety. However, an employer only needs to 'formally' assess the significant risks. He or she is not expected to assess the trivial and minor types of household risks. Staff are expected to read and to act upon these formal risk assessments, and they are unlikely to do so enthusiastically if the file is full of trivia. An assessment of risk is nothing more than a careful examination of what, in your work, could cause harm to people. It is a record that shows whether sufficient precautions have been taken to prevent harm.

The HSE recommends five steps to any risk assessment.

Step 1

Look at what might reasonably be expected to cause harm. Ignore the trivial and concentrate only on significant hazards that could result in serious harm or injury. Manufacturers' data sheets or instructions can also help you spot hazards and put risks in their true perspective.

Step 2

Decide who might be harmed and how. Think about people who might not be in the workplace all the time – cleaners, visitors, contractors or maintenance personnel. Include members of the public or people who share the workplace. Is there a chance that they could be injured by activities taking place in the workplace?

Step 3

Evaluate what is the risk arising from an identified hazard. Is it adequately controlled or should more be done? Even after precautions have been put in place, some risk may remain. What you have to decide, for each significant hazard, is whether this remaining risk is low, medium or high. First of all, ask yourself if you have done all the things that the law says you have got to do. For example, there are legal requirements on the prevention of access to dangerous machinery. Then ask yourself whether generally accepted industry standards are in place, but do not stop there – think for yourself, because the law also says that you must do what is reasonably practicable to keep the workplace safe. Your real aim is to make all risks small by adding precautions, if necessary.

If you find that something needs to be done, ask yourself:

- Can I get rid of this hazard altogether?
- If not, how can I control the risk so that harm is unlikely?

Key fact

Definition

- A hazard is something that might cause harm
- A risk is the chance that harm will be done.

Only use PPE when there is nothing else that you can reasonably do.

If the work that you do varies a lot, or if there is movement between one site and another, select those hazards which you can reasonably foresee, the ones that apply to most jobs and assess the risks for them. After that, if you spot any unusual hazards when you get on-site, take what action seems necessary.

Step 4

Record your findings and say what you are going to do about risks that are not adequately controlled. If there are fewer than five employees you do not need to write anything down, but if there are five or more employees the significant findings of the risk assessment must be recorded. This means writing down the more significant hazards and assessing if they are adequately controlled and recording your most important conclusions. Most employers have a standard risk assessment form which they use, such as that shown in Fig. 1.16, but any format is suitable. The important thing is to make a record.

There is no need to show how the assessment was made, provided you can show that:

1 a proper check was made;
2 you asked those who might be affected;
3 you dealt with all obvious and significant hazards;
4 the precautions are reasonable and the remaining risk is low;
5 you informed your employees about your findings.

Definition

Risk assessments need to be *suitable* and *sufficient*, not perfect.

Risk assessments need to be *suitable* and *sufficient*, not perfect. The two main points are:

1 Are the precautions reasonable?
2 Is there a record to show that a proper check was made?

File away the written assessment in a dedicated file for future reference or use. It can help if an HSE Inspector questions the company's precautions or if the company becomes involved in any legal action. It shows that the company has done what the law requires.

Step 5

Review the assessments from time to time and revise them if necessary.

Method statement

The Construction, Design and Management Regulations and Approved Codes of Practice define a method statement as a written document laying out the work procedure and sequence of operations to ensure health and safety.

If the method statement is written as a result of a risk assessment carried out for a task or operation, then following the prescribed method will reduce the risk.

The safe isolation procedure described in Fig. 1.28 is a method statement. Following this method meets the requirements of the Electricity at Work Regulations, the IET Regulations, and reduces the risk of electric shock to the operative and other people who might be affected by his actions.

Completing a risk assessment

When completing a risk assessment such as that shown in Fig. 1.16, do not be over-complicated. In most firms in the commercial, service and light industrial sector, the hazards are few and simple. Checking them is common sense but necessary.

HAZARD RISK ASSESSMENT	**FLASH-BANG ELECTRICAL CO.**
For Company name or site: .... Address:	Assessment undertaken by: .... Signed: .... Date: ....
STEP 5 Assessment review date:	
STEP 1 List the hazards here	STEP 2 Decide who might be harmed ....
STEP 3 Evaluate (what is) the risk – is it adequately controlled? State risk level as low, medium or high	STEP 4 Further action – what else is required to control any risk identified as medium or high? ....

Figure 1.16 Hazard risk assessment standard form.

Step 1

List only hazards which you could reasonably expect to result in significant harm under the conditions prevailing in your workplace. Use the following examples as a guide:

- slipping or tripping hazards (e.g. from poorly maintained or partly installed floors and stairs);
- fire (e.g. from flammable materials you might be using, such as solvents);
- chemicals (e.g. from battery acid);
- moving parts of machinery (e.g. blades);
- rotating parts of hand tools (e.g. drills);
- accidental discharge of cartridge-operated tools;
- high-pressure air from airlines (e.g. air-powered tools);
- pressure systems (e.g. steam boilers);
- vehicles (e.g. fork lift trucks);
- electricity (e.g. faulty tools and equipment);

- dust (e.g. from grinding operations or thermal insulation);
- fumes (e.g. from welding);
- manual handling (e.g. lifting, moving or supporting loads);
- noise levels too high (e.g. machinery);
- poor lighting levels (e.g. working in temporary or enclosed spaces);
- low temperatures (e.g. working outdoors or in refrigeration plant);
- high temperatures (e.g. working in boiler rooms or furnaces).

Step 2

Decide who might be harmed; do not list individuals by name. Just think about groups of people doing similar work or who might be affected by your work:

- office staff;
- electricians;
- maintenance personnel;
- other contractors on-site;
- operators of equipment;
- cleaners;
- members of the public.

Pay particular attention to those who may be more vulnerable, such as:

- staff with disabilities;
- visitors;
- young or inexperienced staff;
- people working in isolation or enclosed spaces.

Safety first

Safety procedures

- Hazard risk assessment is *an essential part* of any health and safety management system.
- The aim of the planning process is to minimize risk.
- HSE Publication HSG (65).

Step 3

Calculate what is the risk – is it adequately controlled? Have you already taken precautions to protect against the hazards which you have listed in Step 1? For example:

- have you provided adequate information to staff?
- have you provided training or instruction?

Do the precautions already taken

- meet the legal standards required?
- comply with recognized industrial practice?
- represent good practice?
- reduce the risk as far as is reasonably practicable?

If you can answer ‘yes’ to the above points then the risks are adequately controlled, but you need to state the precautions you have put in place. You can refer to company procedures, company rules, company practices, etc. in giving this information. For example, if we consider there might be a risk of electric shock from using electrical power tools, then the risk of a shock will be *less* if the company policy is to portable appliance test (PAT) all power tools each year and to fit a label to the tool showing that it has been tested for electrical safety. If the stated company procedure is to use battery drills whenever possible, or 110V drills when this is not possible, and to *never* use 230V drills, then this again will reduce the risk. If a policy such as this is written down in the company safety policy statement, then you can simply refer to the appropriate section of the safety policy statement and the level of risk will be low. (Note: PAT testing is described in Advanced Electrical Installation Work.)

Step 4

Further action – what more could be done to reduce those risks which were found to be inadequately controlled?

You will need to give priority to those risks that affect large numbers of people or which could result in serious harm. Senior managers should apply the principles below when taking action, if possible in the following order:

1. Remove the risk completely.
2. Try a less risky option.
3. Prevent access to the hazard (e.g. by guarding).
4. Organize work differently in order to reduce exposure to the hazard.
5. Issue PPE.
6. Provide welfare facilities (e.g. washing facilities for removal of contamination and first aid).

Any hazard identified by a risk assessment as *high risk* must be brought to the attention of the person responsible for health and safety within the company. Ideally, in Step 4 of the risk assessment you should be writing 'No further action is required. The risks are under control and identified as low risk.'

The assessor may use as many standard hazard risk assessment forms, such as that shown in Fig. 1.16, as the assessment requires. Upon completion they should be stapled together or placed in a plastic wallet and stored in the dedicated file.

You might like to carry out a risk assessment on a situation you are familiar with at work, using the standard form of Fig. 1.16, or your employer's standard forms.

Safe manual handling

Manual handling is lifting, transporting or supporting loads by hand or by bodily force. The load might be any heavy object, a printer, a VDU, a box of tools or a stepladder. Whatever the heavy object is, it must be moved thoughtfully and carefully, using appropriate lifting techniques if personal pain and injury are to be avoided. *Many people hurt their back, arms and feet, and over one-third of all three-day reported injuries submitted to the HSE each year are the result of manual handling*.

When lifting heavy loads, correct lifting procedures must be adopted to avoid back injuries. Figure 1.17 demonstrates the technique. Do not lift objects from the floor with the back bent and the legs straight as this causes excessive stress on the spine. Always lift with the back straight and the legs bent so that the powerful leg muscles do the lifting work. Bend at the hips and knees to get down to the level of the object being lifted, positioning the body as close to the object as possible. Grasp the object firmly and, keeping the back straight and the head erect, use the leg muscles to raise in a smooth movement. Carry the load close to the body. When putting the object down, keep the back straight and bend at the hips and knees, reversing the lifting procedure. A bad lifting technique will result in sprains, strains and pains. *There have been too many injuries over the years resulting from bad manual handling techniques. The problem has become so serious that the HSE has introduced new legislation* under the Health and Safety at Work Act 1974, the Manual Handling Operations Regulations 1992. Publications such as *Getting to Grips with Manual Handling* can be obtained from HSE Books; the address and Infoline are given in Appendix B.

Definition

Manual handling is lifting, transporting or supporting loads by hand or by bodily force.

Safety first

Lifting

- bend your legs;
- keep your back straight;
- use the leg muscles to raise the weight in a smooth movement.

Visit the companion website for more on this topic.

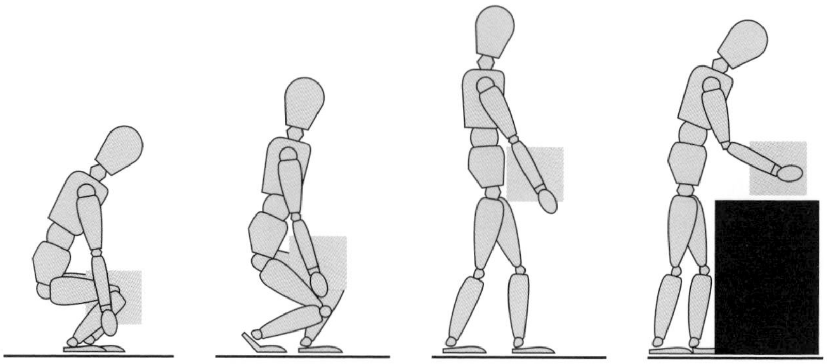

Figure 1.17 Correct manual lifting and carrying procedure.

Figure 1.18 Always use a mechanical aid to transport a load when available.

Where a job involves considerable manual handling, employers must now train employees in the correct lifting procedures and provide the appropriate equipment necessary to promote the safe manual handling of loads.

Consider some 'good practice' when lifting loads:

- Do not lift the load manually if it is more appropriate to use a mechanical aid. Only lift or carry what you can easily manage.
- Always use a trolley, wheelbarrow or truck such as that shown in Fig. 1.18 when these are available.
- Plan ahead to avoid unnecessary or repeated movement of loads.
- Take account of the centre of gravity of the load when lifting – the weight acts through the centre of gravity.
- Never leave a suspended load unsupervised.
- Always lift and lower loads gently.
- Clear obstacles out of the lifting area.
- Use the manual lifting techniques described above and avoid sudden or jerky movements.
- Use gloves when manual handling to avoid injury from rough or sharp edges.
- Take special care when moving loads wrapped in grease or bubble-wrap.
- Never move a load over other people or walk under a suspended load.

Working at height regulations

Working above ground level creates added dangers and slows down the work rate of the electrician. New Work at Height Regulations came into force on 6 April 2005. Every precaution should be taken to ensure that the working platform is appropriate for the purpose and in good condition.

Ladders

The term *ladder* is generally taken to include stepladders and trestles. The use of ladders for working above ground level is only acceptable for access and work of short duration (Work at Height Regulations 2005).

It is advisable to inspect the ladder before climbing it. It should be straight and firm. All rungs and tie rods should be in position and there should be no cracks in the stiles. The ladder should not be painted, since the paint may be hiding defects.

Extension ladders should be erected in the closed position and extended one section at a time. Each section should overlap by at least the number of rungs indicated below:

- ladder up to 4.8 m length – two rungs overlap;
- ladder up to 6.0 m length – three rungs overlap;
- ladder over 6.0 m length – four rungs overlap.

The angle of the ladder to the building should be in the proportion 4 up to 1 out or 75° as shown in Fig. 1.19. The ladder should be lashed at the top and bottom when possible to prevent unwanted movement and placed on firm and level ground. Footing is only considered effective for ladders smaller than 6 m and manufactured securing devices should always be considered. When ladders provide access to a roof or working platform the ladder must extend at least 1.05 m or five rungs above the landing place.

Short ladders may be carried by one person resting the ladder on the shoulder, but longer ladders should be carried by two people, one at each end, to avoid accidents when turning corners.

Long ladders or extension ladders should be erected by two people as shown in Fig. 1.20. One person stands on or 'foots' the ladder, while the other person lifts and walks under the ladder towards the wall. When the ladder is upright it can be positioned in the correct place, at the correct angle and secured before being climbed.

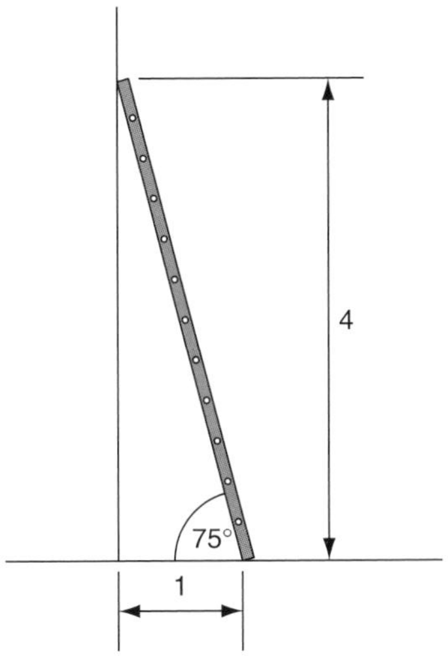

Figure 1.19 A correctly erected ladder.

Safety first

Ladders

New Working at Height Regulations tell us:

- Ladders are only to be used for access.
- Ladders must only be used for work of short duration.

Trestle scaffold

Figure 1.21 shows a trestle scaffold. Two pairs of trestles spanned by scaffolding boards provide a simple working platform. The platform must be at least two boards or 450 mm wide. At least one-third of the trestle must be above the working platform. If the platform is more than 2 m above the ground, toeboards and guardrails must be fitted, and a separate ladder provided for access. The boards which form the working platform should be of equal length and not overhang the trestles by more than four times their own thickness. The maximum span of boards between trestles is:

- 1.3 m for boards 40 mm thick;
- 2.5 m for boards 50 mm thick.

Trestles which are higher than 3.6 m must be tied to the building to give them stability. Where anyone can fall more than 4.5 m from the working platform, trestles may not be used.

Safety first

Scaffold

- Scaffold or mobile towers are always safer than ladders for working above ground.

Visit the companion website for more on this topic.

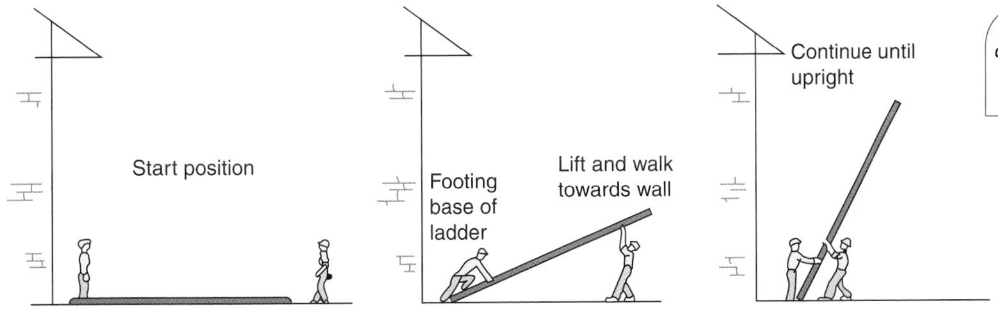

Figure 1.20 Correct procedure for erecting long or extension ladder.

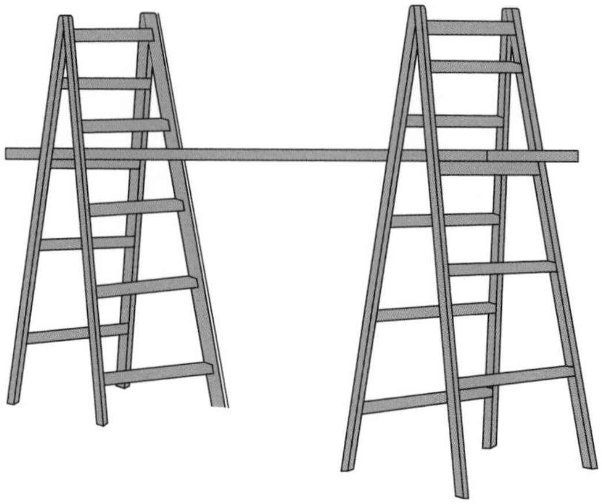

Figure 1.21 A trestle scaffold.

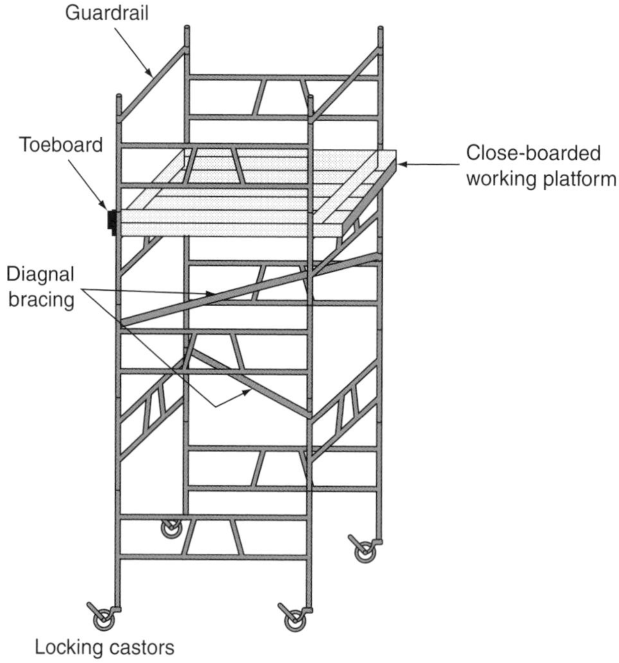

Figure 1.22 A mobile scaffold tower.

Figure 1.23 An example of safe ladder use.

Mobile scaffold towers

Mobile scaffold towers may be constructed of basic scaffold components or made from light alloy tube. The tower is built up by slotting the sections together until the required height is reached. A scaffold tower is shown in Fig. 1.22.

If the working platform is above 2 m from the ground it must be close boarded and fitted with guardrails and toeboards. When the platform is being used, all four wheels must be locked. The platform must not be moved unless it is clear of tools, equipment and workers, and should be pushed at the base of the tower and not at the top.

The stability of the tower depends upon the ratio of the base width to tower height. A ratio of base to height of 1:3 gives good stability. Outriggers can be used to increase stability by effectively increasing the base width. If outriggers are used then they must be fitted diagonally across all four corners of the tower and not on one side only. The tower must not be built more than 12 m high unless it has been specially designed for that purpose. Any tower higher than 9 m should be secured to the structure of the building to increase stability.

Access to the working platform of a scaffold tower should be by a ladder securely fastened vertically to the inside of the tower. Ladders must never be leaned against a tower, since this might push the tower over.

Working in excavations and confined spaces

Working alone

Some working situations are so potentially hazardous that not only must PPE be worn but you must also never work alone and safe working procedures must be in place before your work begins to reduce the risk.

It is unsafe to work in isolation in the following situations:

- when working above ground;
- when working below ground;
- when working in confined spaces;
- when working in excavations;
- when working close to unguarded machinery;
- when a fire risk exists;
- when working close to toxic or corrosive substances.

> **Safety first**
>
> **Working alone**
>
> Never work alone in:
>
> - confined spaces;
> - storage tanks;
> - enclosed ductwork.

Working above ground

We have already looked at this topic as it applies to electrical personnel. The new Work at Height Regulations 2005 tell us that a person is at height if that person could be injured by falling from it. The regulations require that:

- we should avoid working at height if at all possible;
- no work should be done at height which can be done on the ground (e.g. equipment can be assembled on the ground then taken up to height, perhaps for fixing);
- ensure the work at height is properly planned;
- take account of any risk assessments carried out under Regulation 3 of the Management of Health and Safety at Work Regulations.

Working in excavations or below ground

Working below ground might be working in a cellar or an unventilated basement with only one entrance/exit. There is a risk that this entrance/exit might become blocked by materials, fumes or fire. When working in trenches or excavations there is always the risk of the sides collapsing if they are not adequately supported by temporary steel sheets and adequate side bracing. There is also the risk of falling objects, so always:

- wear a hard hat;
- never go into an unsupported excavation;
- erect barriers around the excavation;
- provide good ladder access;
- ensure the work is properly planned;
- take account of the risk assessment before starting work.

Working in confined spaces

When working in confined spaces there is always the risk that you may become trapped or overcome by a lack of oxygen, or by gas, fumes, heat or an accumulation of dust. Examples of confined spaces are:

- storage tanks and silos on farms;
- enclosed sewers and pumping stations;
- furnaces;
- ductwork.

In my experience, electricians spend a lot of time on their knees in confined spaces because many electrical cable systems run out of sight away from public areas of a building.

The Confined Spaces Regulations 1997 require that:

- a risk assessment is carried out before work commences;
- if there is a serious risk of injury in entering the confined space then the work should be done on the outside of the vessel;
- a safe working procedure, such as a 'permit-to-work procedure', is followed and adequate emergency arrangements put in place before work commences.

Working near unguarded machinery

There is an obvious risk in working close to unguarded machinery, and indeed, most machinery will be guarded, but in some production processes and with overhead travelling cranes, this is not always possible. To reduce the risks associated with these hazards:

- have the machinery stopped during your work activity if possible;
- put temporary barriers in place;
- make sure the machine operator knows that you are working on the equipment;
- identify the location of emergency stop buttons;
- take account of the risk assessment before work commences.

A risk of fire

When working in locations containing stored flammable materials such as petrol, paraffin, diesel or bottled gas, there is always the risk of fire. To minimize the risk:

- take account of the risk assessment before work commences;
- keep the area well ventilated;
- locate the fire extinguishers;
- secure your exit from the area;
- locate the nearest fire alarm point;
- follow a safe working procedure and put adequate emergency arrangements in place before work commences.

Definition

The *permit-to-work procedure* is a type of 'safe system to work' procedure used in specialized and potentially dangerous plant process situations.

Permit-to-work system

The **permit-to-work procedure** is a type of 'safe system to work' procedure used in specialized and potentially dangerous plant process situations. The procedure was developed for the chemical industry, but the principle is equally applicable to the management of complex risk in other industries or situations. For example:

- working on part of an assembly line process where goods move through a complex, continuous process from one machine to another (e.g. the food industry);
- repairs to railway tracks, tippers and conveyors;
- working in confined spaces (e.g. vats and storage containers);
- working on or near overhead crane tracks;
- working underground or in deep trenches;
- working on pipelines;
- working near live equipment or unguarded machinery;
- roof work;
- working in hazardous atmospheres (e.g. the petroleum industry);
- working near or with corrosive or toxic substances.

All the above situations are high-risk working situations that should be avoided unless you have received special training and will probably require the completion of a permit-to-work. Permits-to-work must adhere to the following eight principles:

1 Wherever possible the hazard should be eliminated so that the work can be done safely without a permit-to-work.
2 The site manager has overall responsibility for the permit-to-work even though he or she may delegate the responsibility for its issue.
3 The permit must be recognized as the master instruction, which, until it is cancelled, overrides all other instructions.
4 The permit applies to everyone on-site, other trades and subcontractors.
5 The permit must give detailed information; for example: (i) which piece of plant has been isolated and the steps by which this has been achieved; (ii) what work is to be carried out; (iii) the time at which the permit comes into effect.
6 The permit remains in force until the work is completed and is cancelled by the person who issued it.
7 No other work is authorized. If the planned work must be changed, the existing permit must be cancelled and a new one issued.
8 Responsibility for the plant must be clearly defined at all stages because the equipment that is taken out of service is released to those who are to carry out the work.

The people doing the work, the people to whom the permit is given, take on the responsibility of following and maintaining the safeguards set out in the permit, which will define what is to be done (no other work is permitted) and the time-scale in which it is to be carried out.

The permit-to-work system must help communication between everyone involved in the process or type of work. Employers must train staff in the use of such permits and, ideally, training should be designed by the company issuing the permit, so that sufficient emphasis can be given to particular hazards present and the precautions which will be required to be taken. For further details see Permit to Work at www.hse.gov.uk.

Secure electrical isolation

Electric shock occurs when a person becomes part of the electrical circuit. The level or intensity of the shock will depend upon many factors, such as age, fitness and the circumstances in which the shock is received. The lethal level

Safety first

Isolation

- never work 'live';
- isolate;
- secure the isolation;
- prove the supply 'dead' before starting work.

is approximately 50 mA, above which muscles contract, the heart flutters and breathing stops. A shock above the 50 mA level is therefore fatal unless the person is quickly separated from the supply. Below 50 mA only an unpleasant tingling sensation may be experienced or you may be thrown across a room or shocked enough to fall from a roof or ladder, but the resulting fall may lead to serious injury.

To prevent people from receiving an electric shock accidentally, all circuits contain protective devices. All exposed metal is earthed; fuses and miniature circuit-breakers (MCBs) are designed to trip under fault conditions, and residual current devices (RCDs) are designed to trip below the fatal level as described in Chapter 4.

Construction workers and particularly electricians do receive electric shocks, usually as a result of carelessness or unforeseen circumstances. As an electrician working on electrical equipment you must always make sure that the equipment is switched off or electrically isolated before commencing work. Every circuit must be provided with a means of isolation (IET Regulation 132.15). When working on portable equipment or desk top units it is often simply a matter of unplugging the equipment from the adjacent supply. Larger pieces of equipment and electrical machines may require isolating at the local isolator switch before work commences. To deter anyone from re-connecting the supply while work is being carried out on equipment, a sign 'Danger – Electrician at Work' should be displayed on the isolator and the isolation 'secured' with a small padlock or the fuses removed so that no one can re-connect while work is being carried out on that piece of equipment. The Electricity at Work Regulations 1989 are very specific at Regulation 12(1) that we must ensure the disconnection and separation of electrical equipment from every source of supply and that this disconnection and separation is secure. Where a test instrument or voltage indicator is used to prove the supply dead, Regulation 4(3) of the Electricity at Work Regulations 1989 recommends that the following procedure is adopted.

Definition

Electrical isolation – we must ensure the disconnection and separation of electrical equipment from every source of supply and that this disconnection and separation is secure.

1. First, connect the test device such as that shown in Fig. 1.25 to the supply which is to be isolated. The test device should indicate mains voltage.
2. Next, isolate the supply and observe that the test device now reads zero volts.
3. Then connect the same test device to a known live supply or proving unit such as that shown in Fig. 1.26 to 'prove' that the tester is still working correctly.
4. Finally, secure the isolation and place warning signs; only then should work commence.

The test device being used by the electrician must incorporate safe test leads which comply with the Health and Safety Executive Guidance Note 38 on electrical test equipment. These leads should incorporate barriers to prevent the user from touching live terminals when testing and incorporating a protective resistor and be well insulated and robust, such as those shown in Fig. 1.24 and Fig 1.25.

To isolate a piece of equipment or individual circuit successfully, competently, safely and in accordance with all the relevant regulations, we must follow a procedure such as that given by the flow diagram in Fig. 1.28. Start at the top and work down the flow diagram.

When the heavy outlined amber boxes are reached, pause and ask yourself whether everything is satisfactory up until this point. If the answer is 'yes', move on. If the answer is 'no', go back as indicated by the diagram.

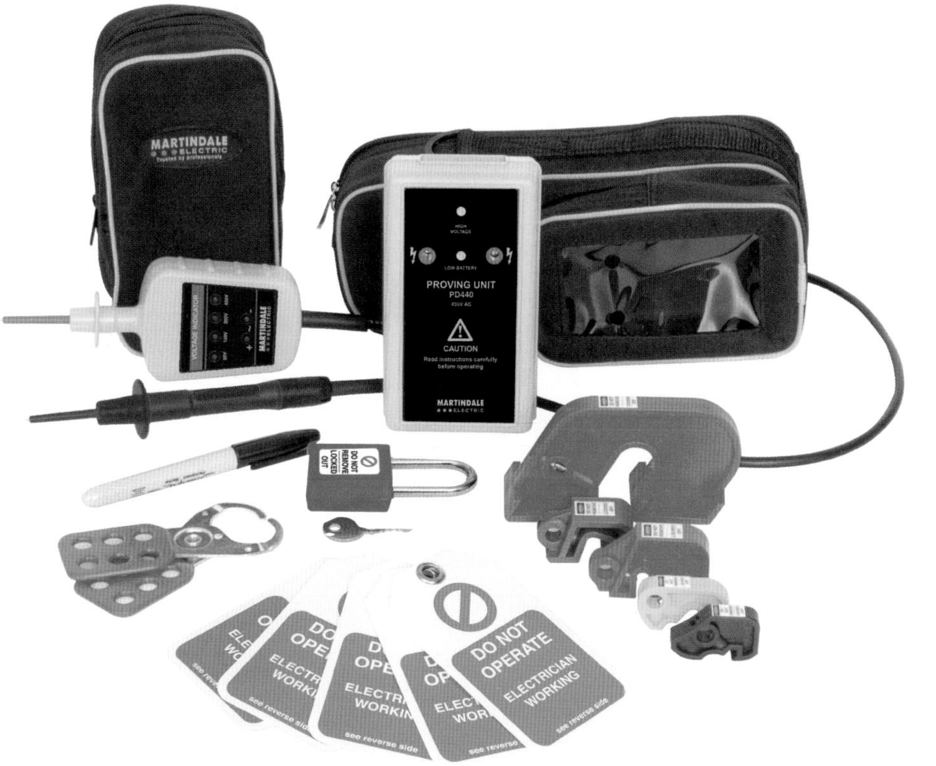

Figure 1.24 Comprehensive lock-out kit for use as part of a safe isolation procedure.

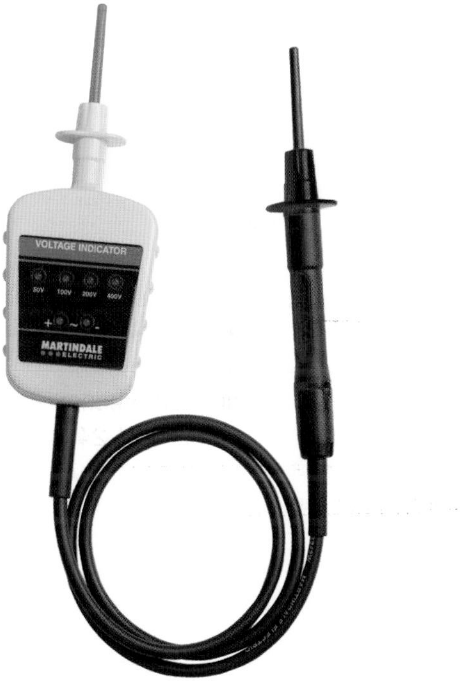

Figure 1.25 Typical voltage indicator.

Figure 1.26 Voltage proving unit.

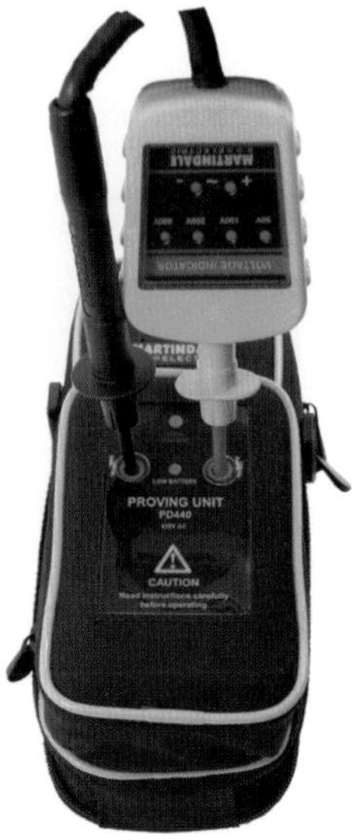

Figure 1.27 Voltage indicator being proved.

Live testing

The Electricity at Work Regulations 1989 at Regulation 4(3) tell us that it is preferable that supplies be made dead before work commences. However, it does acknowledge that some work, such as fault finding and testing, may require the electrical equipment to remain energized. Therefore, if the fault finding and testing can only be successfully carried out live, then the person carrying out the fault diagnosis must:

- be trained so that they understand the equipment and the potential hazards of working live and can, therefore, be deemed 'competent' to carry out that activity;
- only use approved test equipment;
- set up appropriate warning notices and barriers so that the work activity does not create a situation dangerous to others.

While live testing may be required by workers in the electrical industries in order to find the fault, live repair work must not be carried out. The individual circuit or piece of equipment must first be isolated before work commences in order to comply with the Electricity at Work Regulations 1989.

On-site electrical supplies and tools

Temporary electrical supplies on construction sites can save many person-hours of labour by providing energy for fixed and portable tools and lighting. However, as stated previously in this chapter, construction sites are dangerous places and the temporary electrical supplies must be safe. IET Regulation 110.1 tells us that *all* the regulations apply to temporary electrical installations such as construction sites. The frequency of inspection of construction sites is increased to every three months because of the inherent dangers. Regulation 704.313.4 recommends the following voltages for distributing to plant and equipment on construction sites:

400 V – fixed plant such as cranes

230 V – site offices and fixed floodlighting robustly installed

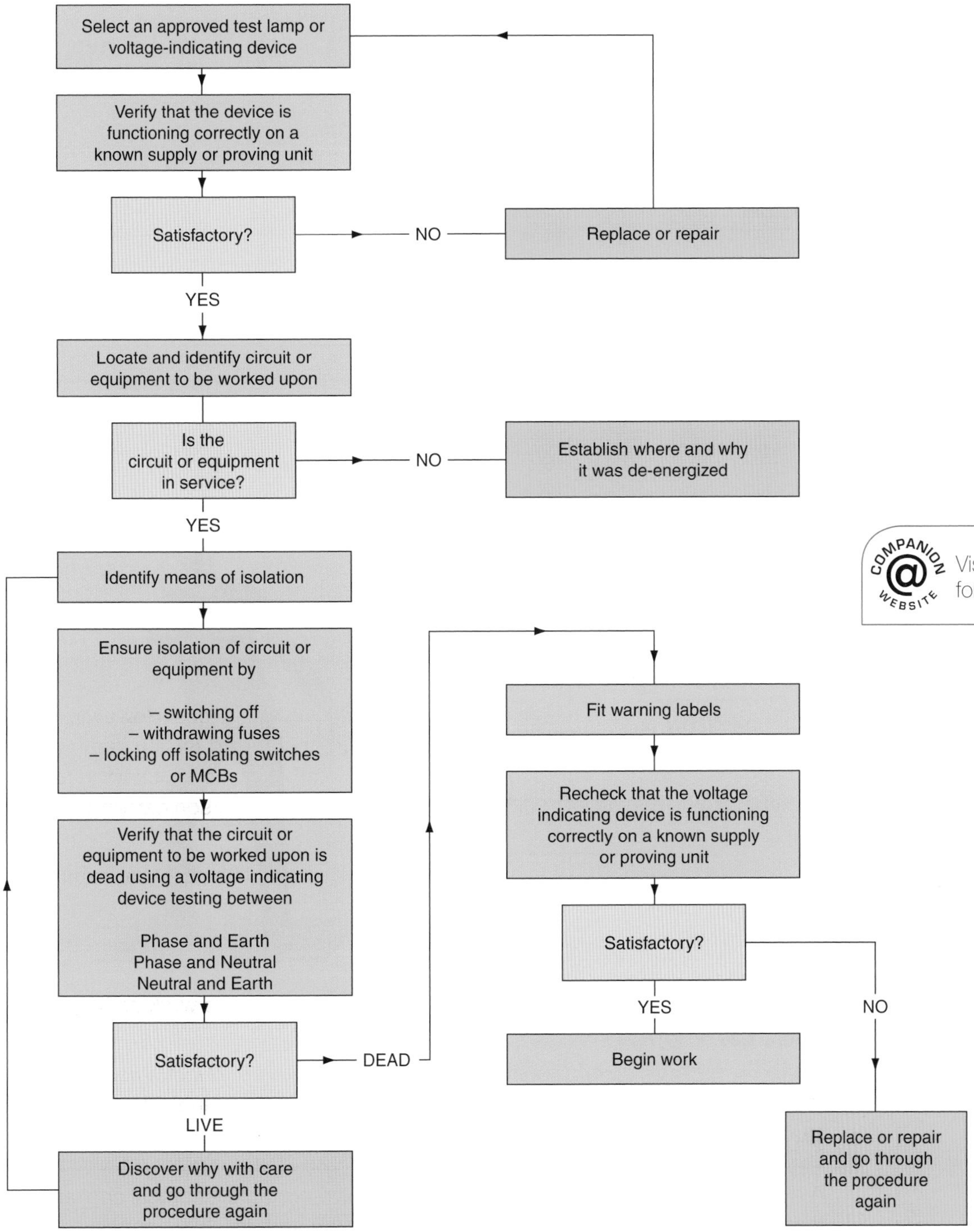

Figure 1.28 Flowchart for a secure isolation procedure.

COMPANION WEBSITE Visit the companion website for more on this topic.

110 V – portable tools and hand lamps

SELV – portable lamps used in damp or confined places.

Portable tools must be fed from a 110 V socket outlet unit (see Fig. 1.29(a)) incorporating splash-proof sockets and plugs with a keyway which prevents a tool from one voltage from being connected to the socket outlet of a different voltage.

Socket outlet and plugs are also colour coded for voltage identification: 25 V violet, 50 V white, 110 V yellow, 230 V blue and 400 V red, as shown in Fig. 1.29(b).

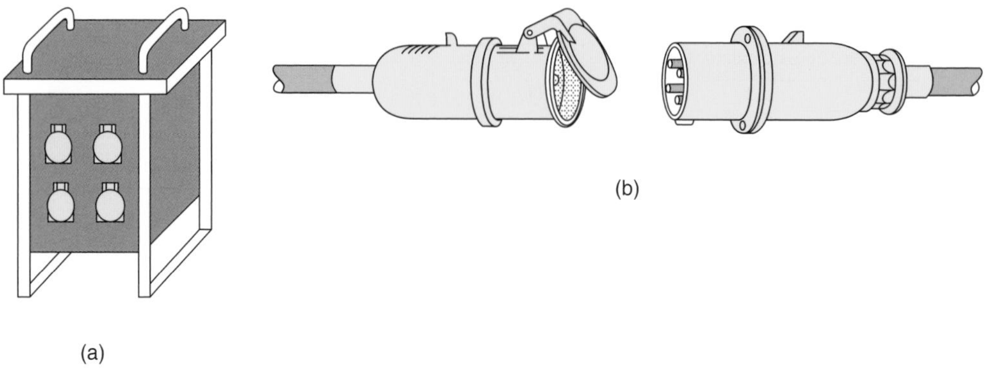

Figure 1.29 110 volts distribution unit and cable connector suitable for construction site electrical supplies: (a) reduced-voltage distribution unit incorporating industrial sockets to BS 4343; (b) industrial plug and connector.

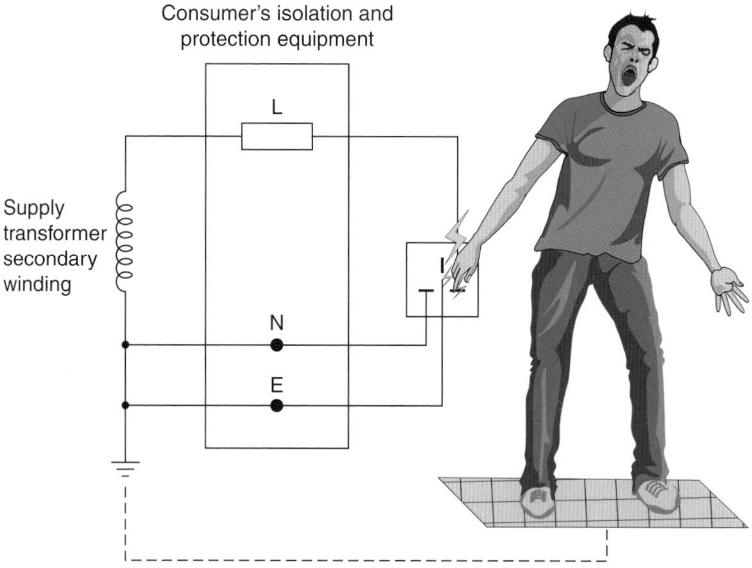

Figure 1.30 Touching live and earth or live and neutral makes a person part of the electrical circuit and can lead to an electric shock.

COMPANION WEBSITE Visit the companion website for more on this topic.

Portable electric appliance testing

A quarter of all serious electrical accidents involve portable electrical appliances; that is, equipment which has a cable lead and plug and which is normally moved around or can easily be moved from place to place. This includes, for example, floor cleaners, kettles, heaters, portable power tools, fans, televisions, desk lamps, photocopiers, fax machines and desktop computers. There is a requirement under the Health and Safety at Work Act for employers to take adequate steps to protect users of portable appliances from the hazards of electric shock and fire. The responsibility for safety applies equally to small as well as large companies. The Electricity at Work Regulations 1989 also place a duty of care upon employers to ensure that the risks associated with the use of electrical equipment are controlled.

Against this background the HSE has produced guidance notes HS(G) 107 *Maintaining Portable and Transportable Electrical Equipment* and leaflets *Maintaining Portable Electrical Equipment in Offices* and *Maintaining*

Portable Electrical Equipment in Hotels and Tourist Accommodation. In these publications the HSE recommends that a three-level system of inspection can give cost-effective maintenance of portable appliances. These are:

1 User checking.
2 Visual inspection by an appointed person.
3 Combined inspection and testing by a competent person or contractor.

A user visually checking the equipment is probably the most important maintenance procedure. About 95% of faults or damage can be identified by just looking. The user should check for obvious damage using common sense. The use of potentially dangerous equipment can then be avoided. Possible dangers to look for are as follows:

- damage to the power cable or lead which exposes the colours of the internal conductors, which are brown, blue and green with a yellow stripe;
- damage to the plug itself. The plug pushes into the wall socket, usually a square pin 13 A socket in the United Kingdom, to make an electrical connection. With the plug removed from the socket the equipment is usually electrically 'dead'. If the bakelite plastic casing of the plug is cracked, broken or burned, or the contact pins are bent, do not use it;
- non-standard joints in the power cable, such as taped joints;
- poor cable retention. The outer sheath of the power cable must be secured and enter the plug at one end and the equipment at the other. The coloured internal conductors must not be visible at either end;
- damage to the casing of the equipment such as cracks, pieces missing, loose or missing screws or signs of melted plastic, burning, scorching or discoloration;
- equipment which has previously been used in unsuitable conditions such as a wet or dusty environment.

If any of the above dangers are present, the equipment should not be used until the person appointed by the company to make a 'visual inspection' has had an opportunity to do so.

A **visual inspection** will be carried out by an appointed person within a company, such person having been trained to carry out this task. In addition to the user checks described above, an inspection could include the removal of the plug top cover to check that:

- a fuse of the correct rating is being used and also that a proper cartridge fuse is being used and not a piece of wire, a nail or silver paper;
- the cord grip is holding the sheath of the cable and not the coloured conductors;
- the wires (conductors) are connected to the correct terminals of the plug top as shown in Fig. 1.31;
- the coloured insulation of each conductor wire goes right up to the terminal so that no bare wire is visible;
- the terminal fixing screws hold the conductor wires securely and the screws are tight;
- all the conductor wires are secured within the terminal;
- there are no internal signs of damage such as overheating, excessive 'blowing' of the cartridge fuse or the intrusion of foreign bodies such as dust, dirt or liquids.

The above inspection cannot apply to 'moulded plugs', which are moulded on to the flexible cable by the manufacturer in order to prevent some of the bad

Definition

The installation must be *visually inspected* before testing begins. The aim of the visual inspection is to confirm that all equipment and accessories are undamaged and comply with the relevant British and European Standards, and also that the installation has been securely and correctly erected.

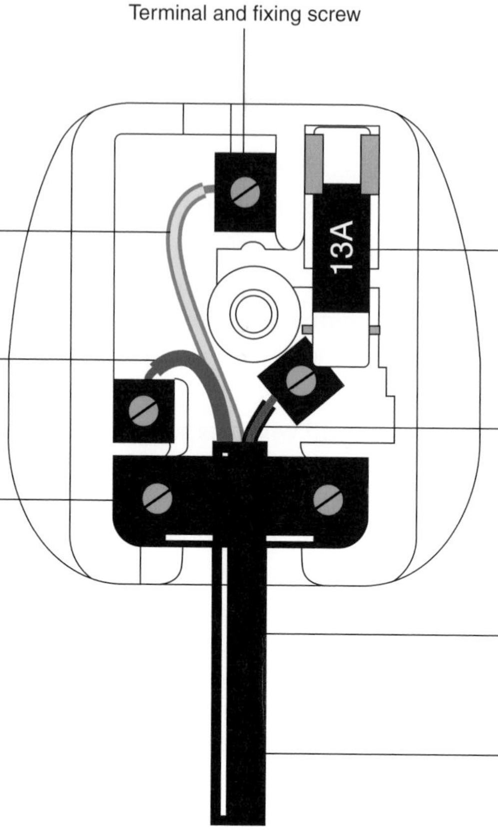

Figure 1.31 Correct connection of plug.

practice described above. In the case of a moulded plug top, only the fuse can be checked. The visual inspection checks described above should also be applied to extension leads and their plugs. The HSE recommends that a simple procedure be written to give guidance to the 'appointed person' carrying out the visual inspection.

Combined inspection and testing is also necessary on some equipment because some faults cannot be seen by just looking – for example, the continuity and effectiveness of earth paths on class I equipment. For some portable appliances the earth is essential to the safe use of the equipment and, therefore, all earthed equipment and most extension leads should be periodically tested and inspected for these faults. All portable appliance test instruments (PAT Testers) will carry out two important tests: earth bonding and insulation resistance.

Earth bonding tests apply a substantial test current, typically about 25 A, down the earth pin of the plug top to an earth probe, which should be connected to any exposed metalwork on the portable appliance being tested. The PAT Tester will then calculate the resistance of the earth bond and either give an actual reading or indicate pass or fail. A satisfactory result for this test would typically be a reading of less than 0.1 Ω. The earth bond test is, of course, not required for double-insulated portable appliances because there will be no earthed metalwork.

Insulation resistance tests apply a substantial test voltage, typically 500 V, between the live and neutral bonded together and the earth. The PAT Tester then calculates the insulation resistance and either gives an actual reading or indicates pass or fail. A satisfactory result for this test would typically be a reading greater than 2 MΩ.

Some PAT Testers offer other tests in addition to the two described above. These are described below.

A flash test tests the insulation resistance at a higher voltage than the 500 V test described above. The flash test uses 1.5 kV for Class 1 portable appliances, that is, earthed appliances, and 3 kV for Class 2 appliances, which are double insulated. The test establishes that the insulation will remain satisfactory under more stringent conditions but must be used with caution, since it may overstress the insulation and will damage electronic equipment. A satisfactory result for this test would typically be less than 3 mA.

A fuse test tests that a fuse is in place and that the portable appliance is switched on prior to carrying out other tests. A visual inspection will be required to establish that the *size* of the fuse is appropriate for that particular portable appliance.

An earth leakage test measures the leakage current to earth through the insulation. It is a useful test to ensure that the portable appliance is not deteriorating and liable to become unsafe. It also ensures that the tested appliances are not responsible for nuisance 'tripping' of RCDs (RCDs – see Chapter 3). A satisfactory reading is typically less than 3 mA.

An operation test proves that the preceding tests were valid (i.e. that the unit was switched on for the tests), that the appliances will work when connected to the appropriate voltage supply and will not draw a dangerously high current from that supply. A satisfactory result for this test would typically be less than 3.2 kW for 230 V equipment and less than 1.8 kW for 110 V equipment.

All PAT Testers are supplied with an operating manual, giving step-by-step instructions for their use and pass and fail scale readings. The HSE suggested intervals for the three levels of checking and inspection of portable appliances in offices and other low-risk environments are given in Table 1.3.

Who does what?

When actual checking, inspecting and testing of portable appliances takes place will depend upon the company's safety policy and risk assessments. In low-risk environments such as offices and schools, the three-level system of checking, inspection and testing recommended by the HSE should be carried out. Everyone can use common sense and carry out the user checks described earlier. Visual inspections must be carried out by a 'competent person' but that person does not need to be an electrician or electronics service engineer. Any sensible member of staff who has received training can carry out this duty. They will need to know what to look for and what to do, but more importantly, they will need to be able to avoid danger to themselves and to others. The HSE recommends that the appointed person follows a simple written procedure for each visual inspection. A simple tick sheet would meet this requirement. For example:

1 Is the correct fuse fitted? Yes/No

2 Is the cord grip holding the cable sheath? Yes/No

The tick sheet should incorporate all the appropriate visual checks and inspections described earlier.

Safety first

Power tools

- Look at the power tools that you use at work.
- Do they have a PAT Test label?
- Is it 'in date'?

Table 1.3 HSE suggested intervals for checking, inspecting and testing of portable appliances in offices and other low-risk environments

Equipment/ environment	User checks	Formal visual inspection	Combined visual inspection and electrical testing
Battery-operated (less than 20V)	No	No	No
Extra low-voltage (less than 50V a.c.): e.g. telephone equipment, low-voltage desk lights	No	No	No
Information technology: e.g. desktop computers, VDU screens	No	Yes, 2–4 years	No if double insulated – otherwise up to 5 years
Photocopiers, fax machines: *not* handheld, rarely moved	No	Yes, 2–4 years	No if double insulated – otherwise up to 5 years
Double-insulated equipment: *not* handheld, moved occasionally, e.g. fans, table lamps, slide projectors	No	Yes, 2–4 years	No
Double-insulated equipment: *handheld*, e.g. power tools	Yes	Yes, 6 months to 1 year	No
Earthed equipment (Class 1): e.g. electric kettles, some floor cleaners, power tools	Yes	Yes, 6 months to 1 year	Yes, 1–2 years
Cables (leads) and plugs connected to the above	Yes	Yes, 6 months to 4 years depending on the type of equipment it is connected to	Yes, 1–5 years depending on the type of equipment it is connected to
Extension leads (mains voltage)	Yes	As above	As above

Testing and inspection require a much greater knowledge than is required for simple checks and visual inspections. This more complex task need not necessarily be carried out by a qualified electrician or electronics service engineer. However, the person carrying out the test must be trained to use the equipment and to interpret the results. In addition, greater knowledge will be required for the inspection of the range of portable appliances which might be tested.

Keeping records

Records of the inspecting and testing of portable appliances are not required by law but within the Electricity at Work Regulations 1989 it is generally accepted

Figure 1.32 Typical PAT Test labels.

that some form of recording of results is required to implement a quality control system. The control system should:

- ensure that someone is nominated to have responsibility for portable appliance inspection and testing;
- maintain a log or register of all portable appliance test results to ensure that equipment is inspected and tested when it is due;
- label tested equipment with the due date for its next inspection and test as shown in Fig. 1.32. If it is out of date, don't use it.

Any piece of equipment which fails a PAT Test should be disabled and taken out of service (usually by cutting off the plug), labelled as faulty and sent for repair.

The register of PAT Test results will help managers to review their maintenance procedures and the frequency of future visual inspections and testing. Combined inspection and testing should be carried out where there is a reason to suspect that the equipment may be faulty, damaged or contaminated but this cannot be verified by visual inspection alone. Inspection and testing should also be carried out after any repair or modification to establish the integrity of the equipment or at the start of a maintenance system, to establish the initial condition of the portable equipment being used by the company.

Electric shock

Definition

Electric shock occurs when a person becomes part of the electrical circuit.

Electric shock occurs when a person becomes part of the electrical circuit, as shown in Fig. 1.30. The level or intensity of the shock will depend upon many factors, such as age, fitness and the circumstances in which the shock is received. The lethal level is approximately 50 mA, above which muscles contract, the heart flutters and breathing stops. A shock above the 50 mA level is therefore fatal unless the person is quickly separated from the supply. Below 50 mA only an unpleasant tingling sensation may be experienced or you may be thrown across a room or fall from a roof or ladder, but the resulting fall may lead to serious injury.

To prevent people from receiving an electric shock accidentally, all circuits must contain protective devices. All exposed metal must be earthed; fuses and miniature circuit-breakers (MCBs) are designed to trip under fault conditions and residual current devices (RCDs) are designed to trip below the fatal level.

Construction workers and particularly electricians do receive electric shocks, usually as a result of carelessness or unforeseen circumstances. When this happens it is necessary to act quickly to prevent the electric shock from becoming fatal. Actions to be taken upon finding a workmate receiving an electric shock are as follows:

- switch off the supply if possible;
- alternatively, remove the person from the supply *without touching him*, e.g. push him off with a piece of wood, pull him off with a scarf, dry towel or coat;
- if breathing or heart has stopped, immediately call professional help by dialling 999 and asking for the ambulance service. Give precise directions to the scene of the accident. The casualty stands the best chance of survival if the emergency services can get a rapid response paramedic team quickly to the scene. They have extensive training and will have specialist equipment with them;
- only then should you apply resuscitation or cardiac massage until the patient recovers, or help arrives;
- treat for shock.

First aid

Despite all the safety precautions taken on construction sites to prevent injury to the workforce, accidents do happen and *you* may be the only other person able to take action to assist a workmate. This section is not intended to replace a first aid course but to give learners the knowledge to understand the types of injuries they may come across in the workplace. If you are not a qualified first aider limit your help to obvious common-sense assistance and call for help, *but* do remember that if a workmate's heart or breathing has stopped as a result of an accident they have only minutes to live unless you act quickly. The Health and Safety (First Aid) Regulations 1981 and relevant approved codes of practice and guidance notes place a duty of care on all employers to provide *adequate* first aid facilities appropriate to the type of work being undertaken. *Adequate* facilities will relate to a number of factors such as:

- How many employees are employed?
- What type of work is being carried out?
- Are there any special or unusual hazards?
- Are employees working in scattered and/or isolated locations?
- Is there shift work or 'out of hours' work being undertaken?
- Is the workplace remote from emergency medical services?
- Are there inexperienced workers on site?
- What were the risks of injury and ill health identified by the company's hazard risk assessment?

The regulations state that:

> *Employers are under a duty to provide such numbers of suitable persons as is adequate and appropriate in the circumstances for rendering first aid to his employees if they are injured or become ill at work. For this purpose a person shall not be suitable unless he or she has undergone such training and has such qualifications as the Health and Safety Executive may approve.*

This is typical of the way in which the Health and Safety Regulations are written. The regulations and codes of practice do not specify numbers, but set out guidelines in respect of the number of first aiders needed, dependent upon the type of company, the hazards present and the number of people employed.

Let us now consider the questions 'what is first aid?' and 'who might become a first aider?' The regulations give the following definitions of first aid. '*First aid* is the treatment of minor injuries which would otherwise receive no treatment or do not need treatment by a doctor or nurse'; *or* 'In cases where a person will require help from a doctor or nurse, first aid is treatment for the purpose of preserving life and minimizing the consequences of an injury or illness until such help is obtained'. A more generally accepted definition of first aid might be as follows: **first aid** is the initial assistance or treatment given to a casualty for any injury or sudden illness before the arrival of an ambulance, doctor or other medically qualified person.

Definition

First aid is the initial assistance or treatment given to a casualty for any injury or sudden illness before the arrival of an ambulance, doctor or other medically qualified person.

Now having defined first aid, who might become a first aider? A **first aider** is someone who has undergone a training course to administer first aid at work and holds a current first aid certificate. The training course and certification must be approved by the HSE. The aims of a first aider are to preserve life, to limit the worsening of the injury or illness and to promote recovery.

Definition

A *first aider* is someone who has undergone a training course to administer first aid at work and holds a current first aid certificate.

A first aider may also undertake the duties of an appointed person. An **appointed person** is someone who is nominated to take charge when someone is injured or becomes ill, including calling an ambulance if required. The appointed person will also look after the first aid equipment, including restocking the first aid box.

Definition

An *appointed person* is someone who is nominated to take charge when someone is injured or becomes ill, including calling an ambulance if required. The appointed person will also look after the first aid equipment, including restocking the first aid box.

Appointed persons should not attempt to give first aid for which they have not been trained, but should limit their help to obvious common-sense assistance and summon professional assistance as required. Suggested numbers of first aid personnel are given in Table 1.1. The actual number of first aid personnel must take into account any special circumstances such as remoteness from medical services, the use of several separate buildings and the company's hazard risk assessment. First aid personnel must be available at all times when people are at work, taking into account shift-working patterns and providing cover for sickness absences.

Every company must have at least one first aid kit under the regulations. The size and contents of the kit will depend upon the nature of the risks involved in the particular working environment and the number of employees. Table 1.2 gives a list of the contents of any first aid box to comply with the HSE Regulations.

There now follows a description of some first aid procedures which should be practised under expert guidance before they are required in an emergency.

Asphyxiation

Asphyxiation is a condition caused by lack of air in the lungs leading to suffocation. Suffocation may cause discomfort by making breathing difficult or it may kill by stopping the breathing. There is a risk of asphyxiation to workers when:

- working in confined spaces;
- working in poorly ventilated spaces;
- working in paint stores and spray booths;
- working in the petrochemical industry;
- working in any environment in which toxic fumes and gases are present.

Definition

Asphyxiation is a condition caused by lack of air in the lungs leading to suffocation. Suffocation may cause discomfort by making breathing difficult or it may kill by stopping the breathing.

Under the Management of Health and Safety at Work Regulations a risk assessment must be made if the environment may be considered hazardous to health. Safety procedures, including respiratory protective equipment, must be in place before work commences.

The treatment for fume inhalation or asphyxia is to get the patient into fresh air but only if you can do this without putting yourself at risk. If the patient is unconscious, proceed with resuscitation as described below.

Bleeding

If the wound is dirty, rinse it under clean running water. Clean the skin around the wound and apply a plaster, pulling the skin together.

If the bleeding is severe, apply direct pressure to reduce the bleeding and raise the limb if possible. Apply a sterile dressing or pad and bandage firmly before obtaining professional advice.

To avoid possible contact with hepatitis or the AIDS virus when dealing with open wounds, first aiders should avoid contact with fresh blood by wearing plastic or rubber protective gloves, or by allowing the casualty to apply pressure to the bleeding wound.

Burns

Remove heat from the burn to relieve the pain by placing the injured part under clean, cold water. Do not remove burnt clothing sticking to the skin. Do not apply lotions or ointments. Do not break blisters or attempt to remove loose skin. Cover the injured area with a clean, dry dressing.

Broken bones

Make the casualty as comfortable as possible by supporting the broken limb either by hand or with padding. Do not move the casualty unless by remaining in that position they are likely to suffer further injury. Obtain professional help as soon as possible.

Contact with chemicals

Wash the affected area very thoroughly with clean, cold water. Remove any contaminated clothing. Cover the affected area with a clean, sterile dressing and seek expert advice. It is a wise precaution to treat all chemical substances as possibly harmful; even commonly used substances can be dangerous if contamination is from concentrated solutions. When handling dangerous substances, it is also good practice to have a neutralizing agent to hand.

Disposal of dangerous substances must not be into the main drains since this can give rise to an environmental hazard, but should be undertaken in accordance with local authority regulations.

Exposure to toxic fumes

Get the casualty into fresh air quickly and encourage deep breathing if he/she is conscious. Resuscitate if breathing has stopped. Obtain expert medical advice as fumes may cause irritation of the lungs.

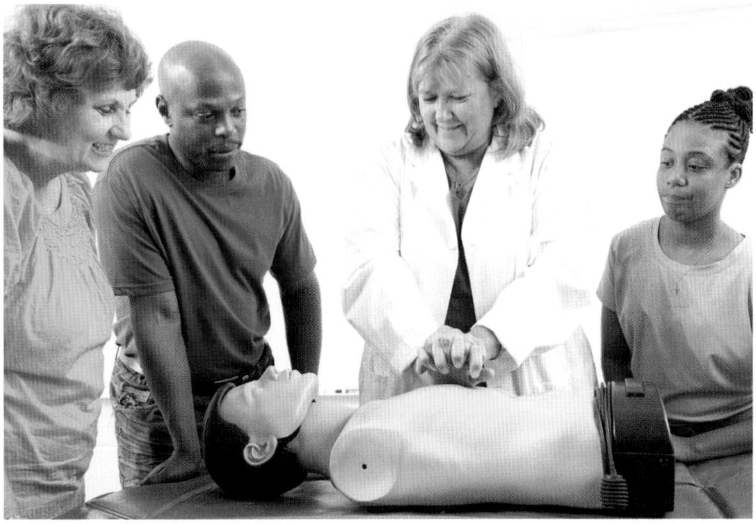

Figure 1.33 Following a few simple rules can save lives.

Sprains and bruising

A cold compress can help to relieve swelling and pain. Soak a towel or cloth in cold water, squeeze it out and place it on the injured part. Renew the compress every few minutes.

Resuscitation – breathing stopped

Remove any restrictions from the face and any vomit, loose or false teeth from the mouth. Loosen tight clothing around the neck, chest and waist. To ensure a good airway, lay the casualty on their back and support the shoulders on some padding. Tilt the head backwards and open the mouth. If the casualty is faintly breathing, lifting the tongue and clearing the airway may be all that is necessary to restore normal breathing.

Heart stopped beating – CPR (Cardio Pulmonary Resuscitation)

This sometimes happens following a severe electric shock. If the casualty's lips are blue, the pupils of their eyes widely dilated and the pulse in their neck cannot be felt, then they may have gone into cardiac arrest. Act quickly and lay the casualty on their back. Kneel down beside them and place the heel of one hand in the centre of their chest. Cover this hand with your other hand and interlace the fingers. Straighten your arms and press down on their chest sharply with the heel of your hands and then release the pressure. Continue to do this 15 times at the rate of one push per second. 'Push hard and fast, where a gold medallion would rest on the chest, to the rhythm of Ha, Ha, Ha, Ha, staying alive, staying alive.' That is, the music from *Saturday Night Fever*, as demonstrated by Vinnie Jones in a recent TV advertisement. Check the casualty's pulse. If none is felt, give a further 15 chest compressions. Continue this procedure until the heartbeat is restored. Pay close attention to the condition of the casualty while giving heart massage. When a pulse is restored the blueness around the mouth will quickly go away and you should stop the heart massage. Treat the casualty for shock, place them in the recovery position and obtain professional help.

Shock

Everyone suffers from shock following an accident. The severity of the shock depends upon the nature and extent of the injury. In cases of severe shock the casualty will become pale and their skin will become clammy from sweating. They may feel faint, have blurred vision, feel sick and complain of thirst. Reassure the casualty that everything that needs to be done is being done. Loosen tight clothing and keep them warm and dry until help arrives. *Do not* move them unnecessarily or give them anything to drink.

Accident reports

Figure 1.34 First aid log-book/accident book with data protection-compliant removable sheets.

Every accident must be reported to an employer and minor accidents reported to a supervisor, safety officer or first aider and the details of the accident and treatment given suitably documented. A first aid log-book or accident book such as that shown in Fig. 1.34 containing first aid treatment record sheets could be used to effectively document accidents which occur in the workplace and the treatment given. Failure to do so may influence the payment of compensation at a later date if an injury leads to permanent disability. To comply with the Data Protection Regulations, from 31 December 2003 all first aid treatment log-books or accident report books must contain perforated sheets which can be removed after completion and filed away for personal security.

If the accident results in death, serious injury or an injury that leads to an absence from work of more than three days, your employer must report the accident to the local office of the HSE. The quickest way to do this is to call the Incident Control Centre on 0845 300 9923. They will require the following information:

- the name of the person injured;
- a summary of what happened;
- a summary of events prior to the accident;
- information about the injury or loss sustained;
- details of witnesses;
- date and time of accident;
- name of the person reporting the incident.

The Incident Control Centre will forward a copy of every report they complete to the employer for them to check and hold on record. However, good practice would recommend an employer or their representative to make an extensive report of any serious accident that occurs in the workplace. In addition to recording the above information, the employer or their representative should:

- sketch diagrams of how the accident occurred, where objects were before and after the accident, where the victim fell, etc.;
- take photographs or video that show how things were after the accident, for example, broken step-ladders, damaged equipment, etc.;
- collect statements from witnesses. Ask them to write down what they saw;
- record the circumstances surrounding the accident. Was the injured person working alone – in the dark – in some other adverse situation or condition – was PPE being worn – was PPE recommended in that area?

The above steps should be taken immediately after the accident has occurred and after the victim has been sent for medical attention. The area should be made safe and the senior management informed so that any actions to prevent a similar occurrence can be put in place. Taking photographs and obtaining witnesses' statements immediately after an accident happens means that evidence may still be around and memories still sharp.

RIDDOR

RIDDOR stands for Reporting of Injuries, Diseases and Dangerous Occurrences Regulation 1995, which is sometimes referred to as RIDDOR 95, or just RIDDOR for short. The HSE requires employers to report some work-related accidents or diseases so that they can identify where and how risks arise, investigate serious accidents and publish statistics and data to help reduce accidents at work.

What needs reporting? Every work-related death, major injury, dangerous occurrence, disease or any injury which results in an absence from work of over three days.

Where an employee or member of the public is killed as a result of an accident at work the employer or his representative must report the accident to the Environmental Health Department of the Local Authority by telephone that day and give brief details. Within 10 days this must be followed up by a completed accident report form (Form No. F2508). Major injuries sustained as a result of an accident at work include amputations, loss of sight (temporary or permanent), fractures to the body other than to fingers, thumbs or toes and any other serious injury. Once again, the Environmental Health Department of the Local Authority must be notified by telephone on the day that the serious injury occurs and the telephone call followed up by a completed Form F2508 within 10 days. Dangerous occurrences are listed in the regulations and include the collapse of a lift, an explosion or injury caused by an explosion, the collapse of a scaffold over 5 m high, the collision of a train with any vehicle, the unintended collapse of a building and the failure of fairground equipment.

Depending upon the seriousness of the event, it may be necessary to immediately report the incident to the Local Authority. However, the incident must be reported within 10 days by completing Form F2508. If a doctor notifies an employer that an employee is suffering from a work-related disease then form F2508A must be completed and sent to the Local Authority. Reportable diseases include certain poisonings, skin diseases, lung disease, infections and occupational cancer. The full list is given within the pad of report forms.

An accident at work resulting in an over-three-day injury (that is, an employee being absent from work for over three days as a result of an accident at work) requires that accident report form F2508 be sent to the Local Authority within 10 days.

An over-three-day injury is one which is not major but results in the injured person being away from work for more than three days not including the day the injury occurred.

Who are the reports sent to? They are sent to the Environmental Health Department of the Local Authority or the area HSE offices (see Appendix B for area office addresses). Accident report forms F2508 can also be obtained from them or by ringing the HSE Infoline, or by ringing the incident contact centre on 0845 300 9923.

For most businesses, a reportable accident, dangerous occurrence or disease is a very rare event. However, if a report is made, the company must keep a record of the occurrence for three years after the date on which the incident happened. The easiest way to do this would probably be to file a photocopy of the completed accident report form F2508, but a record may be kept in any form which is convenient.

Dangerous occurrences and hazardous malfunctions

Definition

Dangerous occurrence – a 'near miss' that could easily have led to serious injury or loss of life. Near-miss accidents occur much more frequently than injury accidents and are, therefore, a good indicator of hazard, which is why the HSE collects these data.

A **dangerous occurrence** is a 'near miss' that could easily have led to serious injury or loss of life. Dangerous occurrences are defined in the Reporting of Injuries, Diseases and Dangerous Occurrences Regulations (RIDDOR) 1995. Near-miss accidents occur much more frequently than injury accidents and are, therefore, a good indicator of hazard, which is why the HSE collects these data. In January 2008 a BA passenger aeroplane lost power to both engines as it prepared to land at Heathrow airport. The pilots glided the plane into a crash landing on the grass just short of the runway. This is one example of a dangerous occurrence which could so easily have been a disaster.

Consider another example – on a wet and windy night a large section of scaffold around a town centre building collapses. Fortunately this happens about midnight when no one is around because of the time and the bad weather. However, if it had occurred at midday, workers would have been using the scaffold and the streets would have been crowded with shoppers. This would be classified as a dangerous occurrence and must be reported to the HSE, who will investigate the cause and, using their wide range of powers, would either:

- stop all work;
- demand the dismantling of the structure;
- issue an Improvement Notice;
- issue a Prohibition Notice;
- prosecute those who have failed in their health and safety duties.

Other reportable dangerous occurrences are:

- the collapse of a lift;
- plant coming into contact with overhead power lines;
- any unexpected collapse which projects material beyond the site boundary;
- the overturning of a road tanker;
- a collision between a car and a train.

Definition

Hazardous malfunction – if a piece of equipment was to fail in its function (that is, fail to do what it is supposed to do) and, as a result of this failure have the potential to cause harm, then this would be defined as a hazardous malfunction.

Hazardous malfunction – if a piece of equipment was to fail in its function (that is, fail to do what it is supposed to do) and, as a result of this failure have the potential to cause harm, then this would be defined as a hazardous malfunction. Consider an example – if a 'materials lift' on a construction site was to collapse when the supply to its motor failed, this would be a hazardous malfunction. All the regulations concerning work equipment state that it must be:

- suitable for its intended use;
- safe in use;
- maintained in a safe condition;
- used only by instructed persons;
- provided with suitable safety measures, protective devices and warning signs.

Disposing of waste

We have said many times in this book so far that having a good attitude to health and safety, working conscientiously and neatly, keeping passageways clear and regularly tidying up the workplace is the sign of a good and competent craftsman. But what do you do with the rubbish that the working environment produces? Well, all the packaging material for electrical fittings and accessories usually goes into either your employer's skip or the skip on-site designated for that purpose. All the off-cuts of conduit, trunking and tray also go into the skip. In fact, most of the general site debris will probably go into the skip and the waste disposal company will take the skip contents to a designated local council land-fill area for safe disposal.

The part coils of cable and any other reusable left-over lengths of conduit, trunking or tray will be taken back to your employer's stores area. Here it will be stored for future use and the returned quantities deducted from the costs allocated to that job.

What goes into the skip for normal disposal into a land-fill site is usually a matter of common sense. However, some substances require special consideration and disposal. We will now look at asbestos and large quantities of used fluorescent tubes.

Control of asbestos at work regulations

In October 2010 the HSE launched a national campaign to raise awareness among electricians and other trades of the risk to their health of coming into contact with asbestos. It is called the 'Hidden Killer Campaign' because asbestos is a killer. It can cause asbestosis, lung cancer and mesothelioma which affect the lining of the lungs and surrounding digestive system.

The British Lung Foundation has calculated that every week on average six electricians, four plumbers and eight joiners die in the UK as a result of exposure to asbestos, making it the greatest cause of work-related deaths. Chief Executive Dr Penny Woods said that 'twice as many people die in Britain each year from asbestos related illnesses than those who die from accidents on the roads'.
For more information about asbestos hazards, visit www.hse.gov.uk/asbestos and the British Lung Foundation www.take5andstayalive.com

Asbestos is a mineral found in many rock formations. When separated it becomes a fluffy, fibrous material with many uses. It was used extensively in the construction industry during the 1960s and 1970s for roofing material, ceiling and floor tiles, fire-resistant board for doors and partitions, for thermal insulation and commercial and industrial pipe lagging.

There are three main types of asbestos:

- chysotile which is white and accounts for about 90% of the asbestos in use today;
- amosite which is brown;
- crocidolite which is blue.

Asbestos cannot be identified by colour alone and a laboratory analysis is required to establish its type. Blue and brown are the two most dangerous forms of asbestos and have been banned from use since 1985. White asbestos was banned from use in 1999.

In the buildings where it was installed some 40 years ago, when left alone, it did not represent a health hazard, but those buildings are increasingly becoming in need of renovation and modernization. It is in the dismantling, drilling and breaking up of these asbestos materials that the health hazard increases. Asbestos is a serious health hazard if the dust is inhaled. The tiny asbestos particles find their way into delicate lung tissue and remain embedded for life, causing constant irritation and eventually, serious lung disease.

Figure 1.35 Hazardous areas should be clearly signposted.

Asbestos materials may be encountered by electricians in decorative finishes such as artex ceiling finishes, plaster and floor tiles. It is also found in control gear such as flash guards and matting in fuse carriers and distribution fuse boards, and in insulation materials in vessels, containers, pipework, ceiling ducts and wall and floor partitions.

Working with asbestos materials is not a job for anyone in the electrical industry. If asbestos is present in situations or buildings where you are expected to work, it should be removed by a specialist contractor before your work commences. Specialist contractors, who will wear fully protective suits and use breathing apparatus, are the only people who can safely and responsibly carry out the removal of asbestos. They will wrap the asbestos in thick plastic bags and store the bags temporarily in a covered and locked skip. This material is then disposed of in a special land-fill site with other toxic industrial waste materials and the site monitored by the local authority for the foreseeable future.

There is a lot of work for electrical contractors in my part of the country, updating and improving the lighting in government buildings and schools. This work often involves removing the old fluorescent fittings hanging on chains or fixed to beams, and installing a suspended ceiling and an appropriate number of recessed modular fluorescent fittings. So what do we do with the old fittings? Well, the fittings are made of sheet steel, a couple of plastic lamp holders, a little cable, a starter and ballast. All of these materials can go into the ordinary skip. However, the fluorescent tubes contain a little mercury and fluorescent powder with toxic elements, which cannot be disposed of in the normal land-fill sites.

Hazardous Waste Regulations 2005

New Hazardous Waste Regulations were introduced in July 2005 and under these regulations lamps and tubes are classified as hazardous. While each lamp contains only a small amount of mercury, vast numbers of lamps and tubes are disposed of in the United Kingdom every year, resulting in a significant environmental threat.

The environmentally responsible way to dispose of fluorescent lamps and tubes is to recycle them.

The process usually goes like this:

- your employer arranges for the local electrical wholesaler to deliver a plastic used lamp waste container of an appropriate size for the job;
- expired lamps and tubes are placed whole into the container, which often has a grating inside to prevent the tubes from breaking when being transported;
- when the container is full of used lamps and tubes, you telephone the electrical wholesaler and ask them to pick up the filled container and deliver it to one of the specialist recycling centres;
- your electrical company will receive a 'Duty of Care Note' and full recycling documents which should be filed safely as proof that the hazardous waste was recycled safely;
- the charge is approximately 50p for each 1800 mm tube and this cost is passed on to the customer through the final account.

Safety first

Waste

- clean up before you leave the job;
- put waste in the correct skip;
- recycle used lamps and tubes;
- get rid of all waste responsibly.

Safe working procedures

The principles which were laid down in the many Acts of Parliament and the regulations that we have already looked at in this chapter control our working environment. They make our workplace safer, but despite all this legislation, workers continue to be injured and killed at work or die as a result of a work-related injury. The number of deaths has consistently averaged about 200 each year for the past eight years. These figures only relate to employees. If you include the self-employed and members of the public killed in work-related accidents, the numbers almost double.

In addition to the deaths, about 28,000 people have major accidents at work and about 130,000 people each year receive minor work-related injuries which keep them off work for more than three days.

It is a mistake to believe that these things only happen in dangerous occupations such as deep-sea diving, mining and quarrying, fishing industry, tunnelling and fire-fighting, or that they only happen in exceptional circumstances such as would never occur in your workplace. This is not the case. Some basic thinking and acting beforehand could have prevented most of these accident statistics from happening.

Causes of accidents

Most accidents are caused by either human error or environmental conditions. **Human errors** include behaving badly or foolishly, being careless and not paying attention to what you should be doing at work, doing things that you are not competent to do or have not been trained to do. You should not work when tired or fatigued and you should never work when you have been drinking alcohol or taking drugs.

Environmental conditions include unguarded or faulty machinery, damaged or faulty tools and equipment, poorly illuminated or ventilated workplaces and untidy, dirty or overcrowded workplaces.

Definition

Human errors include behaving badly or foolishly, being careless and not paying attention to what you should be doing at work.

Definition

Environmental conditions include unguarded or faulty machinery.

Safety first

Safety procedures

- Hazard risk assessment *is an essential part* of any health and safety management system.
- The aim of the planning process is to minimize risk.

HSE publication HSG(65)

The most common causes of accidents

These are:

- slips, trips and falls;
- manual handling, that is, moving objects by hand;
- using equipment, machinery or tools;
- storage of goods and materials which then become unstable;
- fire;
- electricity;
- mechanical handling.

Accident prevention measures

To control the risk of an accident we usually:

- eliminate the cause;
- substitute a procedure or product with less risk;
- enclose the dangerous situation;
- put guards around the hazard;

- use safe systems of work;
- supervise, train and give information to staff;
- if the hazard cannot be removed or minimized, provide PPE.

In other chapters of this book we will look at the application of procedures that make the workplace a safer place to work.

Site visitors

A construction site will have many people working on a building project throughout the construction period. The groundwork people lay the foundations, the steel work is erected, the bricklayers build the walls, the carpenters put on the roof and, only when the building is waterproof do the electricians and mechanical service trades begin to install the electrical and mechanical systems.

If there was an emergency and the site had to be evacuated, how would you know who or how many people were on-site? You can see that there has to be a procedure for logging people in and out, so that the main contractor can identify who is on-site at any one time.

If the architect pops in to make a quick inspection or resolve a problem, he must first 'sign in'. If the managing director of your electrical company drops by to see how work is progressing, he must first 'sign in'. When they leave, they must 'sign out'. How else would you know who is on-site at any one time?

A formal site visitor procedure must be in place to meet the requirements of the health and safety regulations and to maintain site security.

The site visitor procedure must meet the needs of the building project. For example, a high-profile building project, one which pushes the architectural boundaries such as the new Wembley Football Stadium in London, would have attracted many more visitors during its construction period than, say, a food distribution warehouse for a supermarket.

The site visitor procedure is only required to be 'fit for purpose', but whatever system is put in place, here are some suggestions for consideration:

- you will want your visitors to be safe when they are on-site and so you might insist that they wear a hard hat and a high-visibility jacket;
- in some cases, you might want 'ordinary people' visitors to be escorted for their own safety while on-site. Of course this would not apply to professional visitors such as the architect or quantity surveyor;
- the procedure should identify who the visitor is and who they wish to visit;
- you will want to know the time of their arrival and the time of their departure.

Many of these requirements can be met with a simple log-book divided into columns with headings such as date and time in, visitor's name, company name, name of the person to be visited and time out or time of leaving.

Check your understanding

When you have completed the questions, check out the answers at the back of the book.

Note: more than one multiple-choice answer may be correct.

1 For any fire to continue to burn, three components must be present. These are
   a. fuel, wood and cardboard
   b. petrol, oxygen and bottled gas
   c. flames, fuel and heat
   d. fuel, oxygen and heat.

2 A water fire extinguisher is suitable for use on small fires of burning
   a. wood, paper and fabric
   b. flammable liquids
   c. flammable gas
   d. all of the above.

3 A foam fire extinguisher is suitable for use on small fires of burning
   a. wood, paper and fabric
   b. flammable liquids
   c. flammable gas
   d. all of the above.

4 A carbon dioxide gas fire extinguisher is suitable for use on small fires of burning
   a. wood, paper and fabric
   b. flammable liquids
   c. flammable gas
   d. all of the above.

5 A dry powder fire extinguisher is suitable for use on small fires of burning
   a. wood, paper and fabric
   b. flammable liquids
   c. flammable gas
   d. all of the above.

6 A vaporizing foam fire extinguisher is suitable for use on small fires of burning
   a. wood, paper and fabric
   b. flammable liquids
   c. flammable gas
   d. all of the above.

7 You should only attack a fire with a fire extinguisher if
   a. it is burning brightly
   b. you can save someone's property
   c. you can save someone's life
   d. you can do so without putting yourself at risk.

8 A fire extinguisher should only be used to fight
   a. car fires
   b. electrical fires
   c. small fires
   d. big fires.

9 A Statutory Regulation
   a. is the law of the land
   b. must be obeyed
   c. tells us how to comply with the law
   d. is a code of practice.

10 A Non-statutory Regulation
   a. is the law of the land
   b. must be obeyed
   c. tells us how to comply with the law
   d. is a code of practice.

11 Under the Health and Safety at Work Act an employer is responsible for
   a. maintaining plant and equipment
   b. providing PPE
   c. wearing PPE
   d. taking reasonable care to avoid injury.

12 Under the Health and Safety at Work Act an employee is responsible for
   a. maintaining plant and equipment
   b. providing PPE
   c. wearing PPE
   d. taking reasonable care to avoid injury.

13 The IET Wiring Regulations
   a. are statutory regulations
   b. are non-statutory regulations
   c. are codes of good practice
   d. must always be complied with.

14 Before beginning work on a 'live' circuit or piece of equipment you should
   a. only work 'live' if your supervisor is with you
   b. only work 'live' if you feel that you are 'competent' to do so
   c. isolate the circuit or equipment before work commences
   d. secure the isolation before work commences.

15 The initial assistance or treatment given to a casualty for any injury or sudden illness before the arrival of an ambulance or medically qualified person is one definition of
   a. an appointed person
   b. a first aider

c. first aid
d. an adequate first aid facility.

16 Someone who has undergone a training course to administer medical aid at work and holds a current qualification is one definition of
a. a doctor
b. a nurse
c. a first aider
d. a supervisor.

17 Two of the most common categories of risk and causes of accidents at work are
a. slips, trips and falls
b. put guards around the hazard
c. manual handling
d. use safe systems of work.

18 Two of the most common precautions taken to control risks are
a. slips, trips and falls
b. put guards around the hazard
c. manual handling
d. use safe systems of work

19 Something which has the potential to cause harm is one definition of
a. health and safety
b. risk
c. competent person
d. hazard.

20 The chances of harm actually being done is one definition of
a. electricity
b. risk
c. health and safety
d. hazard.

21 A competent person dealing with a hazardous situation
a. must wear appropriate PPE
b. display a health and safety poster
c. reduces the risk
d. increases the risk.

22 Employers of companies employing more than five people must
a. become a member of the NICEIC
b. provide PPE if appropriate
c. carry out a hazard risk assessment
d. display a health and safety poster.

23 There are five parts to a hazard risk assessment procedure. Identify one from the list below:
a. wear appropriate PPE
b. notify the HSE that you intend to carry out a risk assessment
c. list the hazards and who might be harmed
d. substitute a procedure with less risk.

24 Lifting, transporting or supporting heavy objects by hand or bodily force is one definition of
 a. working at height
 b. a mobile scaffold tower
 c. a sack truck
 d. manual handling.

25 When working above ground for long periods of time the most appropriate piece of equipment to use would be
 a. a ladder
 b. a trestle scaffold
 c. a mobile scaffold tower
 d. a pair of sky hooks.

26 The most appropriate piece of equipment to use for gaining access to a permanent scaffold would be
 a. a ladder
 b. a trestle scaffold
 c. a mobile scaffold tower
 d. a pair of sky hooks.

27 The Electricity at Work Regulations tell us that 'we must ensure the disconnection and separation of electrical equipment from every source of supply and the separation must be secure'. A procedure to comply with this regulation is called
 a. work at height
 b. a hazard risk assessment
 c. a safe isolation procedure
 d. a workstation risk assessment.

28 The Electricity at Work Regulations absolutely forbid the following work activity:
 a. working at height
 b. testing live electrical systems
 c. live repair work on electrical circuits
 d. working without the appropriate PPE.

29 'Good housekeeping' at work is about
 a. cleaning up and putting waste in the skip
 b. working safely
 c. making the tea and collecting everyone's lunch
 d. putting tools and equipment away after use.

30 Use bullet points to describe a safe isolation procedure of a 'live' electrical circuit.

31 How does the law enforce the regulations of the Health and Safety at Work Act?

32 List the responsibilities under the Health and Safety at Work Act of
 a. an employer to an employee
 b. an employee to an employer.

33 Safety signs are used in the working environment to give information and warning. Sketch and colour one sign from each of the four categories of signs and state the message given by that sign.

34 State the name of two important Statutory Regulations and one Non-statutory Regulation relevant to the electrical industry.

35 Define what is meant by PPE.

36 State five pieces of PPE which a trainee could be expected to wear at work and the protection given by each piece.

37 Describe the action to be taken upon finding a workmate apparently dead on the floor and connected to an electrical supply.

38 State how the Data Protection Act has changed the way in which we record accident and first aid information at work.

39 List five common categories of risk.

40 List five common precautions which might be taken to control risk.

41 Use bullet points to list the main stages involved in lifting a heavy box from the floor, carrying it across a room and placing it on a worktop, using a safe manual handling technique.

42 Describe a safe manual handling technique for moving a heavy electric motor out of the stores, across a yard and into the back of a van for delivery to site.

43 Use bullet points to list a step-by-step safe electrical isolation procedure for isolating a circuit in a three-phase distribution fuse board.

44 Use bullet points to list each stage in the erection and securing of a long extension ladder. Identify all actions which would make the ladder safe to use.

45 Describe how you would use a mobile scaffold tower to re-lamp all the light fittings in a supermarket. Use bullet points to give a step-by-step account of re-lamping the first two fittings.

46 What is a proving unit used for?

47 The HSE Guidance Note GS 38 tells us about suitable test probe leads. Use a sketch to identify the main recommendations.

48 State how you would deal with the following materials when you are cleaning up at the end of the job:

- pieces of conduit and tray
- cardboard packaging material
- empty cable rolls
- half-full cable rolls
- bending machines for conduit and tray
- your own box of tools
- your employer's power tools
- 100 old fluorescent light fittings
- 200 used fluorescent tubes.

CHAPTER 2

C&G Unit 202

Principles of electrical science

Unit 202 of the City and Guilds 2365–02 syllabus

Learning outcomes – when you have completed this chapter you should:

- understand the principles of electricity;
- understand basic electrical circuits;
- understand the principles of electromagnetism;
- understand basic mechanics;
- know the electrical quantities in star and delta connections;
- describe the operating principle of motors, transformers and luminaires;
- understand a.c. theory.

Basic Electrical Installation Work. 978-1-138-84884-9

This chapter has free associated content, including animations and instructional videos, to support your learning.

When you see the logo, visit the companion website for more on this topic www.routledge.com/cw/linsley

In this chapter we look at the electrical scientific principles and competencies required of a qualified electrician.

Units

Very early units of measurement were based on the things easily available – the length of a stride, the distance from the nose to the outstretched hand, the weight of a stone and the time-lapse of one day. Over the years, new units were introduced and old ones modified. Different branches of science and engineering were working in isolation, using their own units, and the result was an overwhelming variety of units.

Definition

SI units are based upon a small number of fundamental units from which all other units may be derived.

In all branches of science and engineering there is a need for a practical system of units which everyone can use. In 1960, the General Conference of Weights and Measures agreed to an international system called the Système International d'Unités (abbreviated to **SI units**).

SI units are based upon a small number of fundamental units from which all other units may be derived. Table 2.1 describes some of the basic units we shall be using in our electrical studies.

Like all metric systems, SI units have the advantage that prefixes representing various multiples or submultiples may be used to increase or decrease the size of the unit by various powers of 10. Some of the more common prefixes and their symbols are shown in Table 2.2.

Basic circuit theory

Definition

A *conductor* is a material in which the electrons are loosely bound to the central nucleus and are, therefore, free to drift around the material at random from one atom to another, as shown in Fig. 2.1(a). Materials which are good conductors include copper, brass, aluminium and silver.

Definition

An *insulator* is a material in which the outer electrons are tightly bound to the nucleus, so that there are no free electrons to move around the material. Good insulating materials are PVC, rubber, glass and wood.

All matter is made up of atoms which arrange themselves in a regular framework within the material. The atom is made up of a central, positively charged nucleus, surrounded by negatively charged electrons. The electrical properties of a material depend largely upon how tightly these electrons are bound to the central nucleus.

A **conductor** is a material in which the electrons are loosely bound to the central nucleus and are, therefore, free to drift around the material at random from one atom to another, as shown in Fig. 2.1(a). Materials which are good conductors include copper, brass, aluminium and silver.

An **insulator** is a material in which the outer electrons are tightly bound to the nucleus, so that there are no free electrons to move around the material. Good insulating materials are PVC, rubber, glass and wood.

If a battery is attached to a conductor as shown in Fig. 2.1(b), the free electrons drift purposefully in one direction only. The free electrons close to the positive plate of the battery are attracted to it since unlike charges attract, and the free electrons near the negative plate will be repelled from it. For each electron entering the positive terminal of the battery, one will be ejected from the negative terminal, so the number of electrons in the conductor remains constant.

Table 2.1 Basic SI units

Quantity	Measure of	Basic unit	Symbol	Notes
Area	Length × length	Metre squared	m^2	
Current I	Electric current	Ampere	A	
Energy	Ability to do work	Joule	J	Joule is a very small unit 3.6×10^6 J = 1 kWh
Force	The effect on a body	Newton	N	
Frequency	Number of cycles	Hertz	Hz	Mains frequency is 50 Hz
Length	Distance	Metre	m	
Mass	Amount of material	Kilogram	Kg	One metric tonne = 1000 kg
Magnetic flux Φ	Magnetic energy	Weber	Wb	
Magnetic flux density B	Number of lines of magnetic flux	Tesla	T	
Potential or pressure	Voltage	Volt	V	
Period T	Time taken to complete one cycle	Second	s	The 50 Hz mains supply has a period of 20 ms
Power	Rate of doing work	Watt	W	
Resistance	Opposition to current flow	Ohm	Ω	
Resistivity	Resistance of a sample piece of material	Ohm metre	ρ	Resistivity of copper is 17.5×10^{-9} Ωm
Temperature	Hotness or coldness	Kelvin	K	0°C = 273 K. A change of 1 K is the same as 1°C
Time	Time	Second	s	60 s = 1 min 60 min = 1 h
Weight	Force exerted by a mass	Kilogram	kg	1000 kg = 1 tonne

Note: A more detailed description may be found in this chapter.

Table 2.2 Symbols and multiples for use with SI units

Prefix	Symbol	Multiplication factor		
Mega	M	$\times 10^6$	or	×1,000,000
Kilo	k	$\times 10^3$	or	×1000
Hecto	h	$\times 10^2$	or	×100
Deca	da	×10	or	×10
Deci	d	$\times 10^{-1}$	or	÷10
Centi	c	$\times 10^{-2}$	or	÷100
Milli	m	$\times 10^{-3}$	or	÷1000
Micro	μ	$\times 10^{-6}$	or	÷1,000,000

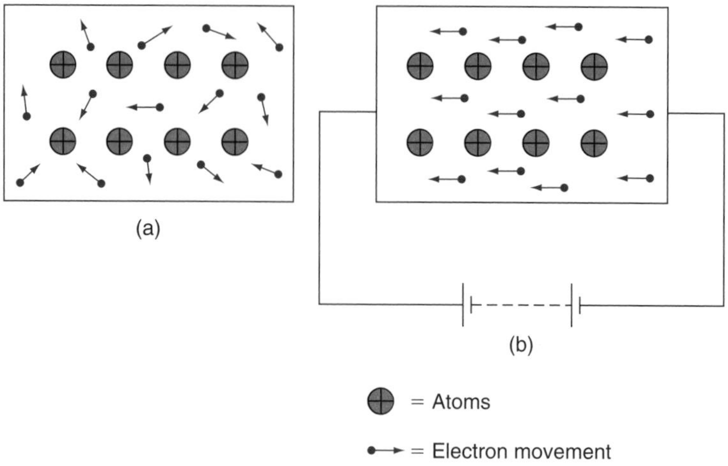

Figure 2.1 Atoms and electrons on a material.

Visit the companion website for more on this topic.

Definition

The drift of electrons within a conductor is known as an *electric current*, measured in amperes and given the symbol *I*.

Definition

The *potential difference* (p.d.) is the change in energy levels measured across the load terminals. This is also called the volt drop or terminal voltage, since e.m.f. and p.d. are both measured in volts.

Definition

Every circuit offers some opposition to current flow, which we call the circuit *resistance,* measured in ohms (symbol Ω).

Definition

Ohm's law, which says that the current passing through a conductor under constant temperature conditions is proportional to the potential difference across the conductor.

Visit the companion website for more on this topic.

The drift of electrons within a conductor is known as an **electric current**, measured in amperes and given the symbol *I*.

For a current to continue to flow, there must be a complete circuit for the electrons to move around. If the circuit is broken by opening a switch, for example, the electron flow and therefore the current will stop immediately.

To cause a current to flow continuously around a circuit, a driving force is required, just as a circulating pump is required to drive water around a central heating system. This driving force is the *electromotive force* (e.m.f.). Each time an electron passes through the source of e.m.f., more energy is provided to send it on its way around the circuit.

An e.m.f. is always associated with energy conversion, such as chemical to electrical in batteries and mechanical to electrical in generators. The energy introduced into the circuit by the e.m.f. is transferred to the load terminals by the circuit conductors.

The **potential difference** (p.d.) is the change in energy levels measured across the load terminals. This is also called the volt drop or terminal voltage, since e.m.f. and p.d. are both measured in volts. **Resistance** in every circuit offers some opposition to current flow, which we call the circuit *resistance*, measured in ohms (symbol Ω), to commemorate the famous German physicist Georg Simon Ohm, who was responsible for the analysis of electrical circuits.

Ohm's law

In 1826, Ohm published details of an experiment he had done to investigate the relationship between the current passing through and the potential difference between the ends of a wire. As a result of this experiment he arrived at a law, now known as **Ohm's law**, which says that the current passing through a conductor under constant temperature conditions is proportional to the potential difference across the conductor. This may be expressed mathematically as

$$V = I \times R (V)$$

Transposing this formula, we also have

$$I = \frac{V}{R} (A) \quad \text{and} \quad R = \frac{V}{I} (\Omega)$$

Example 1

An electric heater, when connected to a 230 V supply, was found to take a current of 4 A. Calculate the element resistance.

$$R = \frac{V}{I}$$

$$\therefore R = \frac{230\ \text{V}}{4\ \text{A}} = 57.5\,\Omega$$

Example 2

The insulation resistance measured between phase conductors on a 400 V supply was found to be 2 MΩ. Calculate the leakage current.

$$I = \frac{V}{R}$$

$$\therefore I = \frac{400\ \text{V}}{2 \times 10^{6}\ \Omega} = 200 \times 10^{-6}\ \text{A} = 200\ \mu\text{A}$$

Example 3

When a 4 Ω resistor was connected across the terminals of an unknown d.c. supply, a current of 3 A flowed. Calculate the supply voltage.

$$V = I \times R$$

$$\therefore V = 3\text{A} \times 4\Omega = 12\text{V}$$

Resistivity

The resistance or opposition to current flow varies for different materials, each having a particular constant value. If we know the resistance of, say, 1 m of a material, then the resistance of 5 m will be five times the resistance of 1 m.

The **resistivity** (symbol ρ – the Greek letter 'rho') of a material is defined as the resistance of a sample of unit length and unit cross-section. Typical values are given in Table 2.3. Using the constants for a particular material we can calculate the resistance of any length and thickness of that material from the equation

$$R = \frac{\rho l}{a}\,(\Omega)$$

where

ρ = the resistivity constant for the material (Ω m)
l = the length of the material (m)
a = the cross-sectional area of the material (m^2).

Table 2.3 gives the resistivity of silver as $16.4 \times 10^{-9}\,\Omega$m, which means that a sample of silver 1 m long and 1 m in cross-section will have a resistance of $16.4 \times 10^{-9}\,\Omega$.

Definition

The *resistivity* (symbol ρ – the Greek letter 'rho') of a material is defined as the resistance of a sample of unit length and unit cross-section.

Table 2.3 Resistivity values

Material	Resistivity (Ω m)
Silver	16.4×10^{-9}
Copper	17.5×10^{-9}
Aluminium	28.5×10^{-9}
Brass	75.0×10^{-9}
Iron	100.0×10^{-9}

Visit the companion website for more on this topic.

Key fact

Resistance: If the length is doubled, resistance will be doubled. If the csa is doubled, the resistance will be halved.

Example 1

Calculate the resistance of 100 m of copper cable of 1.5 mm² cross-sectional area (csa) if the resistivity of copper is taken as $17.5 \times 10^{-9}\,\Omega$ m.

$$R = \frac{\rho l}{a}(\Omega)$$

$$\therefore R = \frac{17.5 \times 10^{-9}\ \Omega\ \text{m} \times 100\ \text{m}}{1.5 \times 10^{-6}\ \text{m}^2} = 1.16\ \Omega$$

Example 2

Calculate the resistance of 100 m of aluminium cable of 1.5 mm² cross-sectional area if the resistivity of aluminium is taken as $28.5 \times 10^{-9}\,\Omega$ m.

$$R = \frac{\rho l}{a}(\Omega)$$

$$\therefore R = \frac{28.5 \times 10^{-9}\ \Omega\ \text{m} \times 100\ \text{m}}{1.5 \times 10^{-6}\ \text{m}^2} = 1.9\ \Omega$$

The above examples show that the resistance of an aluminium cable is some 60% greater than a copper conductor of the same length and cross-section. Therefore, if an aluminium cable is to replace a copper cable, the conductor size must be increased to carry the rated current as given by the tables in Appendix 4 of the IET Regulations and Appendix 6 of the On Site Guide.

The other factor which affects the resistance of a material is the temperature, and we will consider this later.

Try this

Resistance

- Take two 100 m lengths of single cable (2 coils)
- Measure the resistance of 100 m of cable (1 coil)
- Value ______________________________ Ω
- Join the two lengths together (200 m) and again measure the resistance
- Value ______________________________ Ω

- Does this experiment prove resistance is proportional to length?
- If the resistance is doubled, it is proved (QED)! (QED, *quod erat demonstradum*) is the Latin for 'which was to be demonstrated'.)

Basic mechanics and machines

Mechanics is the scientific study of 'machines', where a machine is defined as a device which transmits motion or force from one place to another. An engine is one particular type of machine, an energy-transforming machine, converting fuel energy into a more directly useful form of work.

Most modern machines can be traced back to the five basic machines described by the Greek inventor Hero of Alexandria who lived at about the time of Christ. The machines described by him were the wedge, the screw, the wheel and axle, the pulley, and the lever. Originally they were used for simple purposes, to raise water and to move objects which man alone could not lift, but today their principles are of fundamental importance to our scientific understanding of mechanics. Let us now consider some fundamental mechanical principles and calculations.

Definition

Mechanics is the scientific study of 'machines', where a machine is defined as a device which transmits motion or force from one place to another.

Mass

Mass is a measure of the amount of material in a substance, such as metal, plastic, wood, brick or tissue, which is collectively known as a body. The mass of a body remains constant and can easily be found by comparing it on a set of balance scales with a set of standard masses. The SI unit of mass is the kilogram (kg).

Definition

Mass is a measure of the amount of material in a substance, such as metal, plastic, wood, brick or tissue, which is collectively known as a body. The mass of a body remains constant and can easily be found by comparing it on a set of balance scales with a set of standard masses. The SI unit of mass is the kilogram (kg).

Weight

Weight is a measure of the force which a body exerts on anything which supports it. Normally it exerts this force because it is being attracted towards the Earth by the force of gravity.

For scientific purposes the weight of a body is *not* constant, because gravitational force varies from the Equator to the Poles; in space a body would be 'weightless' but here on Earth under the influence of gravity a 1 kg mass would have a weight of approximately 9.81 N (see also the definition of 'force').

Definition

Weight is a measure of the force which a body exerts on anything which supports it. Normally it exerts this force because it is being attracted towards the Earth by the force of gravity.

Speed

The feeling of speed is something with which we are all familiar. If we travel in a motor vehicle we know that an increase in speed would, excluding accidents, allow us to arrive at our destination more quickly. Therefore, **speed** is concerned with distance travelled and time taken. Suppose we were to travel a distance of 30 miles in one hour; our speed would be an average of 30 miles/h:

$$\text{Speed} = \frac{\text{Distance (m)}}{\text{Time (s)}}$$

Definition

Speed is concerned with distance travelled and time taken.

Velocity

In everyday conversation we often use the word **velocity** to mean the same as speed, and indeed the units are the same. However, for scientific purposes this is

Definition

In everyday conversation we often use the word *velocity* to mean the same as speed, and indeed the units are the same. However, for scientific purposes this is not acceptable since velocity is also concerned with direction.

not acceptable since velocity is also concerned with direction. Velocity is speed in a given direction. For example, the speed of an aircraft might be 200 miles/h, but its velocity would be 200 miles/h in, say, a westerly direction. Speed is a scalar quantity, while velocity is a vector quantity.

$$\text{Velocity} = \frac{\text{Distance (m)}}{\text{Time (s)}}$$

Definition

By definition, *acceleration* is the rate of change in velocity with time.

Acceleration

When an aircraft takes off, it starts from rest and increases its velocity until it can fly. This change in velocity is called its acceleration. By definition, **acceleration** is the rate of change in velocity with time.

$$\text{Acceleration} = \frac{\text{Velocity}}{\text{Time}} = (\text{m/s}^2)$$

Example

If an aircraft accelerates from a velocity of 15 m/s to 35 m/s in 4 s, calculate its average acceleration.

Average velocity = 35 m/s – 15 m/s = 20 m/s

$$\text{Average acceleration} = \frac{\text{Velocity}}{\text{Time}} = \frac{20}{4} = 5\ \text{m/s}^2$$

Thus, the average acceleration is 5 m/s^2.

Definition

The presence of a *force* can only be detected by its effect on a body. A force may cause a stationary object to move or bring a moving body to rest.

Force

The presence of a **force** can only be detected by its effect on a body. A force may cause a stationary object to move or bring a moving body to rest. For example, a number of people pushing a broken-down motor car exert a force which propels it forward, but applying the motor car brakes applies a force on the brake drums which slows down or stops the vehicle. Gravitational force causes objects to fall to the ground. The apple fell from the tree on to Isaac Newton's head as a result of gravitational force. The standard rate of acceleration due to gravity is accepted as 9.81 m/s^2. Therefore, an apple weighing 1 kg will exert a force of 9.81 N, since

$$\text{Force} = \text{Mass} \times \text{Acceleration (N)}$$

The SI unit of force is the newton, symbol N, to commemorate the great English scientist Sir Isaac Newton (1642–1727).

Example

A 50 kg bag of cement falls from a fork lift truck while being lifted on to a storage shelf. Determine the force with which the bag will strike the ground:

$$\text{Force} = \text{Mass} \times \text{Acceleration (N)}$$

$$\text{Force} = 50\ \text{kg} \times 9.81\ \text{m/s}^2 = 490.5\ \text{N}$$

A force can manifest itself in many different ways. Let us consider a few examples:

- 'inertial force' is the force required to get things moving, to change direction or stop, like the motor car discussed above;
- 'cohesive or adhesive force' is the force required to hold things together;
- 'tensile force' is the force pulling things apart;
- 'compressive force' is the force pushing things together;
- 'friction force' is the force which resists or prevents the movement of two surfaces in contact;
- 'shearing force' is the force which moves one face of a material over another;
- 'centripetal force' is the force acting towards the centre when a mass attached to a string is rotated in a circular path;
- 'centrifugal force' is the force acting away from the centre, the opposite to centripetal force;
- 'gravitational force' is the force acting towards the centre of the Earth due to the effect of gravity;
- 'magnetic force' is the force created by a magnetic field;
- 'electrical force' is the force created by an electrical field.

Definition

- 'Inertial force' is the force required to get things moving, to change direction or stop, like the motor car discussed above.
- 'Cohesive or adhesive force' is the force required to hold things together.
- 'Tensile force' is the force pulling things apart.
- 'Compressive force' is the force pushing things together.
- 'Friction force' is the force which resists or prevents the movement of two surfaces in contact.
- 'Shearing force' is the force which moves one face of a material over another.
- 'Centripetal force' is the force acting towards the centre when a mass attached to a string is rotated in a circular path.
- 'Centrifugal force' is the force acting away from the centre, the opposite to centripetal force.
- 'Gravitational force' is the force acting towards the centre of the earth due to the effect of gravity.
- 'Magnetic force' is the force created by a magnetic field.
- 'Electrical force' is the force created by an electrical field.

Pressure or stress

To move a broken-down motor car I might exert a force on the back of the car to propel it forward. My hands would apply a pressure on the body panel at the point of contact with the car. **Pressure** or **stress** is a measure of the force per unit area.

$$\text{Pressure or stress} = \frac{\text{Force}}{\text{Area}} (\text{N/m}^2)$$

Definition

Pressure or *stress* is a measure of the force per unit area.

Example 1

A young woman of mass 60 kg puts all her weight on to the heel of one shoe which has an area of 1 cm^2. Calculate the pressure exerted by the shoe on the floor (assuming the acceleration due to gravity to be 9.81 m/s^2).

$$\text{Pressure} = \frac{\text{Force}}{\text{Area}} (\text{N/m}^2)$$

$$\text{Pressure} = \frac{60\,\text{kg} \times 9.81\,\text{m/s}^2}{1 \times 10^{-4}\,\text{m}^2} = 5886\,\text{kN/m}^2$$

Example 2

A small circus elephant of mass 1 tonne (1000 kg) puts all its weight on to one foot which has a surface area of 400 cm^2. Calculate the pressure exerted by the elephant's foot on the floor, assuming the acceleration due to gravity to be 9.81 m/s^2.

(continued)

Example 2 (continued)

$$\text{Pressure} = \frac{\text{Force}}{\text{Area}} (\text{N/m}^2)$$

$$\text{Pressure} = \frac{1000\,\text{kg} \times 9.81\,\text{m/s}^2}{400 \times 10^{-4}\,\text{m}^2} = 245.3\,\text{kN/m}^2$$

These two examples show that the young woman exerts 24 times more pressure on the ground than the elephant. This is because her mass exerts a force over a much smaller area than the elephant's foot and is the reason why many wooden dance floors are damaged by high-heeled shoes.

Work done

Definition

By definition, *work done* is dependent upon the force applied times the distance moved in the direction of the force.

Suppose a broken-down motor car was to be pushed along a road. Work would be done on the car by applying the force necessary to move it along the road. Heavy breathing and perspiration would be evidence of the work done.

By definition, **work done** is dependent upon the force applied times the distance moved in the direction of the force.

$$\text{Work done} = \text{Force} \times \text{Distance moved in the direction of the force (J)}$$

The SI unit of work done is the newton metre or joule (symbol J). The joule is the preferred unit and it commemorates an English physicist, James Prescott Joule (1818–1889).

Example

A building hoist lifts ten 50 kg bags of cement through a vertical distance of 30 m to the top of a high-rise building. Calculate the work done by the hoist, assuming the acceleration due to gravity to be 9.81 m/s^2.

$$\begin{aligned} \text{Work done} &= \text{Force} \times \text{Distance moved (J)} \\ \text{but force} &= \text{Mass} \times \text{Acceleration (N)} \\ \therefore \text{Work done} &= \text{Mass} \times \text{Acceleration} \times \text{Distance moved (J)} \\ \text{Work done} &= 10 \times 50\,\text{kg} \times 9.81\,\text{m/s}^2 \times 30\,\text{m} \\ \text{Work done} &= 147.15\,\text{kJ} \end{aligned}$$

Power

Definition

By definition, *power* is the rate of doing work.

If one motor car can cover the distance between two points more quickly than another car, we say that the faster car is more powerful. It can do a given amount of work more quickly. By definition, **power** is the rate of doing work.

$$\text{Power} = \frac{\text{Work done}}{\text{Time taken}} (\text{W})$$

The SI unit of power, both electrical and mechanical, is the watt (symbol W). This commemorates the name of James Watt (1736–1819), the inventor of the steam-engine.

Example 1

A building hoist lifts ten 50 kg bags of cement to the top of a 30 m-high building. Calculate the rating (power) of the motor to perform this task in 60 s if the acceleration due to gravity is taken as 9.81 m/s^2.

$$\text{Power} = \frac{\text{Work done}}{\text{Time taken}}(\text{W})$$

$$\text{but work done} = \text{Force} \times \text{Distance moved (J)}$$

$$\text{and force} = \text{Mass} \times \text{Acceleration (N)}$$

By substitution,

$$\text{Power} = \frac{\text{Mass} \times \text{Acceleration} \times \text{Distance moved}}{\text{Time taken}}(\text{W})$$

$$\text{Power} = \frac{10 \times 50\text{ kg} \times 9.81\text{ m/s}^2 \times 30\text{ m}}{60\text{ s}}$$

$$\text{Power} = 2452.5\text{ W}$$

The rating of the building hoist motor will be 2.45 kW.

Example 2

A hydroelectric power-station pump motor working continuously during a seven-hour period raises 856 tonnes of water through a vertical distance of 60 m. Determine the rating (power) of the motor, assuming the acceleration due to gravity is 9.81 m/s^2.

From Example 1,

$$\text{Power} = \frac{\text{Mass} \times \text{Acceleration} \times \text{Distance moved}}{\text{Time taken}}(\text{W})$$

$$\text{Power} = \frac{856 \times 1000\text{ kg} \times 9.81\text{ m/s}^2 \times 60\text{ m}}{7 \times 60 \times 60\text{ s}}$$

$$\text{Power} = 20{,}000\text{ W}$$

The rating of the pump motor is 20 kW.

Example 3

An electric hoist motor raises a load of 500 kg at a velocity of 2 m/s. Calculate the rating (power) of the motor if the acceleration due to gravity is 9.81 m/s^2.

$$\text{Power} = \frac{\text{Mass} \times \text{Acceleration} \times \text{Distance moved}}{\text{Time taken}}(\text{W})$$

$$\text{but Velocity} = \frac{\text{Distance}}{\text{Time}}(\text{m/s})$$

(continued)

Example 3 (continued)

$$\therefore \text{Power} = \text{Mass} \times \text{Acceleration} \times \text{Velocity}$$
$$\text{Power} = 500\,\text{kg} \times 9.81\,\text{m/s}^2 \times 2\,\text{m/s}$$
$$\text{Power} = 9810\,\text{W}.$$

The rating of the hoist motor is 9.81 kW.

Levers and turning force

Definition

A *lever* allows a heavy load to be lifted or moved by a small effort.

Definition

A *lever* is any rigid body which pivots or rotates about a fixed axis or fulcrum.

A **lever** allows a heavy load to be lifted or moved by a small effort. Every time we open a door, turn on a tap or tighten a nut with a spanner, we exert a lever-action turning force. A **lever** is any rigid body which pivots or rotates about a fixed axis or fulcrum. The simplest form of lever is the crowbar, which is useful because it enables a person to lift a load at one end which is greater than the effort applied through his or her arm muscles at the other end. In this way the crowbar is said to provide a 'mechanical advantage'. A washbasin tap and a spanner both provide a mechanical advantage through the simple lever action. The mechanical advantage of a simple lever is dependent upon the length of lever on either side of the fulcrum. Applying the principle of turning forces to a lever, we obtain the formula:

$$\text{Load force} \times \text{Distance from fulcrum} = \text{Effort force} \times \text{Distance from fulcrum}$$

This formula can perhaps better be understood by referring to Fig. 2.3. A small effort at a long distance from the fulcrum can balance a large load at a short distance from the fulcrum. Thus a 'turning force' or 'turning moment' depends upon the distance from the fulcrum and the magnitude of the force.

Figure 2.2 An example of a lever in use.

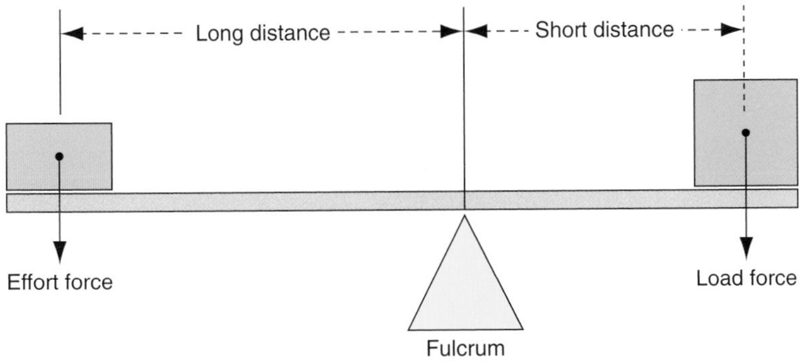

Figure 2.3 Turning forces of a simple lever.

Example

Calculate the effort required to raise a load of 500 kg when the effort is applied at a distance of five times the load distance from the fulcrum (assume the acceleration due to gravity to be 10 m/s^2).

$$\text{Load force} = \text{Mass} \times \text{Acceleration (N)}$$

$$\text{Load force} = 500\text{ kg} \times 10\text{ m/s}^2 = 5000\text{ N}$$

$$\text{Load force} \times \text{Distance from fulcrum} = \text{Effort force} \times \text{Distance from fulcrum}$$

$$5000\text{ N} \times 1\text{ m} = \text{Effort force} \times 5\text{ m}$$

$$\therefore \text{Effort force} = \frac{5000\text{ N} \times 1\text{ m}}{5\text{ m}} = 1000\text{ N}$$

Thus an effort force of 1000 N can overcome a load force of 5000 N using the mechanical advantage of this simple lever.

Levers are grouped into three classes which depend upon the position of the fulcrum or pivot point. Examples of the three classes are shown in Fig. 2.4.

Class 1 lever – the fulcrum is in the middle, between the force and the load. A crowbar and a sack truck are good examples of a class 1 lever. A sack truck is shown in Chapter 1 at Fig 1.18 and in Fig 2.4 on the next page. A pair of plyers or side cutters is an example of two class 1 levers working together.

Class 2 lever – the load is in the middle, between the force and the fulcrum. A wheelbarrow is a good example of a class 2 lever.

Class 3 lever – the force is in the middle, between the load and the fulcrum. You will see many examples of this class 3 lever toning muscles in the gym. A builder’s shovel is another example of a class 3 lever.

Simple machines

Our physical abilities in the field of lifting and moving heavy objects are limited. However, over the centuries we have used our superior intelligence to design tools, mechanisms and machines which have overcome this physical inadequacy. This concept is shown in Fig. 2.5.

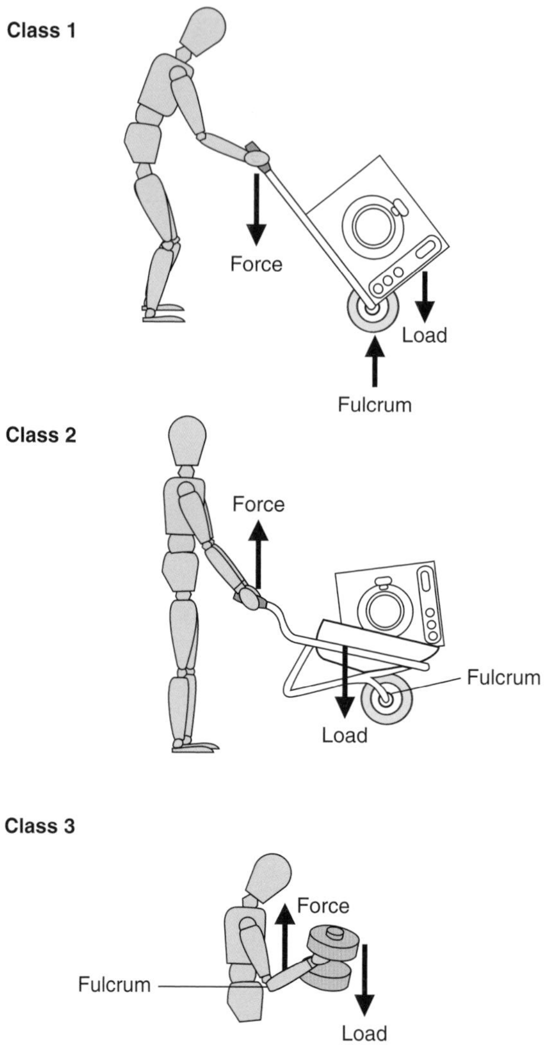

Figure 2.4 Classes of lever.

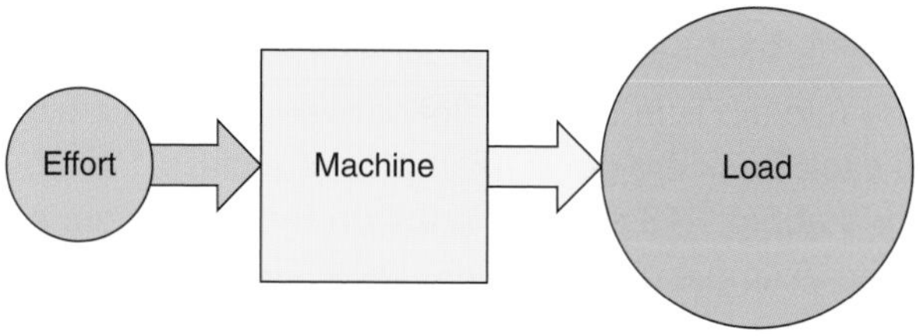

Figure 2.5 Simple machine concept.

Definition

By definition, a *machine* is an assembly of parts, some fixed, others movable, by which motion and force are transmitted. With the aid of a machine we are able to magnify the effort exerted at the input and lift or move large loads at the output.

By definition, a **machine** is an assembly of parts, some fixed, others movable, by which motion and force are transmitted. With the aid of a machine we are able to magnify the effort exerted at the input and lift or move large loads at the output.

Efficiency of any machine

In any machine the power available at the output is less than that which is put in because losses occur in the machine. The losses may result from friction in the bearings, wind resistance to moving parts, heat, noise or vibration.

The ratio of the output power to the input power is known as the **efficiency** of the machine. The symbol for efficiency is the Greek letter 'eta' (η). In general,

$$\eta = \frac{\text{Power output}}{\text{Power input}}$$

Since efficiency is usually expressed as a percentage we modify the general formula as follows:

$$\eta = \frac{\text{Power output}}{\text{Power input}} \times 100$$

Definition

The ratio of the output power to the input power is known as the *efficiency* of the machine. The symbol for efficiency is the Greek letter 'eta' (η). In general,

$$\eta = \frac{\text{Power output}}{\text{Power input}}$$

Example

A transformer feeds the 9.81 kW motor driving the mechanical hoist of the previous example. The input power to the transformer was found to be 10.9 kW. Find the efficiency of the transformer.

$$\eta = \frac{\text{Power output}}{\text{Power input}} \times 100$$

$$\eta = \frac{9.81\,\text{kW}}{10.9\,\text{kW}} \times 100 = 90\%$$

Thus the transformer is 90% efficient. Note that efficiency has no units, but is simply expressed as a percentage.

Electrical machines

Electrical machines are energy converters. If the machine input is mechanical energy and the output electrical energy then that machine is a generator, as shown in Fig. 2.6(a). Alternatively, if the machine input is electrical energy and the output mechanical energy then the machine is a motor, as shown in Fig. 2.6(b).

An electrical machine may be used as a motor or a generator, although in practice the machine will work more efficiently when operated in the mode for which it was designed.

Simple a.c. generator or alternator

If a simple loop of wire is rotated between the poles of a permanent magnet, as shown in Fig. 2.7, the loop of wire will cut the lines of magnetic flux between the north and south poles. This flux cutting will induce an e.m.f. in the wire by **Faraday's law** which states that *when a conductor cuts or is cut by a magnetic field, an e.m.f. is induced in that conductor*. If the generated e.m.f. is collected by carbon brushes at the slip rings and displayed on the screen of a cathode ray oscilloscope, the waveform will be seen to be approximately sinusoidal. Alternately changing, first positive and then negative, then positive again, gives an alternating output.

Definition

Faraday's law which states that *when a conductor cuts or is cut by a magnetic field, an e.m.f. is induced in that conductor*.

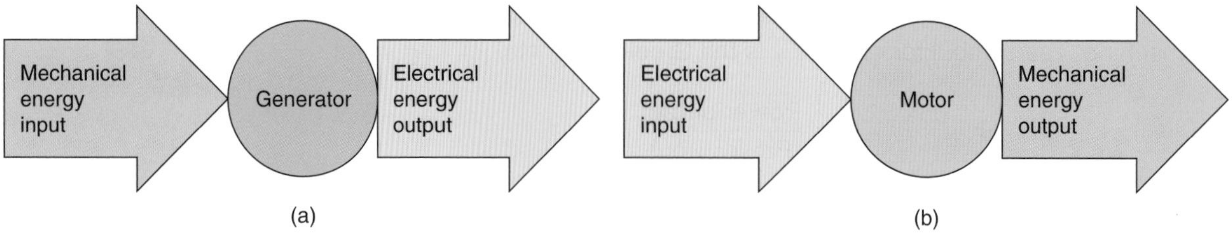

Figure 2.6 Electrical machines as energy converters.

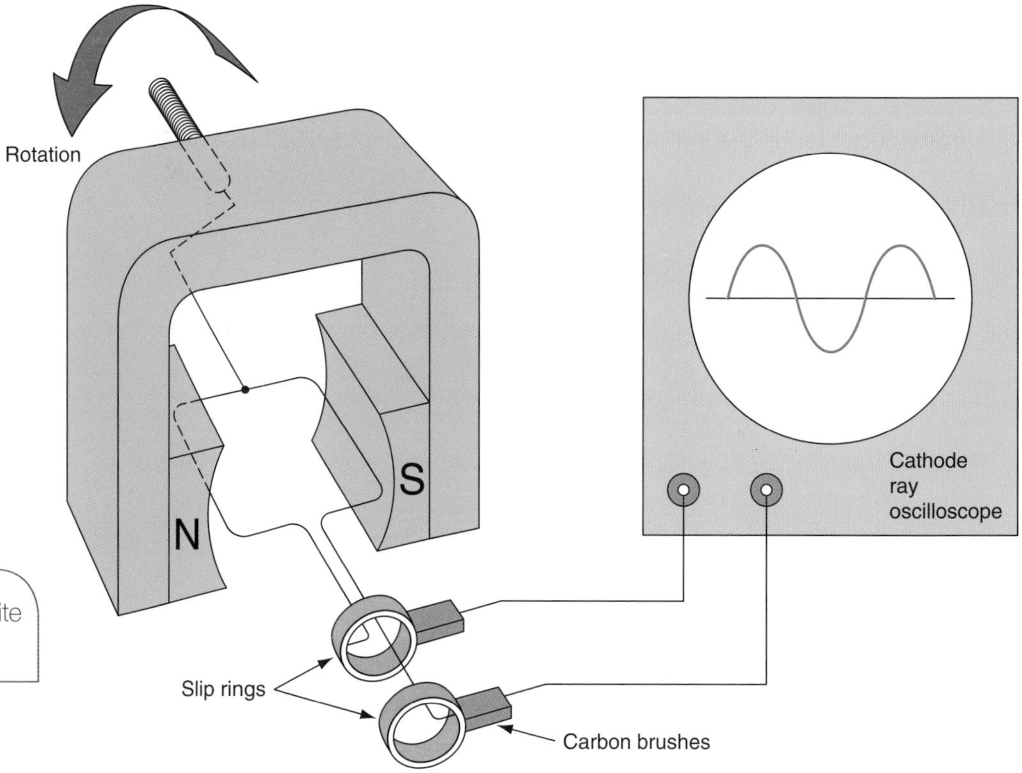

Figure 2.7 Simple a.c. generator or alternator.

Visit the companion website for more on this topic.

Simple d.c. generator or dynamo

If the slip rings of Fig. 2.7 are replaced by a single split ring, called a commutator, the generated e.m.f. will be seen to be in one direction, as shown in Fig. 2.8. The action of the commutator is to reverse the generated e.m.f. every half-cycle, rather like an automatic change-over switch. However, this simple arrangement produces a very bumpy d.c. output. In a practical machine, the commutator would contain many segments and many windings to produce a smoother d.c. output similar to the unidirectional battery supply shown in Fig. 2.9.

Alternating current theory

The supply which we obtain from a car battery is a unidirectional or d.c. supply, whereas the mains electricity supply is alternating or a.c. (see Fig. 2.9).

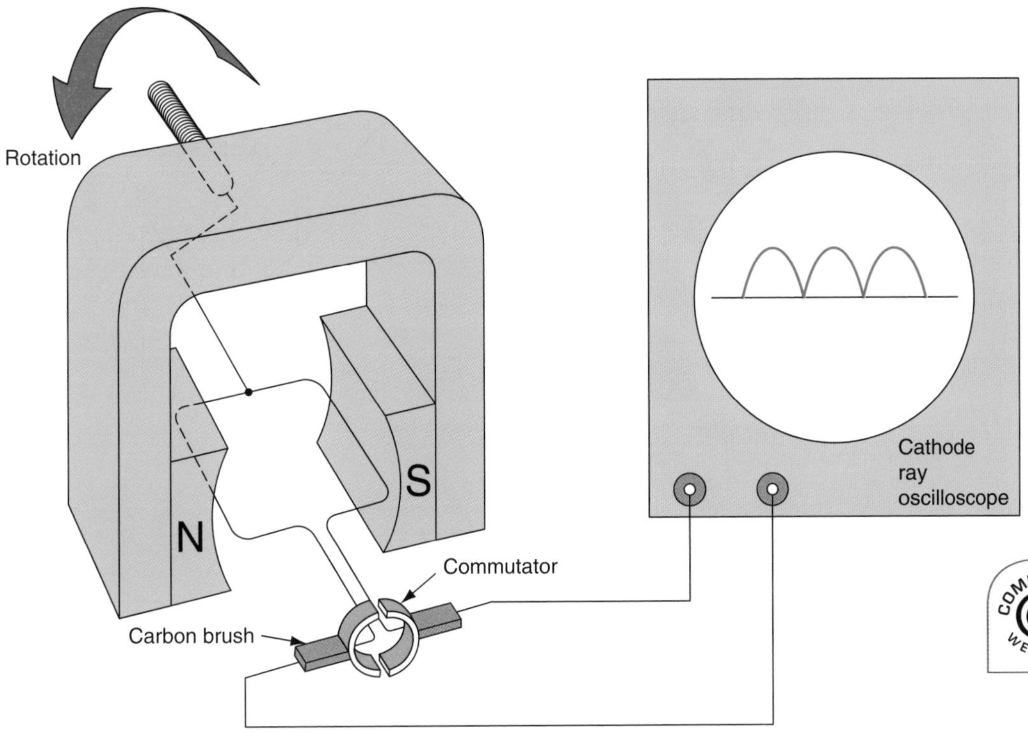

Figure 2.8 Simple d.c. generator or dynamo.

One of the reasons for using alternating supplies for the electricity mains supply is because we can very easily change the voltage levels by using a transformer which will only work on an a.c. supply.

The generated alternating supply at the power-station is transformed up to 132,000 V, or more, for efficient transmission along the national grid conductors.

Most electrical equipment makes use of alternating current supplies, and for this reason a knowledge of alternating waveforms is necessary for all practising electricians.

When a coil of wire is rotated inside a magnetic field as shown in Fig. 2.7, a voltage is induced in the coil. The induced voltage follows a mathematical law known as the sinusoidal law and, therefore, we can say that a sine wave has been generated. Such a waveform has the characteristics displayed in Fig. 2.10.

In the United Kingdom we generate electricity at a frequency of 50 Hz and the time taken to complete each cycle is given by

$$T = \frac{1}{f}$$

$$\therefore T = \frac{1}{50\,\text{Hz}} = 0.02\,\text{s}$$

An alternating waveform is constantly changing from zero to a maximum, first in one direction, then in the opposite direction, and so the instantaneous values of the generated voltage are always changing. A useful description of the electrical effects of an a.c. waveform can be given by the maximum, average and r.m.s. values of the waveform.

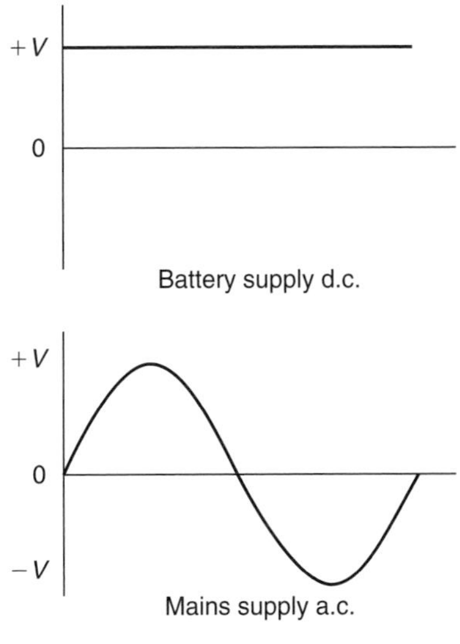

Figure 2.9 Unidirectional and alternating supply.

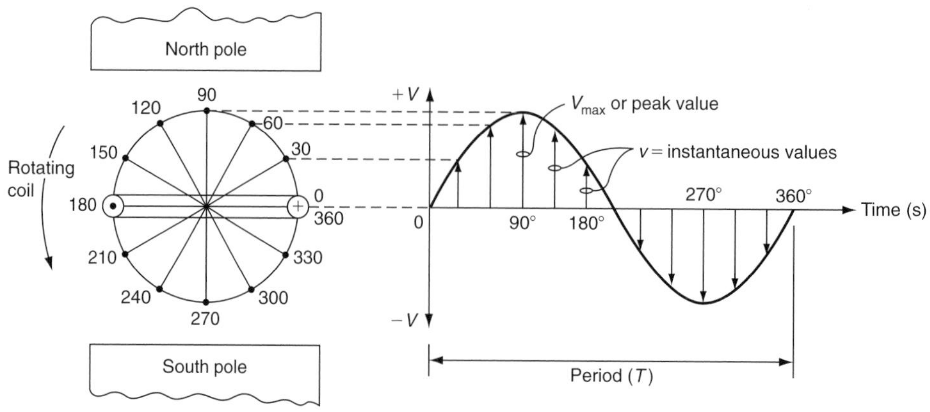

Figure 2.10 Characteristics of a sine wave.

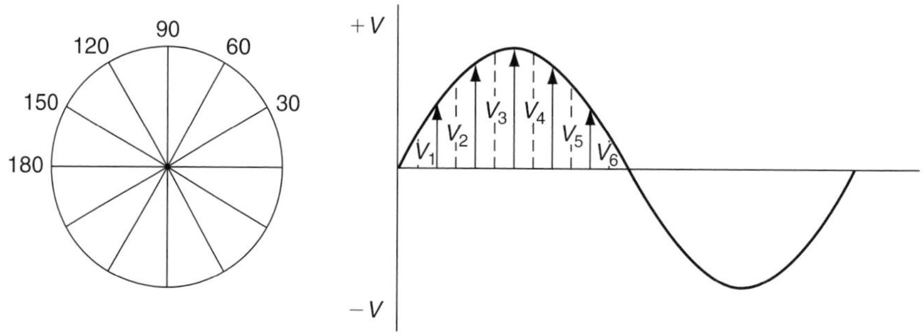

Figure 2.11 Sinusoidal waveform showing instantaneous values of voltage.

The maximum or peak value is the greatest instantaneous value reached by the generated waveform. Cable and equipment insulation levels must be equal to or greater than this value.

The average value is the average over one half-cycle of the instantaneous values as they change from zero to a maximum and can be found from the following formula applied to the sinusoidal waveform shown in Fig. 2.11.

$$V_{av} = \frac{V_1 + V_2 + V_3 + V_4 + V_5 + V_6}{6} = 0.637 V_{max}$$

For any sinusoidal waveform the average value is equal to 0.637 of the maximum value.

The r.m.s. value is the square root of the mean of the individual squared values and is the value of an a.c. voltage which produces the same heating effect as a d.c. voltage. The value can be found from the following formula applied to the sinusoidal waveform shown in Fig. 2.11.

$$V_{r.m.s.} = \sqrt{\frac{V_1^2 + V_2^2 + V_3^2 + V_4^2 + V_5^2 + V_6^2}{6}}$$
$$= 0.7071 V_{max}$$

For any sinusoidal waveform the r.m.s. value is equal to 0.7071 of the maximum value.

Example

The sinusoidal waveform applied to a particular circuit has a maximum value of 325.3 V. Calculate the average and r.m.s. value of the waveform.

$$\text{Average value } V_{av} = 0.637 \times V_{max}$$
$$\therefore V_{av} = 0\ 637 \times 325\ 3 = 207.2\text{V}$$
$$\text{r.m.s. value } V_{r.m.s.} = 0.7071 \times V_{max}$$
$$V_{r.m.s.} = 0.7071 \times 325.3 = 230\text{V}$$

When we say that the main supply to a domestic property is 230 V, we really mean 230 $V_{r.m.s}$. Such a waveform has an average value of about 207.2 V and a maximum value of almost 325.3 V but because the r.m.s. value gives the d.c. equivalent value we almost always give the r.m.s. value without identifying it as such.

Figure 2.12 Both conductors and insulators are used in electrical cable.

Properties of conductors and insulators

In Fig. 2.1 earlier in this chapter we looked at the atomic structure of materials. All materials are made up of atoms and electrons. What makes them different materials is the way in which the atoms and electrons are arranged and how strongly the electrons are attracted to the atoms.

A **conductor** is a material, usually a metal, in which the electrons are loosely bound to the central nucleus. These electrons can easily become 'free electrons' which allows heat and electricity to pass easily through the material.

An **insulator** is a material, usually a non-metal, in which the electrons are very firmly bound to the nucleus and, therefore, will not allow heat or electricity to pass through it.

Let us now define the terms and properties of some of the materials used in the electrical industry.

Ferrous A word used to describe all metals in which the main constituent is iron. The word 'ferrous' comes from the Latin word *ferrum*, meaning iron. Ferrous metals have magnetic properties. Cast iron, wrought iron and steel are all ferrous metals.

Non-ferrous Metals which *do not* contain iron are called non-ferrous. They are non-magnetic and resist rusting. Copper, aluminium, tin, lead, zinc and brass are examples of non-ferrous metals.

Alloy An alloy is a mixture of two or more metals. Brass is an alloy of copper and zinc, usually in the ratio 70–30% or 60–40%.

Corrosion The destruction of a metal by chemical action. Most corrosion takes place when a metal is in contact with moisture (see also mild steel and zinc).

Thermoplastic polymers These may be repeatedly warmed and cooled without appreciable changes occurring in the properties of the material. They are good insulators, but give off toxic fumes when burned. They have a flexible quality when operated up to a maximum temperature of 70°C but should not be flexed when the air temperature is near 0°C, otherwise they may crack.

Polyvinylchloride (PVC) used for cable insulation is a thermoplastic polymer.

Definitions

A *conductor* is a material, usually a metal, in which the electrons are loosely bound to the central nucleus. These electrons can easily become 'free electrons' which allows heat and electricity to pass easily through the material.

An *insulator* is a material, usually a non-metal, in which the electrons are very firmly bound to the nucleus and, therefore, will not allow heat or electricity to pass through it.

Ferrous A word used to describe all metals in which the main constituent is iron.

Non-ferrous Metals which *do not* contain iron are called non-ferrous.

Alloy An alloy is a mixture of two or more metals.

Corrosion The destruction of a metal by chemical action.

Thermoplastic polymers These may be repeatedly warmed and cooled without appreciable changes occurring in the properties of the material.

Polyvinylchloride (PVC) used for cable insulation is a thermoplastic polymer.

Definitions

Thermosetting polymers Once heated and formed, products made from thermosetting polymers are fixed rigidly. Plug tops, socket outlets and switch plates are made from this material.

Rubber is a tough elastic substance made from the sap of tropical plants.

Synthetic rubber is manufactured, as opposed to being produced naturally.

Silicon rubber Introducing organic compounds into synthetic rubber produces a good insulating material such as FP 200 cables.

Magnesium oxide The conductors of mineral-insulated metal-sheathed (MICC) cables are insulated with compressed magnesium oxide.

Thermosetting polymers Once heated and formed, products made from thermosetting polymers are fixed rigidly. Plug tops, socket outlets and switch plates are made from this material.

Rubber is a tough elastic substance made from the sap of tropical plants. It is a good insulator, but degrades and becomes brittle when exposed to sunlight.

Synthetic rubber is manufactured, as opposed to being produced naturally. Synthetic or artificial rubber is carefully manufactured to have all the good qualities of natural rubber – flexibility, good insulation and suitability for use over a wide range of temperatures.

Silicon rubber Introducing organic compounds into synthetic rubber produces a good insulating material which is flexible over a wide range of temperatures and which retains its insulating properties even when burned. These properties make it ideal for cables used in fire alarm installations such as FP200 cables.

Magnesium oxide The conductors of mineral-insulated metal-sheathed (MICC) cables are insulated with compressed magnesium oxide, a white chalk-like substance which is heat resistant and a good insulator, and lasts for many years. The magnesium oxide insulation, copper conductors and sheath, often additionally manufactured with various external sheaths to provide further protection from corrosion and weather, produce a cable designed for long-life and high-temperature installations. However, the magnesium oxide is very hygroscopic, which means that it attracts moisture and, therefore, the cable must be terminated with a special moisture-excluding seal, as shown in Fig. 3.40.

Copper

Copper is extracted from an ore which is mined in South Africa, North America, Australia and Chile. For electrical purposes it is refined to about 98.8% pure copper, the impurities being extracted from the ore by smelting and electrolysis. It is a very good conductor, is non-magnetic and offers considerable resistance to atmospheric corrosion. Copper toughens with work, but may be annealed, or softened, by heating to dull red before quenching.

Copper forms the largest portion of the alloy brass, and is used in the manufacture of electrical cables, domestic heating systems, refrigerator tubes and vehicle radiators. An attractive soft reddish-brown metal, copper is easily worked and is also used to manufacture decorative articles and jewellery.

Aluminium

Aluminium is a grey-white metal obtained from the mineral bauxite which is found in the United States, Germany and the Russian Federation. It is a very good conductor, is non-magnetic, offers very good resistance to atmospheric corrosion and is notable for its extreme softness and lightness. It is used in the manufacture of power cables. The overhead cables of the National Grid are made of an aluminium conductor reinforced by a core of steel. Copper conductors would be too heavy to support themselves between the pylons. Lightness and resistance to corrosion make aluminium an ideal metal for the manufacture of cooking pots and food containers.

Aluminium alloys retain the corrosion resistance properties of pure aluminium with an increase in strength. The alloys are cast into cylinder heads and gearboxes for motor cars, and switch-boxes and luminaires for electrical installations. Special processes and fluxes have now been developed which allow aluminium to be welded and soldered.

Try this

Materials

List five 'good insulator' materials being used at your place of work.

Brass

Brass is a non-ferrous alloy of copper and zinc which is easily cast. Because it is harder than copper or aluminium it is easily machined. It is a good conductor and is highly resistant to corrosion. For these reasons it is often used in the electrical and plumbing trades. Taps, valves, pipes, electrical terminals, plug top pins and terminal glands for steel wire armour (SWA) and MI cables are some of the many applications.

Brass is an attractive yellow metal which is also used for decorative household articles and jewellery. The combined properties of being an attractive metal which is highly resistant to corrosion make it a popular metal for ships' furnishings.

Cast steel

Cast steel is also called tool steel or high-carbon steel. It is an alloy of iron and carbon which is melted in airtight crucibles and then poured into moulds to form ingots. These ingots are then rolled or pressed into various shapes from which the finished products are made. Cast steel can be hardened and tempered and is therefore ideal for manufacturing tools. Hammer heads, pliers, wire cutters, chisels, files and many machine parts are also made from cast steel.

Mild steel

Mild steel is also an alloy of iron and carbon but contains much less carbon than cast steel. It can be filed, drilled or sawn quite easily and may be bent when hot or cold, but repeated cold bending may cause it to fracture. In moist conditions corrosion takes place rapidly unless the metal is protected. Mild steel is the most widely used metal in the world, having considerable strength and rigidity without being brittle. Ships, bridges, girders, motor car bodies, bicycles, nails, screws, conduit, trunking, tray and SWA are all made of mild steel.

Try this

Materials

List five 'good conductor' materials being used at your place of work.

Zinc

Zinc is a non-ferrous metal which is used mainly to protect steel against corrosion and in making the alloy brass. Mild steel coated with zinc is sometimes called *galvanized steel*, and this coating considerably improves steel's resistance to corrosion. Conduit, trunking, tray, SWA, outside luminaires and electricity pylons are made of galvanized steel.

Resistors in series and in parallel

In an electrical circuit resistors may be connected in series, in parallel, or in various combinations of series and parallel connections.

Series-connected resistors

Top tip

Series resistance

The total value of resistance in a series circuit is always greater than the largest individual value.

In any series circuit a current I will flow through all parts of the circuit as a result of the potential difference supplied by a battery V_T. Therefore, we say that in a series circuit the current is common throughout that circuit.

When the current flows through each resistor in the circuit, for example, R_1, R_2 and R_3 in Fig. 2.13, there will be a voltage drop across that resistor whose value will be determined by the values of I and R, since from Ohm's law $V = I \times R$. The sum of the individual voltage drops, for example, V_1, V_2 and V_3 in Fig. 2.13, will be equal to the total voltage V_T.

We can summarize these statements as follows. For any series circuit, I is common throughout the circuit and

$$V_T = V_1 + V_2 + V_3 \qquad \text{(Equation 1)}$$

Let us call the total circuit resistance R_T. From Ohm's law we know that $V = I \times R$ and therefore

$$\left.\begin{aligned} \text{Total voltage } V_T &= I \times R_T \\ \text{Voltage drop across } R_1 \text{ is } V_1 &= I \times R_1 \\ \text{Voltage drop across } R_2 \text{ is } V_2 &= I \times R_2 \\ \text{Voltage drop across } R_3 \text{ is } V_3 &= I \times R_3 \end{aligned}\right\} \qquad \text{(Equation 2)}$$

We are looking for an expression for the total resistance in any series circuit and, if we substitute equation (2) into equation (1), we have:

$$\begin{aligned} V_T &= V_1 + V_2 + V_3 \\ \therefore I \times R_T &= I \times R_1 + I \times R_2 + I \times R_3 \end{aligned}$$

Now, since I is common to all terms in the equation, we can divide both sides of the equation by I. This will cancel out I to leave us with an expression for the circuit resistance:

$$R_T = R_1 + R_2 + R_3$$

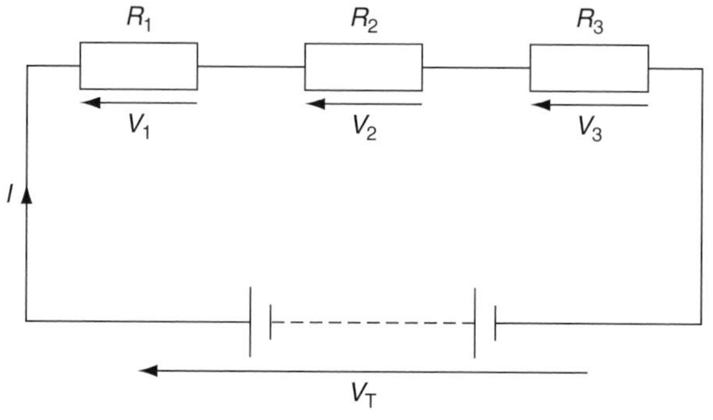

Figure 2.13 A series circuit.

Note that the derivation of this formula is given for information only. Craft students need only state the expression $R_T = R_1 + R_2 + R_3$ for series connections.

Parallel-connected resistors

In any parallel circuit, as shown in Fig. 2.14, the same voltage acts across all branches of the circuit. The total current will divide when it reaches a resistor junction, part of it flowing in each resistor. The sum of the individual currents, for example, I_1, I_2 and I_3 in Fig. 2.14, will be equal to the total current I_T.

We can summarize these statements as follows. For any parallel circuit, V is common to all branches of the circuit and

$$I_T = I_1 + I_2 + I_3 \qquad \text{(Equation 3)}$$

Let us call the total resistance R_T.

From Ohm's law we know, that $I = \frac{V}{R}$, and therefore

$$\text{the total current } I_T = \frac{V}{R_T}$$

$$\text{the current through } R_1 \text{ is } I_1 = \frac{V}{R_1}$$

$$\text{the current through } R_2 \text{ is } I_2 = \frac{V}{R_2}$$

$$\text{the current through } R_3 \text{ is } I_3 = \frac{V}{R_3} \qquad \text{(Equation 4)}$$

We are looking for an expression for the equivalent resistance R_T in any *parallel* circuit and, if we substitute equations (4) into equation (3), we have:

$$I_T = I_1 + I_2 + I_3$$

$$\therefore \frac{V}{R_T} = \frac{V}{R_1} + \frac{V}{R_2} + \frac{V}{R_3}$$

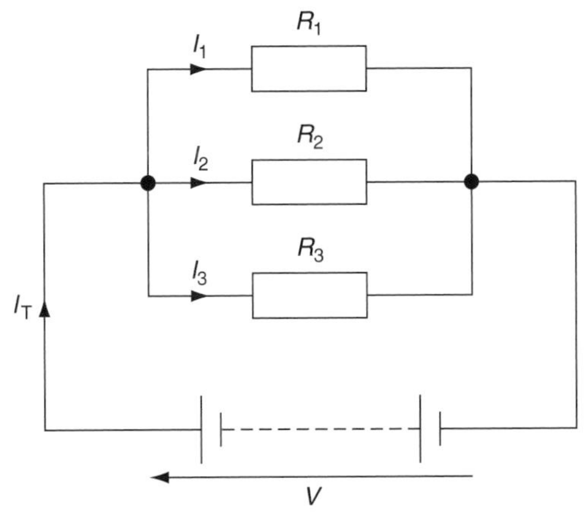

Figure 2.14 A parallel circuit.

Key fact

Resistors

In a series circuit, total resistance $R_T = R_1 + R_2 + R_3$ ohms.

Top tip

Parallel resistance

The total value of resistance in a parallel circuit is always less than the smallest individual value.

Key fact

Resistors

In a parallel circuit, total resistance can be found from

$$\frac{1}{R_T} = \frac{1}{R_1} + \frac{1}{R_2} + \frac{1}{R_3}$$

Now, since V is common to all terms in the equation, we can divide both sides by V, leaving us with an expression for the circuit resistance:

$$\frac{1}{R_T} = \frac{1}{R_1} + \frac{1}{R_2} + \frac{1}{R_3}$$

Note that the derivation of this formula is given for information only. Craft students need only state the expression $1/R_T = 1/R_1 + 1/R_2 + 1/R_3$ for parallel connections.

Example 1

Three 6 Ω resistors are connected (a) in series (see Fig. 2.15), and (b) in parallel (see Fig. 2.16), across a 12 V battery. For each method of connection, find the total resistance and the values of all currents and voltages.

For any series connection

$$R_T = R_1 + R_2 + R_3$$
$$\therefore R_T = 6\,\Omega + 6\,\Omega + 6\,\Omega = 18\,\Omega$$

Total current $I_T = \frac{V_T}{R_T}$

$$\therefore I_T = \frac{12\text{ V}}{18\,\Omega} = 0.67\text{ A}$$

The voltage drop across R_1 is

$$V_1 = I_T \times R_1$$
$$\therefore V_1 = 0.67\text{ A} \times 6\,\Omega = 4\text{ V}$$

The voltage drop across R_2 is

$$V_2 = I_T \times R_2$$
$$\therefore V_2 = 0.67\text{ A} \times 6\,\Omega = 4\text{ V}$$

The voltage drop across R_3 is

$$V_3 = I_T \times R_3$$
$$\therefore V_3 = 0.67\text{ A} \times 6\,\Omega = 4\text{ V}$$

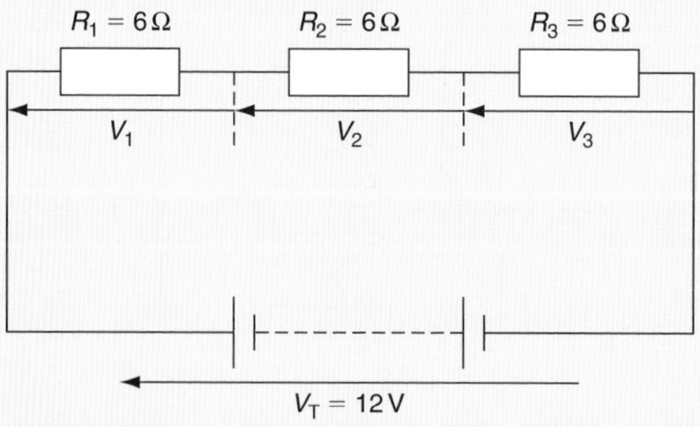

Figure 2.15 Resistors in series.

(continued)

Example 1 (continued)

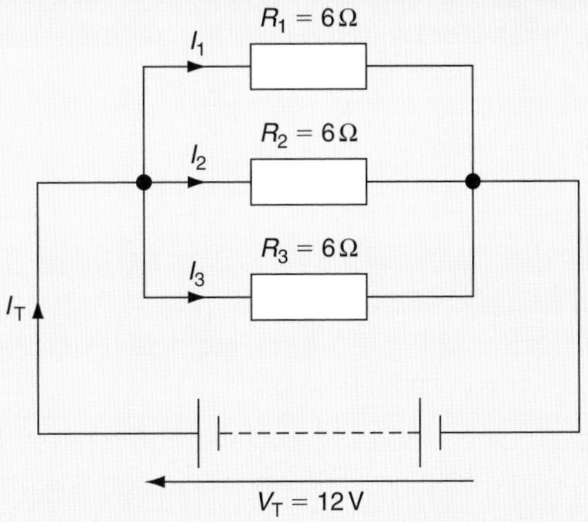

Figure 2.16 Resistors in parallel.

For any parallel connection

$$\frac{1}{R_T} = \frac{1}{R_1} + \frac{1}{R_2} + \frac{1}{R_3}$$

$$\therefore \frac{1}{R_T} = \frac{1}{6\,\Omega} + \frac{1}{6\,\Omega} + \frac{1}{6\,\Omega}$$

$$\frac{1}{R_T} = \frac{1+1+1}{6\,\Omega} = \frac{3}{6\,\Omega}$$

$$R_T = \frac{6\,\Omega}{3} = 2\,\Omega$$

Total current $I_T = \frac{V_T}{R_T}$

$$\therefore I_T = \frac{12\text{ V}}{2\,\Omega} = 6\text{ A}$$

The current flowing through R_1 is

$$I_1 = \frac{V_T}{R_1}$$

$$\therefore I_1 = \frac{12\text{ V}}{6\,\Omega} = 2\text{ A}$$

The current flowing through R_2 is

$$I_2 = \frac{V_T}{R_2}$$

$$\therefore I_2 = \frac{12\text{ V}}{6\,\Omega} = 2\text{ A}$$

The current flowing through R_3 is

$$I_3 = \frac{V_T}{R_3}$$

$$\therefore I_3 = \frac{12\text{ V}}{6\,\Omega} = 2\text{ A}$$

Series and parallel combinations

The most complex arrangement of series and parallel resistors can be simplified into a single equivalent resistor by combining the separate rules for series and parallel resistors.

Example 2

Resolve the circuit shown in Fig. 2.17 into a single resistor and calculate the potential difference across each resistor.

By inspection, the circuit contains a parallel group consisting of R_3, R_4 and R_5 and a series group consisting of R_1 and R_2 in series with the equivalent resistor for the parallel branch.

Consider the parallel group. We will label this group R_P. Then

$$\frac{1}{R_P} = \frac{1}{R_3} + \frac{1}{R_4} + \frac{1}{R_5}$$

$$\frac{1}{R_P} = \frac{1}{2\,\Omega} + \frac{1}{3\,\Omega} + \frac{1}{6\,\Omega}$$

$$\frac{1}{R_P} = \frac{3+2+1}{6\,\Omega} = \frac{6}{6\,\Omega}$$

$$R_P = \frac{6\,\Omega}{6} = 1\,\Omega$$

Figure 2.17 may now be represented by the more simple equivalent shown in Fig. 2.18.

Since all resistors are now in series,

$$R_T = R_1 + R_2 + R_P$$

$$\therefore R_T = 3\,\Omega + 6\,\Omega + 1\,\Omega = 10\,\Omega$$

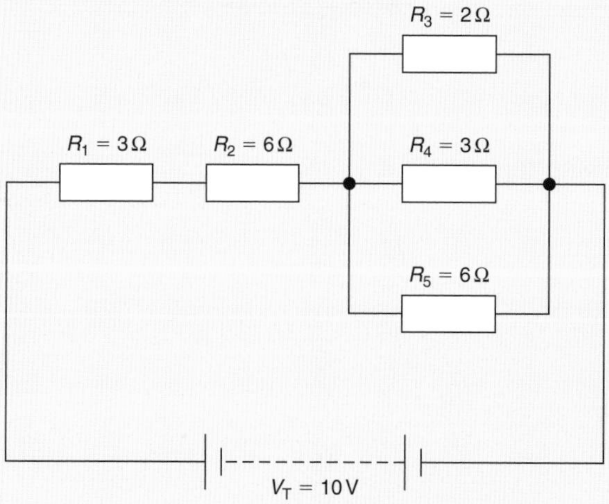

Figure 2.17 A series/parallel circuit.

(continued)

Example 2 (continued)

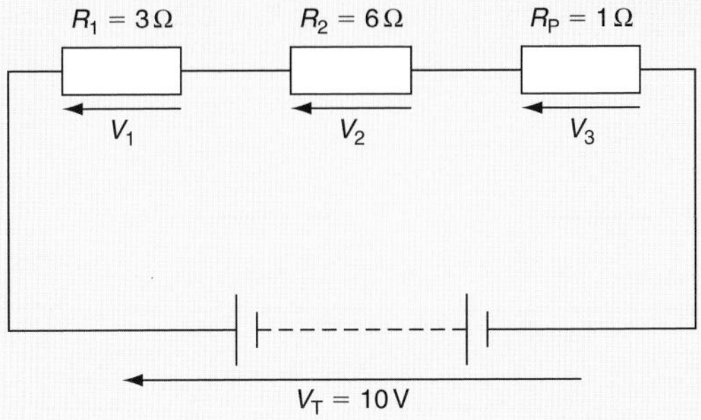

Figure 2.18 Equivalent series circuit.

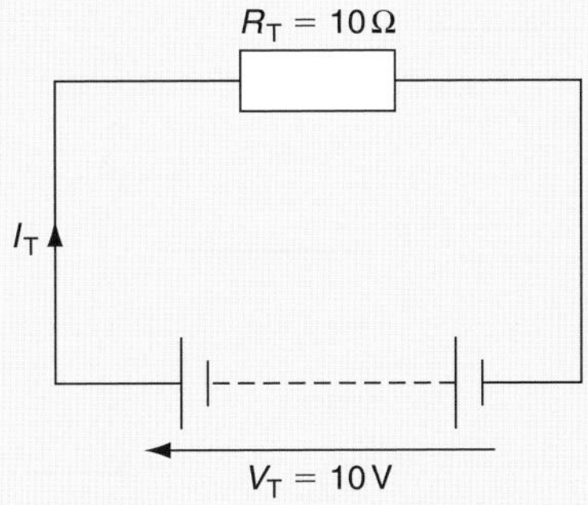

Figure 2.19 Single equivalent resistor for Fig. 2.17.

Thus, the circuit may be represented by a single equivalent resistor of value $10\,\Omega$ as shown in Fig. 2.19. The total current flowing in the circuit may be found by using Ohm's law:

$$I_T = \frac{V_T}{R_T} = \frac{10\,V}{10\,\Omega} = 1\,A$$

The potential differences across the individual resistors are

$$V_1 = I_T \times R_1 = 1\,A \times 3\,\Omega = 3\,V$$
$$V_2 = I_T \times R_2 = 1\,A \times 6\,\Omega = 6\,V$$
$$V_P = I_T \times R_P = 1\,A \times 1\Omega = 1V$$

Since the same voltage acts across all branches of a parallel circuit the same p.d. of 1 V will exist across each resistor in the parallel branch R_3, R_4 and R_5.

Example 3

Determine the total resistance and the current flowing through each resistor for the circuit shown in Fig. 2.20.

By inspection, it can be seen that R_1 and R_2 are connected in series while R_3 is connected in parallel across R_1 and R_2. The circuit may be more easily understood if we redraw it as in Fig. 2.21.

(*continued*)

Example 3 (continued)

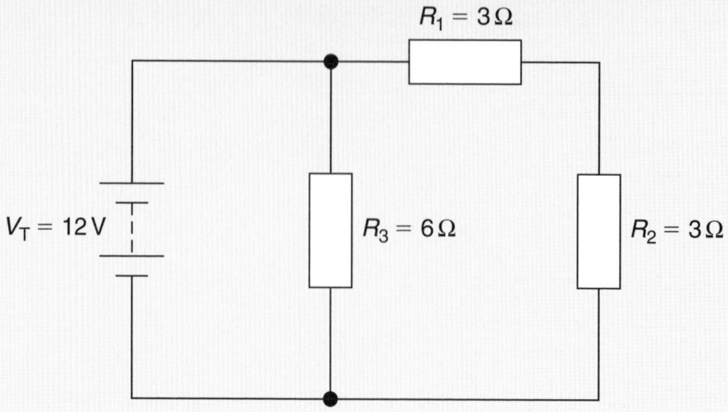

Figure 2.20 A series/parallel circuit for Example 3.

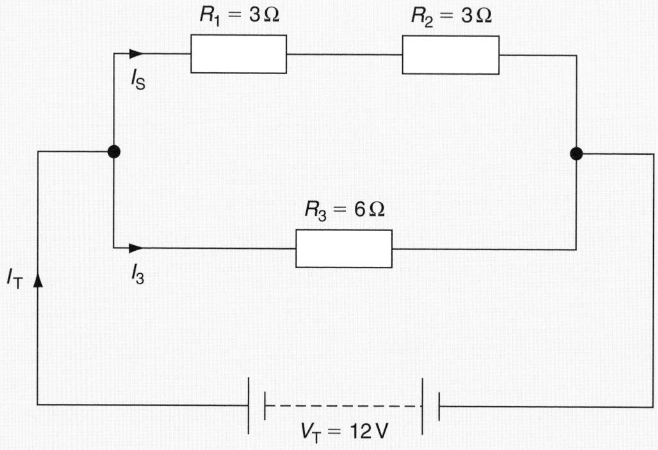

Figure 2.21 Equivalent circuit for Example 3.

For the series branch, the equivalent resistor can be found from

$$R_S = R_1 + R_2$$
$$\therefore R_S = 3\,\Omega + 3\,\Omega = 6\,\Omega$$

Figure 2.21 may now be represented by a more simple equivalent circuit, as in Fig. 2.22.

Since the resistors are now in parallel, the equivalent resistance may be found from

$$\frac{1}{R_T} = \frac{1}{R_S} + \frac{1}{R_3}$$
$$\therefore \frac{1}{R_T} = \frac{1}{6\,\Omega} + \frac{1}{6\,\Omega}$$
$$\frac{1}{R_T} = \frac{1+1}{6\,\Omega} = \frac{2}{6\,\Omega}$$
$$R_T = \frac{6\,\Omega}{2} = 3\,\Omega$$

(continued)

Example 3 (continued)

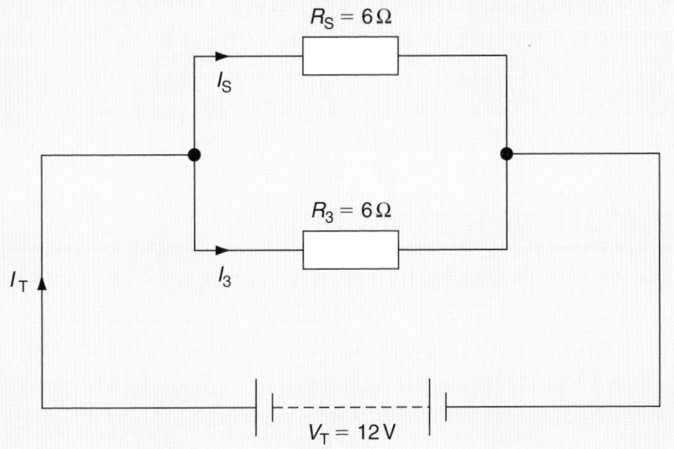

Figure 2.22 Simplified equivalent circuit for Example 2.

The total current is

$$I_T = \frac{V_T}{R_T} = \frac{12\ \text{V}}{3\ \Omega} = 4\ \text{A}$$

Let us call the current flowing through resistor $R_3 I_3$

$$\therefore\ I_3 = \frac{V_T}{R_3} = \frac{12\ \text{V}}{6\ \Omega} = 2\ \text{A}$$

Let us call the current flowing through both resistors R_1 and R_2, as shown in Fig. 2.21, I_S

$$\therefore\ I_S = \frac{V_T}{R_S} = \frac{12\ \text{V}}{6\ \Omega} = 2\ \text{A}$$

Measuring volts and amps

The type of instrument to be purchased for general use in the electrical industries is a difficult choice because there are so many different types on the market and every manufacturer's representative is convinced that his company's product is the best. However, most instruments can be broadly grouped under two general headings: those having *analogue* and those with *digital* displays.

Analogue meters or instruments

Analogue meters have a pointer moving across a calibrated scale. They are the only choice when a general trend or variation in value is to be observed. Hi-fi equipment often uses analogue displays to indicate how power levels vary with time, which is more informative than a specific value. Red or danger zones can be indicated on industrial instruments. The fuel gauge on a motor car often indicates full, half-full or danger on an analogue display which is much more informative than an indication of the exact number of litres of petrol remaining in the tank.

Definition

Analogue meters have a pointer moving across a calibrated scale.

These meters are only accurate when used in the calibrated position – usually horizontally.

Most meters using an analogue scale incorporate a mirror to eliminate parallax error. The user must look straight at the pointer on the scale when taking readings and the correct position is indicated when the pointer image in the mirror is hidden behind the actual pointer. That is the point at which a reading should be taken from the appropriate scale of the instrument.

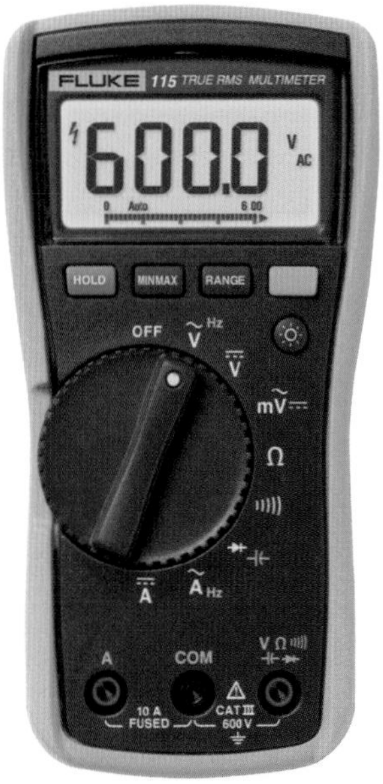

Figure 2.23 Digital multimeter suitable for testing electrical and electronic circuits.

Digital meters or instruments

Digital meters provide the same functions as analogue meters but they display the indicated value using a seven-segment LED to give a numerical value of the measurement. Modern digital meters use semiconductor technology to give the instrument a very high-input impedance, typically about 10 MΩ, and, therefore, they are ideal for testing most electrical or electronic circuits.

The choice between an analogue and a digital display is a difficult one and must be dictated by specific circumstances. However, if you are an electrician or service engineer intending to purchase a new instrument, I think on balance that a good-quality digital multimeter such as that shown in Fig. 2.23 would be best. Having no moving parts, digital meters tend to be more rugged and, having a very high-input impedance, they are ideally suited to testing all circuits that an electrician might work on in their daily work.

Definition

Digital meters provide the same functions as analogue meters but they display the indicated value using a seven-segment LED to give a numerical value of the measurement.

The multimeter

Multimeters are designed to measure voltage, current or resistance. Before taking measurements the appropriate volt, ampere or ohm scale should be selected. To avoid damaging the instrument it is good practice first to switch to the highest value on a particular scale range. For example, if the 10 A scale is first selected and a reading of 2.5 A is displayed, we then know that a more appropriate scale would be the 3 A or 5 A range. This will give a more accurate reading which might be, say, 2.49 A. When the multimeter is used as an ammeter to measure current it must be connected in series with the test circuit, as shown in Fig. 2.24(a). When used as a voltmeter the multimeter must be connected in parallel with the component, as shown in Fig. 2.24(b).

When using a commercial multirange meter as an ohmmeter for testing electronic components, care must be exercised in identifying the positive terminal. The red terminal of the meter, identifying the positive input for testing voltage and current, usually becomes the negative terminal when the meter is used as an ohmmeter because of the way the internal battery is connected to the meter movement. Check the meter manufacturer's handbook before using a multimeter to test electronic components.

Safety first

Voltmeters and ammeters

Electrical test meters are expensive and easy to steal:

- protect them by placing in their protective case;
- keep them safe, out of sight and locked up.

The three effects of an electric current

When an electric current flows in a circuit it can have one or more of the following three effects: **heating**, **magnetic** or **chemical**.

Definition

When an electric current flows in a circuit it can have one or more of the following three effects: *heating, magnetic* or *chemical*.

Heating effect

The movement of electrons within a conductor, which is the flow of an electric current, causes an increase in the temperature of the conductor. The amount of heat generated by this current flow depends upon the type and dimensions of the conductor and the quantity of current flowing. By changing these variables, a conductor may be operated hot and used as the heating element of a fire, or be operated cool and used as an electrical installation conductor.

The heating effect of an electric current is also the principle upon which a fuse gives protection to a circuit. The fuse element is made of a metal with a low melting point and forms a part of the electrical circuit. If an excessive current flows, the fuse element overheats and melts, breaking the circuit.

Magnetic effect

Whenever a current flows in a conductor a magnetic field is set up around the conductor like an extension of the insulation. The magnetic field increases with the current and collapses if the current is switched off. A conductor carrying current and wound into a solenoid produces a magnetic field very similar to a permanent magnet, but has the advantage of being switched on and off by any switch which controls the circuit current.

The magnetic effect of an electric current is the principle upon which electric bells, relays, instruments, motors and generators work.

Chemical effect

When an electric current flows through a conducting liquid, the liquid is separated into its chemical parts. The conductors which make contact with the liquid are called the anode and cathode. The liquid itself is called the electrolyte, and the process is called *electrolysis*.

Electrolysis is an industrial process used in the refining of metals and electroplating. It was one of the earliest industrial applications of electric current. Most of the aluminium produced today is extracted from its ore by electrochemical methods. Electroplating serves a double purpose by protecting a base metal from atmospheric erosion and also giving it a more expensive and attractive appearance. Silver and nickel plating has long been used to enhance the appearance of cutlery, candlesticks and sporting trophies.

An anode and cathode of dissimilar metal placed in an electrolyte can react chemically and produce an e.m.f. When a load is connected across the anode and cathode, a current is drawn from this arrangement, which is called a cell. A battery is made up of a number of cells. It has many useful applications in providing portable electrical power, but electrochemical action can also be undesirable since it is the basis of electrochemical corrosion which rots our motor cars, industrial containers and bridges.

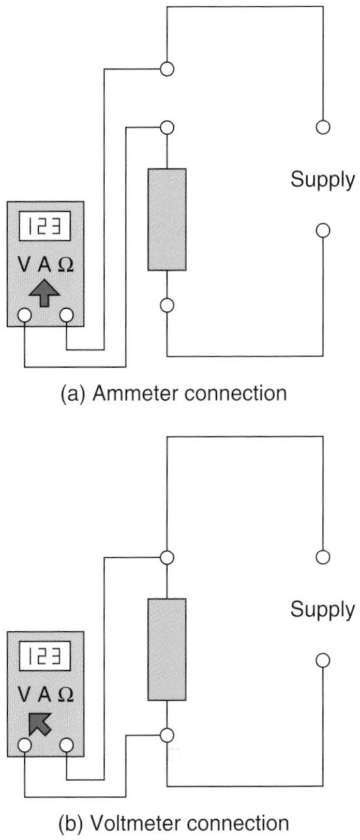

Figure 2.24 Using a multimeter (a) as an ammeter and (b) as a voltmeter.

Magnetism

The Greeks knew as early as 600 BC that a certain form of iron ore, now known as magnetite or lodestone, had the property of attracting small pieces of iron. Later, during the Middle Ages, navigational compasses were made using the magnetic properties of lodestone. Small pieces of lodestone attached to wooden splints floating in a bowl of water always came to rest pointing in a north–south direction. The word lodestone is derived from an old English word meaning 'the way', and the word magnetism is derived from Magnesia, the place where magnetic ore was first discovered.

Iron, nickel and cobalt are the only elements which are attracted strongly by a magnet. These materials are said to be *ferromagnetic*. Copper, brass, wood, PVC and glass are not attracted by a magnet and are, therefore, described as *non-magnetic*.

COMPANION WEBSITE Visit the companion website for more on this topic.

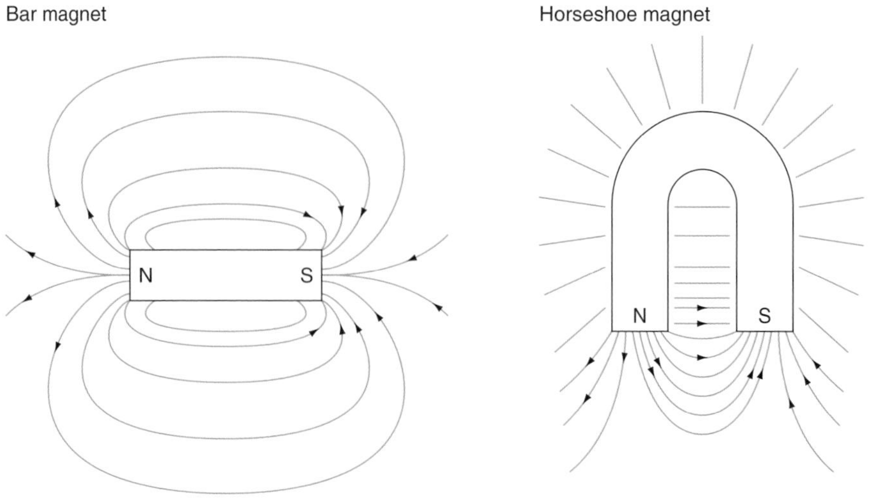

Figure 2.25 Magnetic fields around a permanent magnet.

Some basic rules of magnetism

1. Lines of magnetic flux have no physical existence, but they were introduced by Michael Faraday (1791–1867) as a way of explaining the magnetic energy existing in space or in a material. They help us to visualize and explain the magnetic effects. The symbol used for magnetic flux is the Greek letter Φ (phi) and the unit of magnetic flux is the weber (symbol Wb), pronounced 'veber', to commemorate the work of the German physicist Wilhelm Weber (1804–1891).
2. Lines of magnetic flux always form closed loops.
3. Lines of magnetic flux behave like stretched elastic bands, always trying to shorten themselves.
4. Lines of magnetic flux never cross over each other.
5. Lines of magnetic flux travel along a magnetic material and always emerge out of the 'north pole' end of the magnet.
6. Lines of magnetic flux pass through space and non-magnetic materials undisturbed.
7. The region of space through which the influence of a magnet can be detected is called the **magnetic field** of that magnet.
8. The number of lines of magnetic flux within a magnetic field is a measure of the flux density. Strong magnetic fields have a high-flux density. The symbol used for flux density is *B*, and the unit of flux density is the tesla (symbol T), to commemorate the work of the Croatian-born American physicist Nikola Tesla (1856–1943).
9. The places on a magnetic material where the lines of flux are concentrated are called the **magnetic poles**.
10. Like poles repel; unlike poles attract. These two statements are sometimes called the **'first laws of magnetism'** and are shown in Fig. 2.26.

Definition

The region of space through which the influence of a magnet can be detected is called the *magnetic field* of that magnet.

Definition

The places on a magnetic material where the lines of flux are concentrated are called the *magnetic poles*.

Definition

Like poles repel; unlike poles attract. These two statements are sometimes called the *'first laws of magnetism'* and are shown in Fig. 2.26.

Magnetic fields

If a permanent magnet is placed on a surface and covered by a piece of paper, iron filings can be shaken on to the paper from a dispenser. Gently tapping the paper then causes the filings to take up the shape of the magnetic field surrounding the permanent magnet. The magnetic fields around a permanent magnet are shown in Figs 2.25 and 2.26.

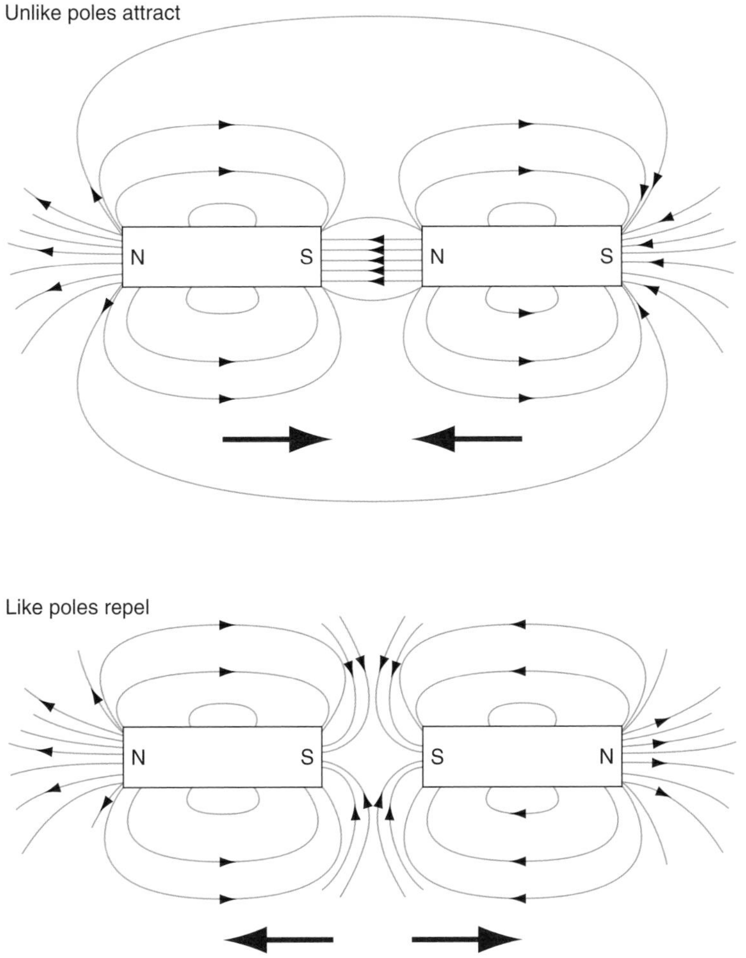

Figure 2.26 The first laws of magnetism.

Visit the companion website for more on this topic.

Electromagnetism

Electricity and magnetism have been inseparably connected since the experiments by Oersted and Faraday in the early nineteenth century. An electric current flowing in a conductor produces a magnetic field 'around' the conductor which is proportional to the current. Thus a small current produces a weak magnetic field, while a large current will produce a strong magnetic field. The magnetic field 'spirals' around the conductor, as shown in Fig. 2.27, and its direction can be determined by the 'dot' or 'cross' notation and the 'screw rule'. To do this, we think of the current as being represented by a dart or arrow inside the conductor. The dot represents current coming towards us when we would see the point of the arrow or dart inside the conductor. The cross represents current going away from us when we would see the flights of the dart or arrow. Imagine a corkscrew or screw being turned so that it will move in the direction of the current. Therefore, if the current was coming out of the paper, as shown in Fig. 2.27(a), the magnetic field would be spiralling anticlockwise around the conductor. If the current was going into the paper, as shown by Fig. 2.27(b), the magnetic field would spiral clockwise around the conductor.

A current flowing in a *coil* of wire or solenoid establishes a magnetic field which is very similar to that of a bar magnet. Winding the coil around a soft iron core

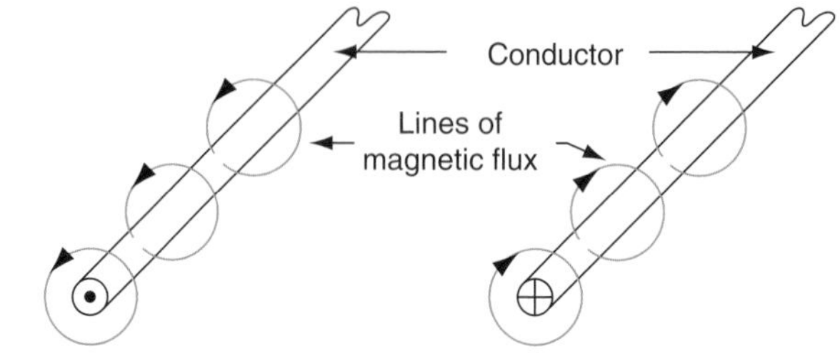

Figure 2.27 Magnetic fields around a current-carrying conductor.

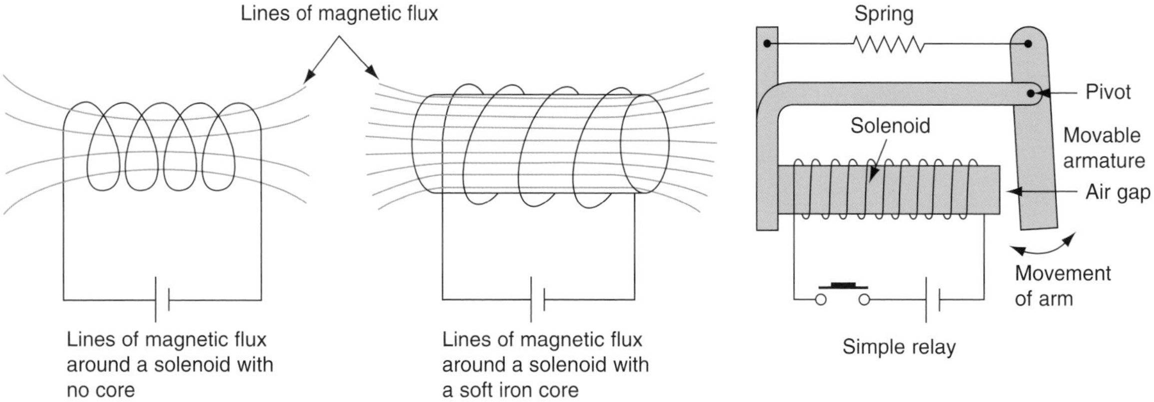

Figure 2.28 The solenoid and one practical application: the relay.

Visit the companion website for more on this topic.

increases the flux density because the lines of magnetic flux concentrate on the magnetic material. The advantage of the electromagnet when compared with the permanent magnet is that the magnetism of the electromagnet can be switched on and off by a functional switch controlling the coil current. This effect is put to practical use in the electrical relay as used in a motor starter or alarm circuit. Figure 2.28 shows the structure and one application of the solenoid.

The electrical relay

Definition

A *relay* is an electromagnetic switch operated by a solenoid.

A **relay** is an electromagnetic switch operated by a solenoid. We looked at the action of a solenoid in Fig. 2.28. The solenoid in a relay operates a number of switch contacts as it moves under the electromagnetic forces. Relays can be used to switch circuits on or off at a distance remotely. The energizing circuit, the solenoid, is completely separate to the switch contacts and, therefore, the relay can switch high-voltage, high-power circuits from a low-voltage switching circuit. This gives the relay many applications in motor control circuits, electronics and instrumentation systems. Figure 2.29 shows a simple relay.

Electrical transformers

A transformer is an electrical machine without moving parts, which is used to change the value of an alternating voltage.

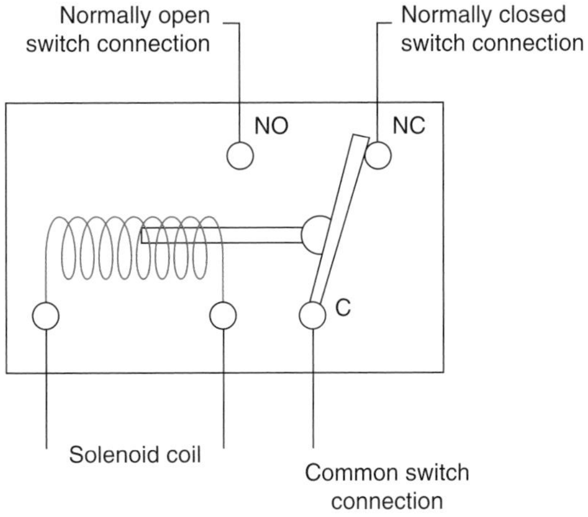

Figure 2.29 A simple relay.

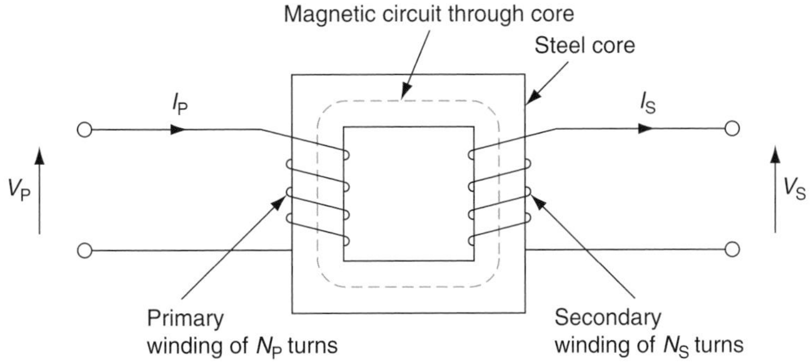

Figure 2.30 A simple transformer.

A transformer will only work on an alternating supply, it will not normally work from a d.c. supply such as a battery.

- A transformer such as that shown in Fig. 2.30 consists of two coils called the primary and secondary coils or windings, wound on to a common core. The iron core of the transformer is not solid but made up of very thin sheets called laminations, to improve efficiency.
- An alternating voltage applied to the primary winding establishes an alternating magnetic flux in the core.
- The magnetic flux in the core causes a voltage to be induced in the secondary winding of the transformers.
- The voltage in both the primary and secondary windings is proportional to the number of turns.
- This means that if you increase the number of secondary turns you will increase the output voltage. This has an application in power distribution.
- Alternatively, reducing the number of secondary turns will reduce the output voltage. This is useful for low-voltage supplies such as domestic bell transformers. Because it has no moving parts, a transformer can have a very high efficiency. Large power transformers, used on electrical distribution systems, can have an efficiency of better than 90%.

Large power transformers need cooling to take the heat generated away from the core. This is often achieved by totally immersing the core and windings in insulating oil. A sketch of an oil-filled transformer can be seen in Fig. 2.31.

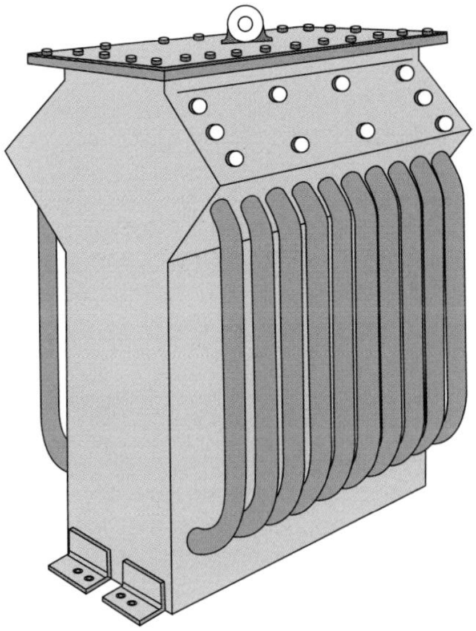

Figure 2.31 Typical oil-filled power transformer.

Figure 2.32 Transmission tower.

Very small transformers are used in electronic applications. Small transformers are used as isolating transformers in shaver sockets and can also be used to supply separated extra-low-voltage (SELV) sources. Equipment supplied from a SELV source may be installed in a bathroom or shower room, provided that it is suitably enclosed and protected from the ingress of moisture. This includes equipment such as water heaters, pumps for showers and whirlpool baths.

Try this

Have you seen any transformers in action? Were they big or small – what were they being used for? Have you been close up to a transmission tower, perhaps when you were walking in the countryside? Make a note in the margin.

Key fact

Definitions

- Try to remember these a.c. circuit definitions.
- Write them down if it helps.

Definition

In any circuit, *resistance* is defined as opposition to current flow.

Visit the companion website for more on this topic.

Definition

Inductive reactance (X_L) is the opposition to an a.c. current in an inductive circuit. It causes the current in the circuit to lag behind the applied voltage.

Alternating current theory

Commercial quantities of electricity for industry, commerce and domestic use are generated as a.c. in large power-stations and distributed around the United Kingdom on the National Grid to the end user. The d.c. electricity has many applications where portability or an emergency stand-by supply is important but for large quantities of power it has to be an a.c. supply because it is so easy to change the voltage levels using a transformer.

Rotating a simple loop of wire or coils of wire between the poles of a magnet, such as that shown simplified in Fig. 2.33, will cut the north–south lines of magnetic flux and induce an a.c. voltage in the loop or coils of wire as shown by the display on a cathode ray oscilloscope.

This is an a.c. supply, an alternating current supply. The basic principle of the a.c. supply generated in a power-station is exactly the same as Fig. 2.33 except that powerful electromagnets are used and the power for rotation comes from a steam turbine.

In this section we will first of all consider the theoretical circuits of pure resistance, inductance and capacitance acting alone in an a.c. circuit before going on to consider the practical circuits of resistance, inductance and capacitance acting together. Let us first define some of our terms of reference.

Resistance

In any circuit, **resistance** is defined as opposition to current flow. From Ohm's law:

$$R = \frac{V_R}{I_R} (\Omega)$$

However, in an a.c. circuit, resistance is only part of the opposition to current flow. The inductance and capacitance of an a.c. circuit also cause an opposition to current flow, which we call *reactance*.

Inductive reactance (X_L) is the opposition to an a.c. current in an inductive circuit. It causes the current in the circuit to lag behind the applied voltage, as shown in Fig. 2.34. It is given by the formula:

$$X_L = 2\pi f L \ (\Omega)$$

where

$\pi = 3.142$ (a constant)
F = the frequency of the supply
L = the inductance of the circuit or by

$$X_L = \frac{V_L}{I_L}$$

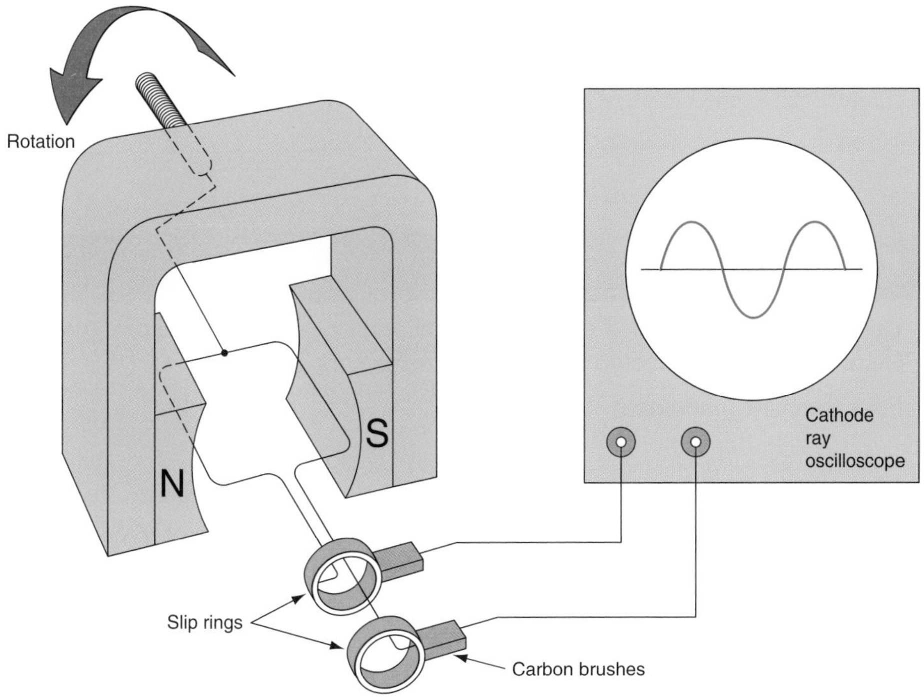

Figure 2.33 Simple a.c. generator or alternator.

Visit the companion website for more on this topic.

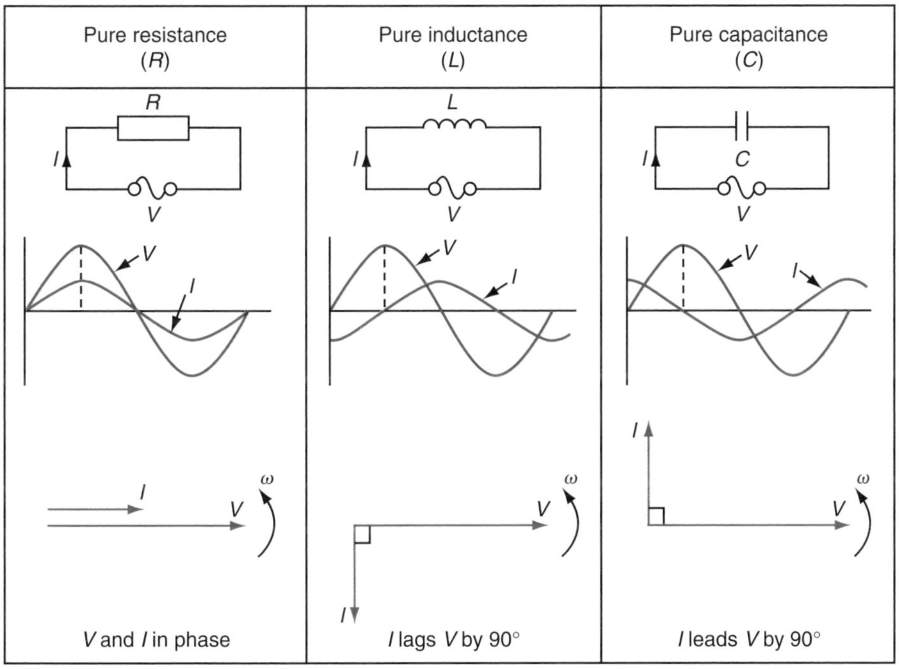

Figure 2.34 Voltage and current relationships in resistive, capacitive and inductive circuits.

Definition

Capacitive reactance (X_C) is the opposition to an a.c. current in a capacitive circuit. It causes the current in the circuit to lead ahead of the voltage.

Capacitive reactance (X_C) is the opposition to an a.c. current in a capacitive circuit. It causes the current in the circuit to lead ahead of the voltage, as shown in Fig. 2.34. It is given by the formula:

$$X_C = \frac{1}{2\pi fC} \ (\Omega)$$

where π and f are defined as before and C is the capacitance of the circuit. It can also be expressed as:

$$X_C = \frac{V_C}{I_C}$$

Example

Calculate the reactance of a 150 μF capacitor and a 0.05 H inductor if they were separately connected to the 50 Hz mains supply.

For capacitive reactance:

$$X_C = \frac{1}{2\pi fC}$$

where $f = 50\,\text{Hz}$ and $C = 150\,\mu\text{F} = 150 \times 10^{-6}\,\text{F}$

$$\therefore X_C = \frac{1}{2 \times 3.142 \times 50\,\text{Hz} \times 150 \times 10^{-6}\,\text{F}} = 21.2\,\Omega$$

For inductive reactance:

$$X_L = 2\pi fL$$

where $f = 50\,\text{Hz}$ and $L = 0.05\,\text{H}$

$$\therefore X_l = 2 \times 3.142 \times 50\,\text{Hz} \times 0.05\,\text{H} = 15.7\,\Omega$$

Impedance

Definition

The total opposition to current flow in an a.c. circuit is called *impedance* and given the symbol Z.

The total opposition to current flow in an a.c. circuit is called **impedance** and given the symbol Z. Thus impedance is the combined opposition to current flow of the resistance, inductive reactance and capacitive reactance of the circuit and can be calculated from the formula:

$$Z = \sqrt{R^2 + X^2} \ (\Omega)$$

or

$$Z = \frac{V_T}{I_T}$$

Example 1

Calculate the impedance when a 5 Ω resistor is connected in series with a 12 Ω inductive reactance.

$$Z = \sqrt{R^2 + X_L^2}\ (\Omega)$$
$$\therefore Z = \sqrt{5^2 + 12^2}$$
$$Z = \sqrt{25 + 144}$$
$$Z = \sqrt{169}$$
$$Z = 13\Omega$$

Example 2

Calculate the impedance when a 48 Ω resistor is connected in series with a 55 Ω capacitive reactance.

$$Z = \sqrt{R^2 + X_C^2}\ (\Omega)$$
$$\therefore Z = \sqrt{48^2 + 55^2}$$
$$Z = \sqrt{2304 + 3025}$$
$$Z = \sqrt{5329}$$
$$Z = 73\Omega$$

Resistance, inductance and capacitance in an a.c. circuit

When a resistor only is connected to an a.c. circuit the current and voltage waveforms remain together, starting and finishing at the same time. We say that the waveforms are *in phase*.

When a pure inductor is connected to an a.c. circuit the current lags behind the voltage waveform by an angle of 90°. We say that the current *lags* the voltage by 90°. When a pure capacitor is connected to an a.c. circuit the current *leads* the voltage by an angle of 90°. These various effects can be observed on an oscilloscope, but the circuit diagram, waveform diagram and phasor diagram for each circuit are shown in Fig. 2.34.

Phasor diagrams

Phasor diagrams and a.c. circuits are an inseparable combination. Phasor diagrams allow us to produce a model or picture of the circuit under consideration which helps us to understand the circuit. A phasor is a straight line, having definite length and direction, which represents to scale the magnitude and direction of a quantity such as a current, voltage or impedance.

To find the combined effect of two quantities we combine their phasors by adding the beginning of the second phasor to the end of the first. The combined effect of the two quantities is shown by the resultant phasor, which is measured from the original zero position to the end of the last phasor.

Definition

A *phasor* is a straight line, having definite length and direction, which represents to scale the magnitude and direction of a quantity such as a current, voltage or impedance.

Example

Find by phasor addition the combined effect of currents *A* and *B* acting in a circuit. Current *A* has a value of 4 A, and current *B* a value of 3 A, leading *A* by 90°.We usually assume phasors to rotate anticlockwise and so the complete diagram will be as shown in Fig. 2.32. Choose a scale of, for example, 1 A = 1 cm and draw the phasors to scale, that is, *A* = 4 cm and *B* = 3 cm, leading *A* by 90°.

The magnitude of the resultant phasor can be measured from the phasor diagram and is found to be 5 A acting at a phase angle ϕ of about 37° leading *A*. We therefore say that the combined effect of currents *A* and *B* is a current of 5 A at an angle of 37° leading *A*.

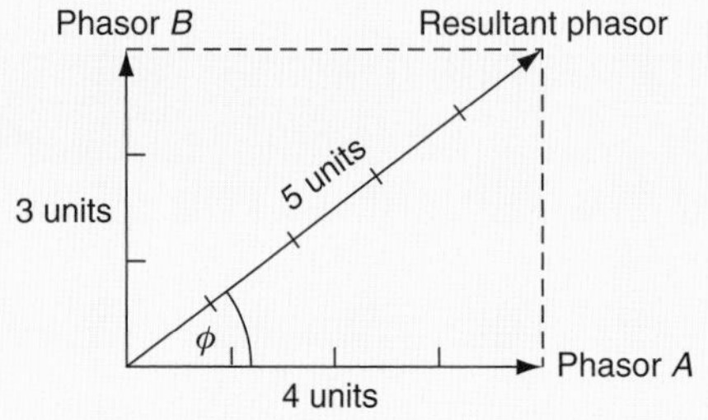

Figure 2.35 The phasor addition of currents A and B.

Phase angle ϕ

In an a.c. circuit containing resistance only, such as a heating circuit, the voltage and current are in phase, which means that they reach their peak and zero values together, as shown in Fig. 2.36(a).

In an a.c. circuit containing inductance, such as a motor or discharge lighting circuit, the current often reaches its maximum value after the voltage, which means that the current and voltage are out of phase with each other, as shown in Fig. 2.36(b). The phase difference, measured in degrees between the current and voltage, is called the phase angle of the circuit, and is denoted by the symbol ϕ, the lower-case Greek letter phi.

When circuits contain two or more separate elements, such as RL, RC or RLC, the phase angle between the total voltage and total current will be neither 0° nor 90° but will be determined by the relative values of resistance and reactance in the circuit. In Fig. 2.37 the phase angle between applied voltage and current is some angle ϕ.

Alternating current series circuits

In a circuit containing a resistor and inductor connected in series as shown in Fig. 2.37, the current *I* will flow through the resistor and the inductor causing the voltage V_R to be dropped across the resistor and V_L to be dropped across the inductor. The sum of these voltages will be equal to the total voltage V_T but because this is an a.c. circuit the voltages must be added by phasor addition. The result is shown in Fig. 2.37, where V_R is drawn to scale and in phase with

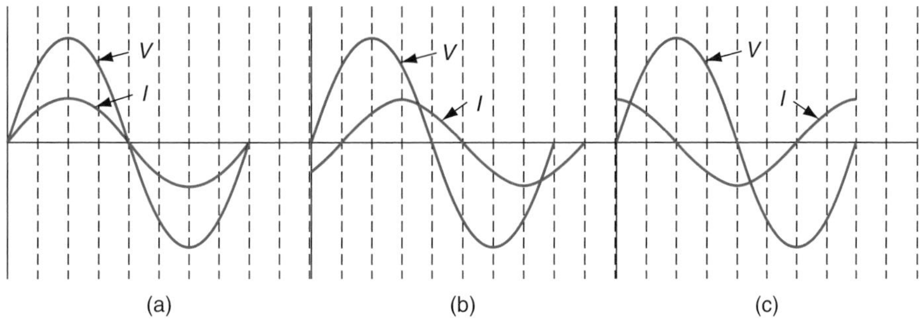

Figure 2.36 Phase relationship of a.c. waveform: (a) V and I in phase, phase angle $\phi = 0°$ and power factor = $\cos \phi = 1$; (b) V and I displaced by 45°, $\phi = 45°$ and p.f. = 0.707; and (c) V and I displaced by 90°, $\phi = 90°$ and p.f. = 0.

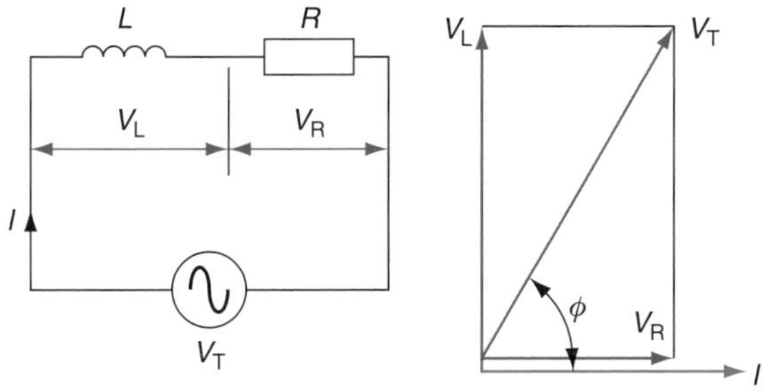

Figure 2.37 A series RL circuit and phasor diagram.

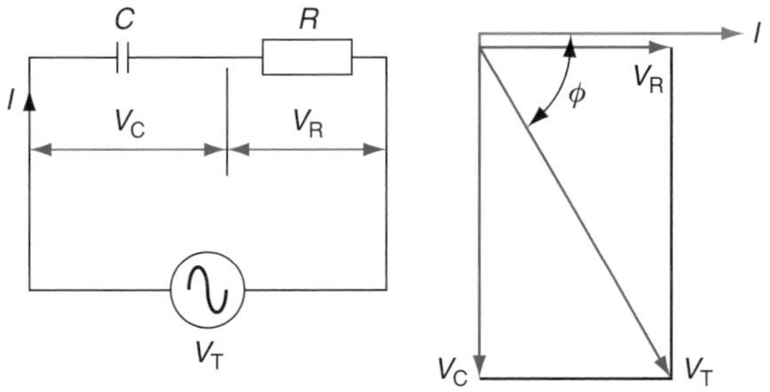

Figure 2.38 A series RC circuit and phasor diagram.

the current and V_L is drawn to scale and leading the current by 90°. The phasor addition of these two voltages gives us the magnitude and direction of V_T, which leads the current by some angle ϕ.

In a circuit containing a resistor and capacitor connected in series as shown in Fig. 2.38, the current I will flow through the resistor and capacitor causing voltage drops V_R and V_C. The voltage V_R will be in phase with the current and V_C will lag the current by 90°. The phasor addition of these voltages is equal to the total voltage V_T which, as can be seen in Fig. 2.38, is lagging the current by some angle ϕ.

The impedance triangle

We have now established the general shape of the phasor diagram for a series a.c. circuit. Figures 2.37 and 2.38 show the voltage phasors, but we know that $V_R = IR$, $V_L = IX_L$, $V_C = IX_C$ and $V_T = IZ$, and therefore the phasor diagrams

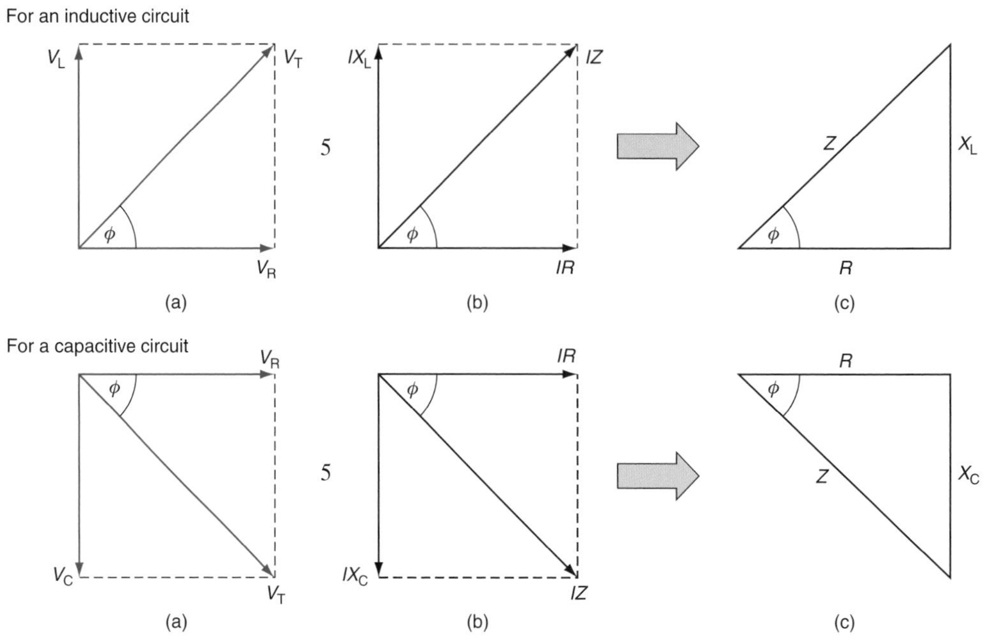

Figure 2.39 Phasor diagram and impedance triangle.

(a) and (b) in Fig. 2.39 must be equal. From Fig. 2.39(b), by the theorem of Pythagoras, we have:

$$(IZ)^2 = (IR)^2 + (IX)^2$$
$$I^2Z^2 = I^2R^2 + I^2X^2$$

If we now divide throughout by I^2 we have:

$$Z^2 = R^2 + X^2$$
$$\text{or } Z = \sqrt{R^2 + X^2}\ \Omega$$

The phasor diagram can be simplified to the impedance triangle given in Fig. 2.39(c).

Example 1

A coil of 0.15 H is connected in series with a 50 Ω resistor across a 100 V 50 Hz supply. Calculate (a) the reactance of the coil, (b) the impedance of the circuit and (c) the current.

For (a)

$$X_L = 2\pi fL\ (\Omega)$$
$$\therefore X_L = 2 \times 3.142 \times 50\,\text{Hz} \times 0.15\,\text{H} = 47.1\,\Omega$$

For (b)

$$Z = \sqrt{R^2 + X^2}\ (\Omega)$$
$$\therefore Z = \sqrt{(50\,\Omega)^2 + (47.1\,\Omega)^2} = 68.69\,\Omega$$

For (c)

$$I = \frac{V}{Z}\ (\text{A})$$
$$\therefore I = \frac{100V}{68.69\,\Omega} = 1.46\,\text{A}$$

Example 2

A 60 μF capacitor is connected in series with a 100 Ω resistor across a 230 V 50 Hz supply. Calculate (a) the reactance of the capacitor, (b) the impedance of the circuit and (c) the current.

For (a)

$$X_C = \frac{1}{2\pi fC}\ (\Omega)$$

$$\therefore X_C = \frac{1}{2\pi \times 50\,\text{Hz} \times 60 \times 10^{-6}\,\text{F}} = 53.05\,\Omega$$

For (b)

$$Z = \sqrt{R^2 + X^2}\ (\Omega)$$

$$\therefore Z = \sqrt{(100\,\Omega)^2 + (53.05\,\Omega)^2} = 113.2\,\Omega$$

For (c)

$$I = \frac{V}{Z}\ (\text{A})$$

$$\therefore I = \frac{230\,\text{V}}{113.2\,\Omega} = 2.03\,\text{A}$$

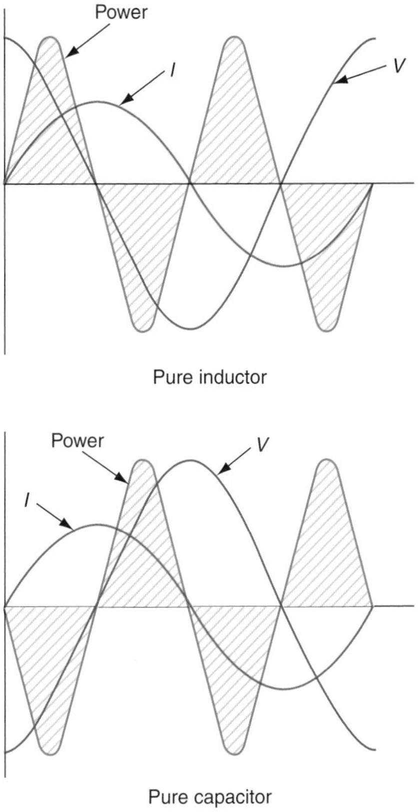

Figure 2.40 Waveform for the a.c. power in purely inductive and purely capacitive circuits.

Power and power factor

Power factor (p.f.) is defined as the cosine of the phase angle between the current and voltage:

$$\text{p.f.} = \cos\phi$$

If the current lags the voltage as shown in Fig. 2.37, we say that the p.f. is lagging, and if the current leads the voltage as shown in Fig. 2.38, the p.f. is said to be leading. From the trigonometry of the impedance triangle shown in Fig. 2.39, p.f. is also equal to:

$$\text{p.f.} = \cos\phi = \frac{R}{Z} = \frac{V_R}{V_T}$$

The electrical power in a circuit is the product of the instantaneous values of the voltage and current. Figure 2.40 shows the voltage and current waveform for a pure inductor and pure capacitor. The power waveform is obtained from the product of V and I at every instant in the cycle. It can be seen that the power waveform reverses every quarter-cycle, indicating that energy is alternately being fed into and taken out of the inductor and capacitor. When considered over one complete cycle, the positive and negative portions are equal, showing that the average power consumed by a pure inductor or capacitor is zero. This shows that inductors and capacitors store energy during one part of the voltage cycle and feed it back into the supply later in the cycle. Inductors store energy as a magnetic field and capacitors as an electric field.

In an electric circuit more power is taken from the supply than is fed back into it, since some power is dissipated by the resistance of the circuit, and therefore:

$$P = I^2R\ (\text{W})$$

Definition

Power factor (p.f.) is defined as the cosine of the phase angle between the current and voltage.

In any d.c. circuit the power consumed is given by the product of the voltage and current, because in a d.c. circuit voltage and current are in phase. In an a.c. circuit the power consumed is given by the product of the current and that part of the voltage which is in phase with the current. The in-phase component of the voltage is given by $V \cos \phi$, and so power can also be given by the equation:

$$P = VI \cos \phi \text{ (W)}$$

Example 1

A coil has a resistance of 30 Ω and a reactance of 40 Ω when connected to a 250 V supply. Calculate (a) the impedance, (b) the current, (c) the p.f. and (d) the power.

For (a)

$$Z = \sqrt{R^2 + X^2} \ (\Omega)$$

$$\therefore Z = \sqrt{(30\,\Omega)^2 + (40\,\Omega)^2} = 50\,\Omega$$

For (b)

$$I = \frac{V}{Z} \text{ (A)}$$

$$\therefore I = \frac{250\text{ V}}{50\,\Omega} = 5\text{ A}$$

For (c)

$$\text{p.f.} = \cos \phi = \frac{R}{Z}$$

$$\therefore \text{p.f.} = \frac{30\,\Omega}{50\,\Omega} = 0.6 \text{ lagging}$$

For (d)

$$P = VI \cos \phi \text{ (W)}$$

$$\therefore P = 250\text{ V} \times 5\text{ A} \times 0.6 = 750\text{ W}$$

Example 2

A capacitor of reactance 12 Ω is connected in series with a 9 Ω resistor across a 150 V supply. Calculate (a) the impedance of the circuit, (b) the current, (c) the p.f. and (d) the power.

For (a)

$$Z = \sqrt{R^2 + X^2} \ (\Omega)$$

$$\therefore Z = \sqrt{(9\,\Omega)^2 + (12\,\Omega)^2} = 15\,\Omega$$

For (b)

$$I = \frac{V}{Z} \text{ (A)}$$

$$\therefore I = \frac{150\text{ V}}{15\,\Omega} = 10\text{ A}$$

For (c)

$$\text{p.f.} = \cos \phi = \frac{R}{Z}$$

$$\therefore \text{p.f.} = \frac{9\,\Omega}{15\,\Omega} = 0.6 \text{ leading}$$

For (d)

$$P = VI \cos \phi \text{ (W)}$$

$$\therefore P = 150\text{ V} \times 10\text{ A} \times 0.6 = 900\text{ W}$$

The power factor of most industrial loads is lagging because the machines and discharge lighting used in industry are mostly inductive. This causes an additional magnetizing current to be drawn from the supply, which does not produce power, but does need to be supplied, making supply cables larger.

Example 3

A 230 V supply feeds three 1.84 kW loads with power factors of 1, 0.8 and 0.4. Calculate the current at each power factor.

The current is given by:

$$I = \frac{P}{V \cos \phi}$$

where $P = 1.84$ kW $= 1840$ W and $V = 230$ V. If the p.f. is 1, then:

$$I = \frac{1840\,\text{W}}{230\,\text{V} \times 1} = 8\,\text{A}$$

For a p.f. of 0.8:

$$I = \frac{1840\,\text{W}}{230\,\text{V} \times 0.8} = 10\,\text{A}$$

For a p.f. of 0.4:

$$I = \frac{1840\,\text{W}}{230\,\text{V} \times 0.4} = 20\,\text{A}$$

It can be seen from these calculations that a 1.84 kW load supplied at a power factor of 0.4 would require a 20 A cable, while the same load at unity power factor could be supplied with an 8 A cable. There may also be the problem of higher voltage drops in the supply cables. As a result, the supply companies encourage installation engineers to improve their power factor to a value close to 1 and sometimes charge penalties if the power factor falls below 0.8.

Power factor correction

Most installations have a low or bad power factor because of the inductive nature of the load. A capacitor has the opposite effect of an inductor, and so it seems reasonable to add a capacitor to a load which is known to have a lower or bad power factor, for example, a motor.

Figure 2.41(a) shows an industrial load with a low power factor. If a capacitor is connected in parallel with the load, the capacitor current I_C leads the applied voltage by 90°. When this capacitor current is added to the load current, as shown in Fig. 2.41(b), the resultant load current has a much improved power factor. However, using a slightly bigger capacitor, the load current can be pushed up until it is 'in phase' with the voltage, as can be seen in Fig. 2.41(c).

Capacitors may be connected across the main busbars of industrial loads in order to provide power factor improvement, but smaller capacitors may also be connected across an individual piece of equipment, as is the case for fluorescent light fittings.

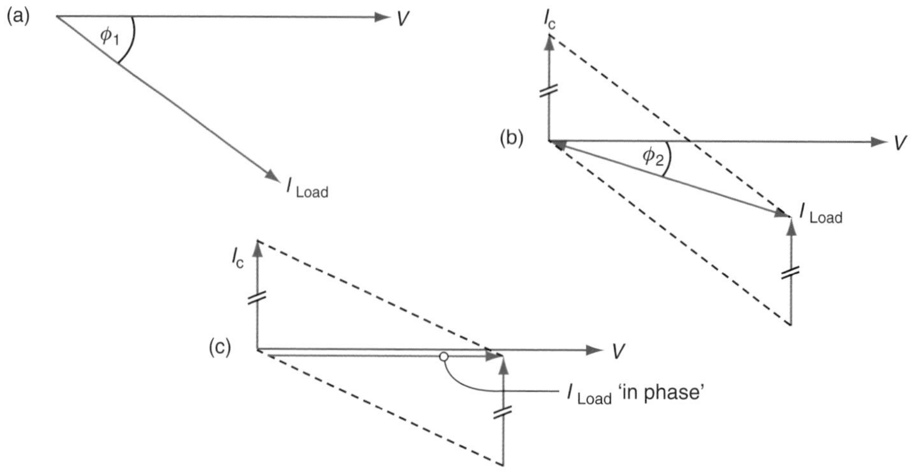

Figure 2.41 Power factor improvement using capacitors.

Electrical machines

All electrical machines operate on the principles of magnetism. The basic rules of magnetism were laid down earlier in this chapter. Here we will look at some of the laws of magnetism as they apply to electrical machines, such as generators, motors and transformers.

A current-carrying conductor maintains a magnetic field around the conductor which is proportional to the current flowing. When this magnetic field interacts with another magnetic field, forces are exerted which describe the basic principles of electric motors.

Michael Faraday demonstrated on 29 August 1831 that electricity could be produced by magnetism. He stated that 'when a conductor cuts or is cut by a magnetic field an e.m.f. is induced in that conductor. The amount of induced e.m.f. is proportional to the rate or speed at which the magnetic field cuts the conductor.' This basic principle laid down the laws of present-day electricity generation where a strong magnetic field is rotated inside a coil of wire to generate electricity.

Self- and mutual inductance

If a coil of wire is wound on to an iron core, as shown in Fig. 2.42, a magnetic field will become established in the core when a current flows in the coil due to the switch being closed.

When the switch is opened the current stops flowing and, therefore, the magnetic flux collapses. The collapsing magnetic flux induces an e.m.f. into the coil and this voltage appears across the switch contacts. The effect is known as *self-inductance*, or just *inductance*, and is one property of any coil. The unit of inductance is the henry (symbol H), to commemorate the work of the American physicist Joseph Henry (1797–1878), and a circuit is said to possess an inductance of 1 henry when an e.m.f. of 1 volt is induced in the circuit by a current changing at the rate of 1 ampere per second.

Fluorescent light fittings contain a choke or inductive coil in series with the tube and starter lamp. The starter lamp switches on and off very quickly, causing

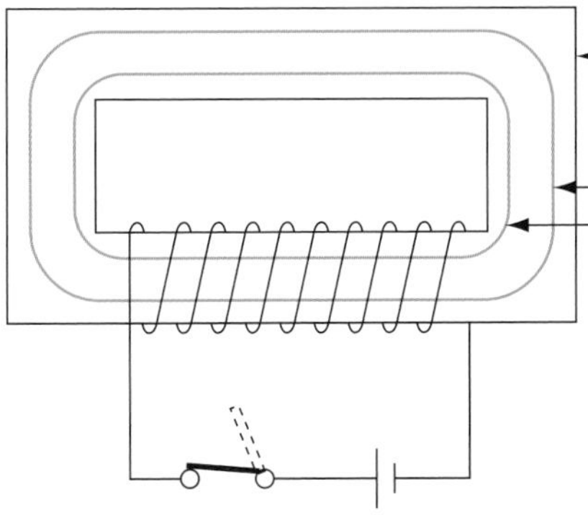

Figure 2.42 An inductive coil or choke.

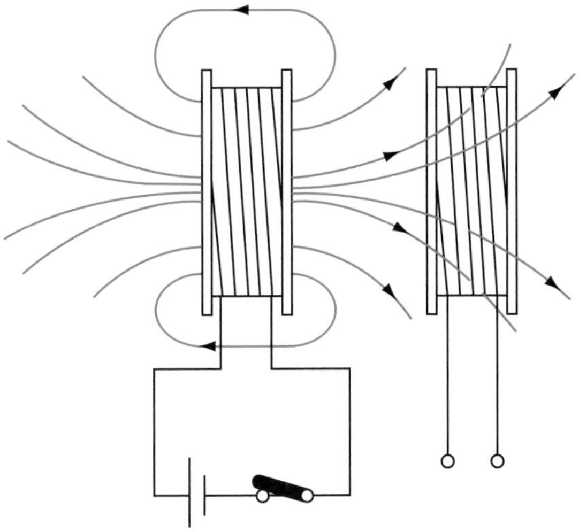

Figure 2.43 Mutual inductance between two coils.

rapid current changes which induce a large voltage across the tube electrodes sufficient to strike an arc in the tube.

When two separate coils are placed close together, as they are in a transformer, a current in one coil produces a magnetic flux which links with the second coil. This induces a voltage in the second coil and is the basic principle of the transformer action which is described later in this chapter. The two coils in this case are said to possess **mutual inductance**, as shown in Fig. 2.43. A mutual inductance of 1 henry exists between two coils when a uniformly varying current of 1 ampere per second in one coil produces an e.m.f. of 1 volt in the other coil.

Definition

A *mutual inductance* of 1 henry exists between two coils when a uniformly varying current of 1 ampere per second in one coil produces an e.m.f. of 1 volt in the other coil.

The e.m.f. induced in a coil such as that shown on the right-hand side in Fig. 2.43 is dependent upon the rate of change of magnetic flux and the number of turns on the coil.

Energy stored in a magnetic field

When we open the switch of an inductive circuit such as an electric motor or fluorescent light circuit the magnetic flux collapses and produces an arc across the switch contacts. The arc is produced by the stored magnetic energy being discharged across the switch contacts.

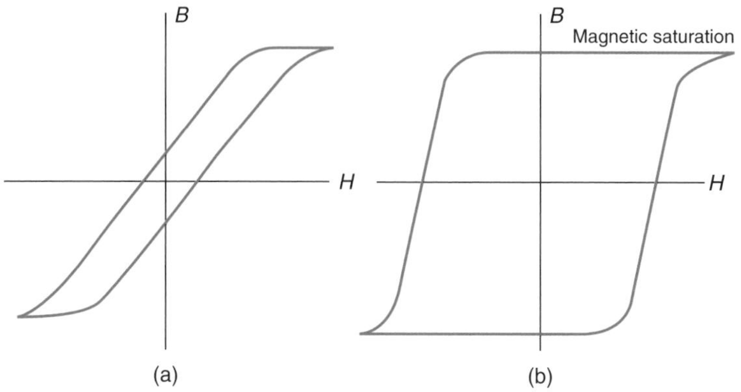

Figure 2.44 Magnetic hysteresis loops: (a) electromagnetic material, and (b) permanent magnetic material.

Magnetic hysteresis

There are many different types of magnetic material and they all respond differently to being magnetized. Some materials magnetize easily, and some are difficult to magnetize. Some materials retain their magnetism, while others lose it. The result will look like the graphs shown in Fig. 2.44 and are called hysteresis loops.

Definition

Magnetic hysteresis loops describe the way in which different materials respond to being magnetized.

Magnetic hysteresis loops describe the way in which different materials respond to being magnetized.

Materials from which permanent magnets are made should display a wide hysteresis loop, as shown by loop (b) in Fig. 2.44.

The core of an electromagnet is required to magnetize easily, and to lose its magnetism equally easily when switched off. Suitable materials will, therefore, display a narrow hysteresis loop, as shown by loop (a) in Fig. 2.44.

When an iron core is subjected to alternating magnetization, as in a transformer, the energy loss occurs at every cycle and so constitutes a continuous power loss, and, therefore, for applications such as transformers, a material with a narrow hysteresis loop is required.

Direct current motors

COMPANION @ WEBSITE Visit the companion website for more on this topic.

All electric motors work on the principle that when a current-carrying conductor is placed in a magnetic field it will experience a force. An electric motor uses this magnetic force to turn the shaft of the electric motor. Let us try to understand this action. If a current-carrying conductor is placed into the field of a permanent magnet, as shown in Fig. 2.45(c), a force *F* will be exerted on the conductor to push it out of the magnetic field.

To understand the force, let us consider each magnetic field acting alone. Figure 2.45(a) shows the magnetic field due to the current-carrying conductor only. Figure 2.45(b) shows the magnetic field due to the permanent magnet in which is placed the conductor carrying no current. Figure 2.45(c) shows the effect of the combined magnetic fields which are distorted and, because lines of magnetic flux never cross but behave like stretched elastic bands, always trying to find the shorter distance between a north and south pole, the force *F* is exerted on the conductor, pushing it out of the permanent magnetic field.

This is the basic motor principle, and the force F is dependent upon the strength of the magnetic field B, the magnitude of the current flowing in the conductor I and the length of conductor within the magnetic field l. The following equation expresses this relationship:

$$F = BlI \text{ (N)}$$

where B is in tesla, l is in metres, I is in amperes and F is in newtons.

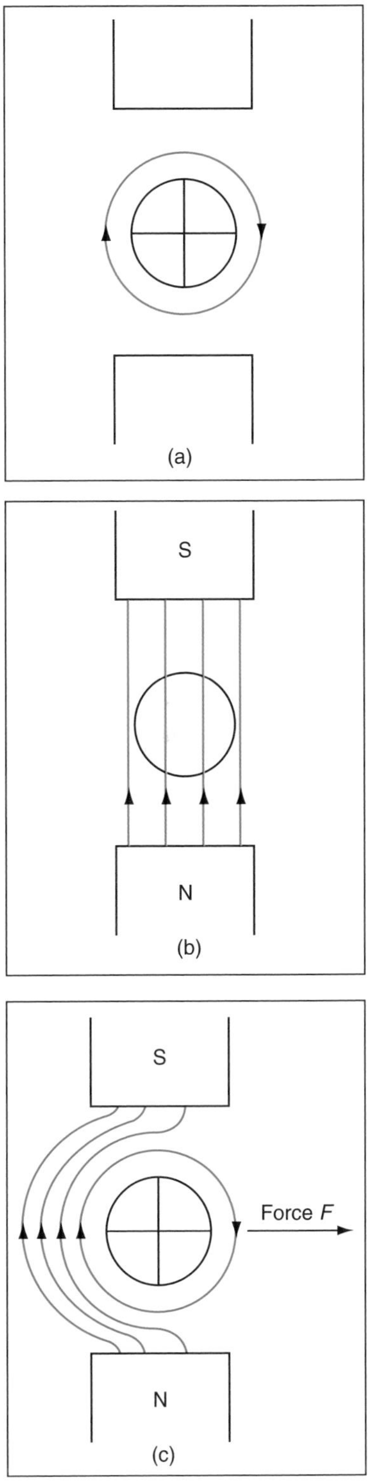

Figure 2.45 Force on a conductor in a magnetic field.

Example

A coil, which is made up of a conductor some 15 m in length, lies at right-angles to a magnetic field of strength 5 T. Calculate the force on the conductor when 15 A flows in the coil.

$$F = BlI \text{ (N)}$$

$$F = 5\text{T} \times 15\text{m} \times 15\text{A} = 1125\text{N}$$

Practical d.c. motors

Practical motors are constructed as shown in Fig. 2.46. All d.c. motors contain a field winding wound on pole pieces attached to a steel yoke. The armature winding rotates between the poles and is connected to the commutator. Contact with the external circuit is made through carbon brushes rubbing on the commutator segments. Direct current motors are classified by the way in which the field and armature windings are connected, which may be in series or in parallel.

Series motor

The field and armature windings are connected in series and consequently share the same current. The series motor has the characteristics of a high starting torque but a speed which varies with load. Figure 2.47 shows series motor connections and characteristics. For this reason the motor is only suitable for direct coupling to a load, except in very small motors, such as vacuum cleaners and hand drills, and is ideally suited for applications where the machine must start on load, such as electric trains, cranes and hoists.

Reversal of rotation may be achieved by reversing the connections of either the field or armature windings but not both. This characteristic means that the machine will run on both a.c. or d.c. and is, therefore, sometimes referred to as a 'universal' motor.

Three-phase a.c. motors

If a three-phase supply is connected to three separate windings equally distributed around the stationary part or stator of an electrical machine, an alternating current circulates in the coils and establishes a magnetic flux. The magnetic field established by the three-phase currents travels around the stator, establishing a rotating magnetic flux, creating magnetic forces on the rotor which turns the shaft on the motor.

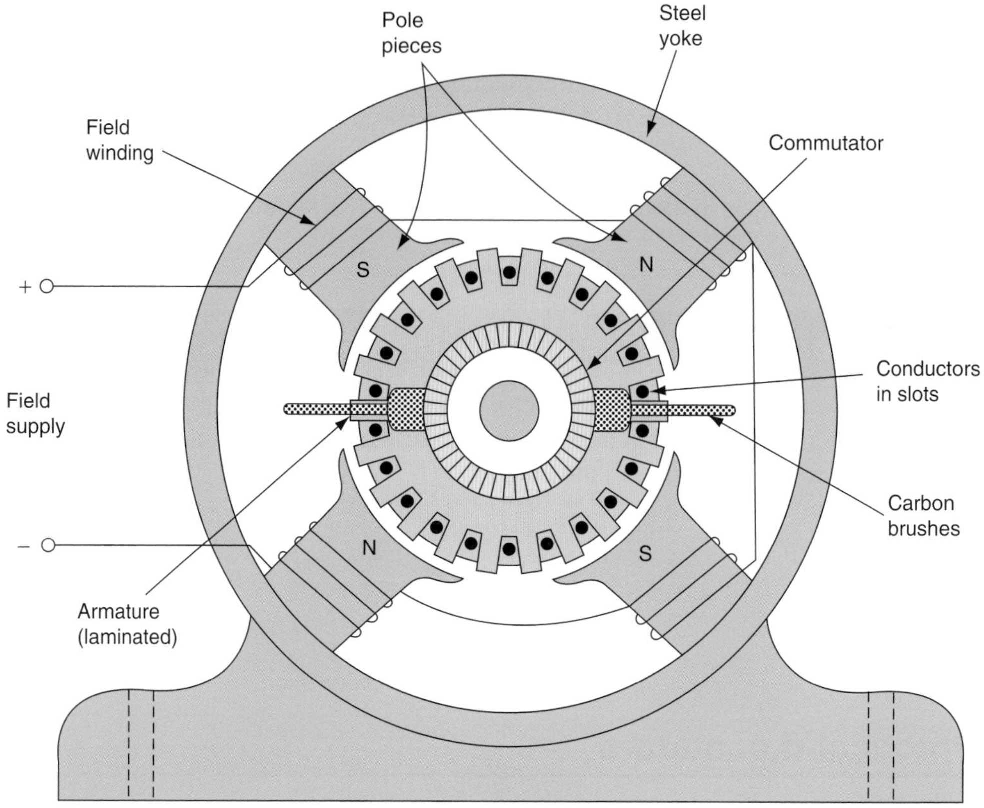

Figure 2.46 Showing d.c. machine construction.

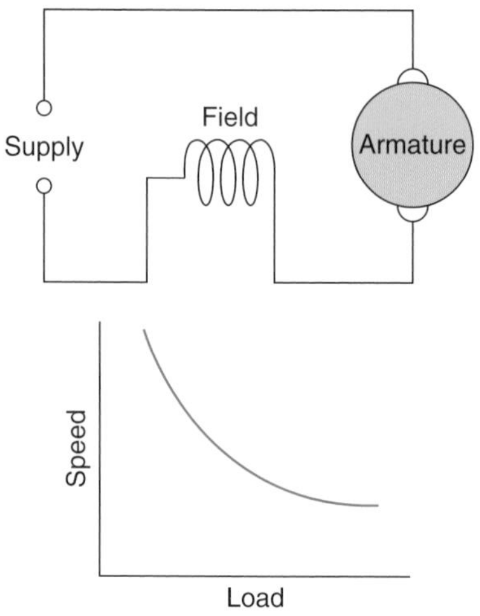

Figure 2.47 Series motor connections and characteristics.

Three-phase induction motor

When a three-phase supply is connected to insulated coils set into slots in the inner surface of the stator or stationary part of an induction motor, as shown in Fig. 2.49(a), a rotating magnetic flux is produced. The rotating magnetic flux cuts the conductors of the rotor and induces an e.m.f. in the rotor conductors by Faraday's law, which states that when a conductor cuts or is cut by a magnetic field an e.m.f. is induced in that conductor, the magnitude of which is proportional to the *rate* at which the conductor cuts or is cut by the magnetic flux. This induced e.m.f. causes rotor currents to flow and establish a magnetic flux which reacts with the stator flux and causes a force to be exerted on the rotor conductors, turning the rotor, as shown in Fig. 2.49(b).

The turning force or torque experienced by the rotor is produced by inducing an e.m.f. into the rotor conductors due to the *relative* motion between the conductors and the rotating field. The torque produces rotation in the same direction as the rotating magnetic field.

Rotor construction

There are two types of induction motor rotor – the wound rotor and the cage rotor. The cage rotor consists of a laminated cylinder of silicon steel with copper or aluminium bars slotted in holes around the circumference and short-circuited at each end of the cylinder, as shown in Fig. 2.50. In small motors the rotor is cast in aluminium. Better starting and quieter running are achieved if the bars are slightly skewed. This type of rotor is extremely robust and since there are no

Figure 2.48 (a) The internal workings of an induction motor; (b) The external shell of an induction motor.

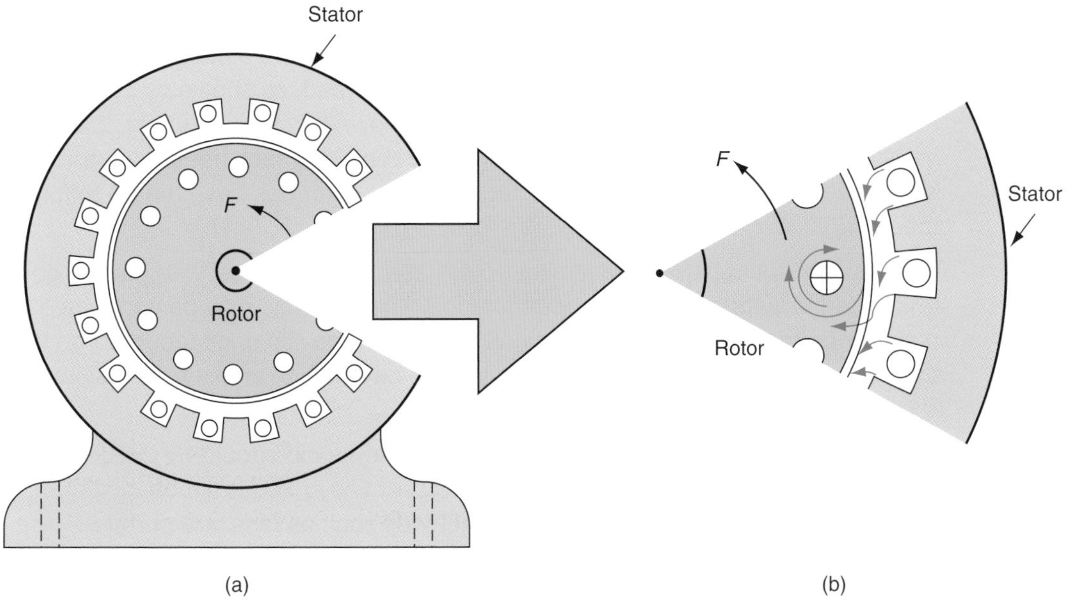

Figure 2.49 Segment taken out of an induction motor to show turning force: (a) construction of an induction motor, and (b) production of torque by magnetic fields.

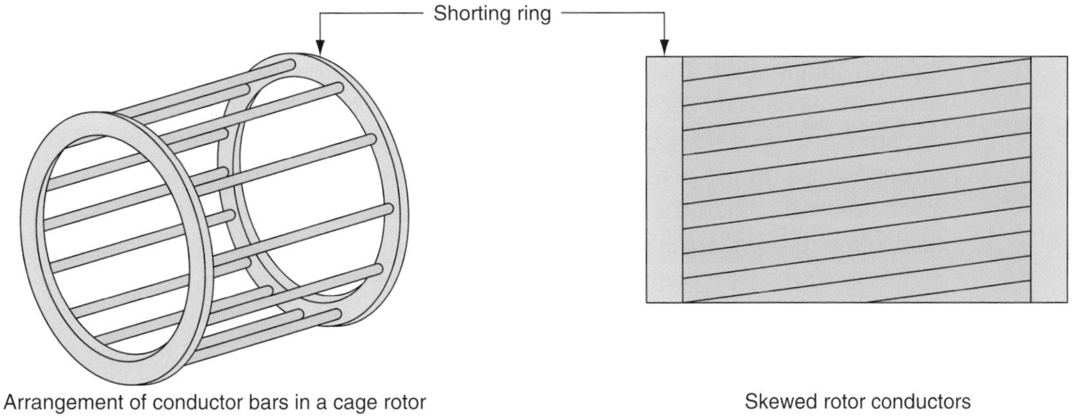

Figure 2.50 Construction of a cage rotor.

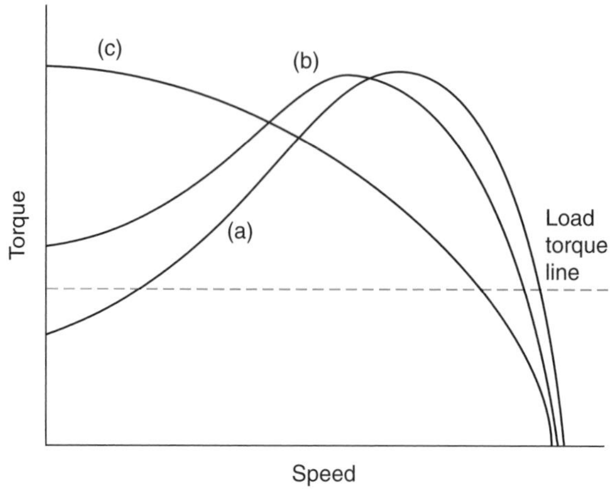

Figure 2.51 Various speed–torque characteristics for an induction motor.

external connections there is no need for slip rings or brushes. A machine fitted with a cage rotor does suffer from a low starting torque and a machine must be chosen which has a higher starting torque than the load, as shown by curve (b) in Fig. 2.51. A machine with the characteristic shown by curve (a) in Fig. 2.51 would not start since the load torque is greater than the machine-starting torque. Alternatively the load may be connected after the motor has been run up to full speed.

The wound rotor consists of a laminated cylinder of silicon steel with copper coils embedded in slots around the circumference. The windings may be connected in star or delta and the end connections brought out to slip rings mounted on the shaft. Connection by carbon brushes can then be made to an external resistance to improve starting.

The cage induction motor has a small starting torque and should be used with light loads or started with the load disconnected. The speed is almost constant. Its applications are for constant speed machines such as fans and pumps. Reversal of rotation is achieved by reversing any two of the stator winding connections.

Single-phase a.c. motors

A single-phase a.c. supply produces a pulsating magnetic field, not the rotating magnetic field produced by a three-phase supply. All a.c. motors require a rotating field to start. Therefore, single-phase a.c. motors have two windings which are electrically separated by about 90°. The two windings are known as the start and run windings. The magnetic fields produced by currents flowing through these out-of-phase windings create the rotating field and turning force required to start the motor. Once rotation is established, the pulsating field in the run winding is sufficient to maintain rotation and the start winding is disconnected by a centrifugal switch which operates when the motor has reached about 80% of the full load speed.

A cage rotor is used on single-phase a.c. motors, the turning force being produced in the way described previously for three-phase induction motors and shown in Fig. 2.49. Because both windings carry currents which are out of phase with each other, the motor is known as a 'split-phase' motor. The phase

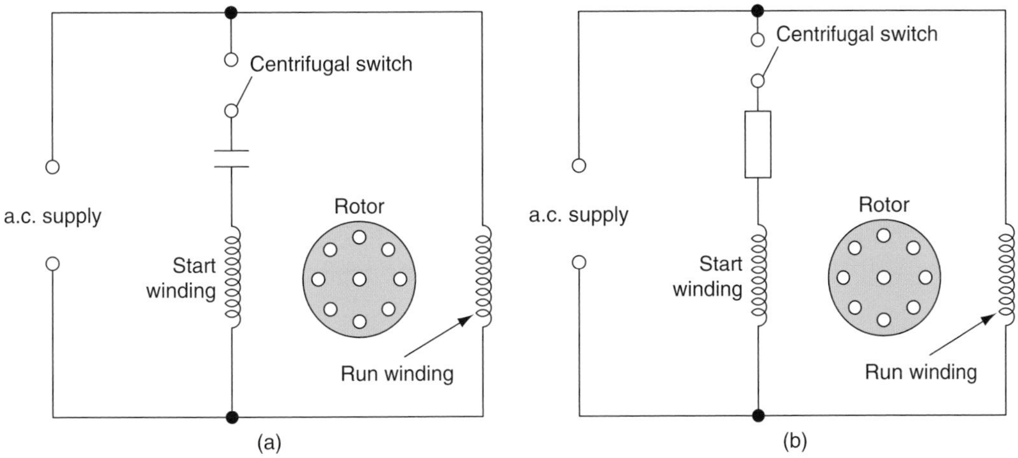

Figure 2.52 Circuit diagram of: (a) capacitor split-phase motors, and (b) resistance split-phase motors.

displacement between the currents in the windings is achieved in one of two ways:

- by connecting a capacitor in series with the start winding, as shown in Fig. 2.52(a), which gives a 90° phase difference between the currents in the start and run windings;
- by designing the start winding to have a high resistance and the run winding a high inductance, once again creating a 90° phase shift between the currents in each winding, as shown in Fig. 2.52(b).

When the motor is first switched on, the centrifugal switch is closed and the magnetic fields from the two coils produce the turning force required to run the rotor up to full speed. When the motor reaches about 80% of full speed, the centrifugal switch clicks open and the machine continues to run on the magnetic flux created by the run winding only.

Split-phase motors are constant speed machines with a low starting torque and are used on light loads such as fans, pumps, refrigerators and washing machines. Reversal of rotation may be achieved by reversing the connections to the start or run windings, but not both.

Shaded pole motors

The shaded pole motor is a simple, robust, single-phase motor, which is suitable for very small machines with a rating of less than about 50 W. Figure 2.53 shows a shaded pole motor. It has a cage rotor and the moving field is produced by enclosing one side of each stator pole in a solid copper or brass ring, called a shading ring, which displaces the magnetic field and creates an artificial phase shift.

Shaded pole motors are constant speed machines with a very low starting torque and are used on very light loads such as oven fans, record turntable motors and electric fan heaters. Reversal of rotation is theoretically possible by moving the shading rings to the opposite side of the stator pole face. However, in practice this is often not a simple process, but the motors are symmetrical and it is sometimes easier to reverse the rotor by removing the fixing bolts and reversing the whole motor.

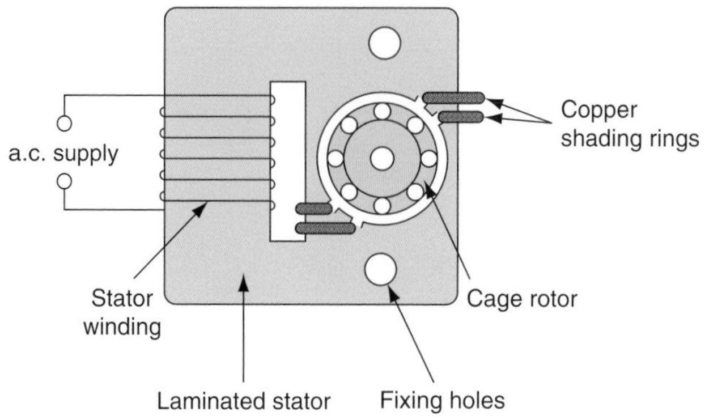

Figure 2.53 Shaded pole motor.

There are more motors operating from single-phase supplies than all other types of motor added together. Most of them operate as very small motors in domestic and business machines where single-phase supplies are most common.

Balancing single-phase loads

A three-phase load such as a motor has equally balanced phases since the resistance of each phase winding will be the same. Therefore, the current taken by each phase will be equal. When connecting single-phase loads to a three-phase supply, care should be taken to distribute the single-phase loads equally across the three phases so that each phase carries approximately the same current. Equally distributing the single-phase loads across the three-phase supply is known as 'balancing' the load. A lighting load of 18 luminaires would be 'balanced' if six luminaires were connected to each of the three phases.

Star and delta connections

The three-phase windings of an a.c. generator may be star connected or delta connected as shown in Fig. 2.54. The important relationship between phase and line currents and voltages is also shown. The square root of 3 ($\sqrt{3}$) is simply a constant for three-phase circuits, and has a value of 1.732. The delta connection is used for electrical power transmission because only three conductors are required. Delta connection is also used to connect the windings of most three-phase motors because the phase windings are perfectly balanced and, therefore, do not require a neutral connection.

Making a star connection at the local substation has the advantage that two voltages become available – a line voltage of 400 V between any two phases, and a phase voltage of 230 V between line and neutral which is connected to the star point.

In any star-connected system currents flow along the lines (I_L), through the load and return by the neutral conductor connected to the star point. In a *balanced* three-phase system all currents have the same value and when they are added up by phasor addition, we find the resultant current is zero. Therefore, no current flows in the neutral and the star point is at zero volts. The star point of the distribution transformer is earthed because earth is also at zero potential.

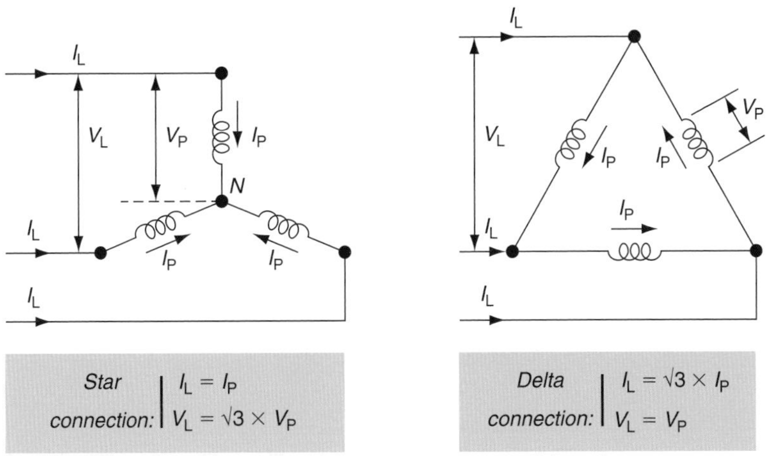

Figure 2.54 Star and delta connections.

A star-connected system is also called a three-phase four-wire system and allows us to connect single-phase loads to a three-phase system.

Three-phase power

We know from our single-phase alternating current theory earlier in this chapter that power can be found from the following formula:

$$\text{Power} = VI \cos\phi \text{ (W)}$$

In any balanced three-phase system, the total power is equal to three times the power in any one phase.

$$\therefore \text{ Total three-phase power} = 3V_P I_P \cos\phi \text{ (W)} \qquad \text{(Equation 1)}$$

Now for a star connection:

$$V_P = \frac{V_L}{\sqrt{3}} \text{ and } I_L = I_P \qquad \text{(Equation 2)}$$

Substituting Equation (2) into Equation (1), we have:

$$\text{Total three-phase power} = \sqrt{3}\, V_L L_L \cos\phi \text{ (W)}$$

Now consider a delta connection:

$$V_P = V_L \text{ and } I_P = \frac{I_L}{\sqrt{3}} \qquad \text{(Equation 3)}$$

Substituting Equation (3) into Equation (1) we have, for any balanced three-phase load,

$$\text{Total three-phase power} = \sqrt{3}\, V_L L_L \cos\phi \text{ (W)}$$

So, a general equation for three-phase power is:

$$\text{Power} = \sqrt{3}\, V_L I_L \cos\phi$$

Example 1

A balanced star-connected three-phase load of 10 Ω per phase is supplied from a 400 V, 50 Hz mains supply at unity power factor. Calculate (a) the phase voltage, (b) the line current and (c) the total power consumed.

For a star connection

$$V_L = \sqrt{3}\,V_P \text{ and } I_L = I_P$$

For (a)

$$V_P = \frac{V_L}{\sqrt{3}}\,(\text{V})$$

$$V_P = \frac{400\,\text{V}}{1.732} = 230.9\,\text{V}$$

For (b)

$$I_L = I_P = \frac{V_P}{R_P}\,(\text{A})$$

$$I_L = I_P = \frac{230.9\,\text{V}}{10\,\Omega} = 23.09\,\text{A}$$

For (c)

$$\text{Power} = \sqrt{3}\,V_L\,I_L\cos\phi\,(\text{W})$$

$$\therefore \text{Power} = 1.732 \times 400\,\text{V} \times 23.09\,\text{A} \times 1 = 16\,\text{kW}$$

Example 2

A 20 kW, 400 V balanced delta-connected load has a power factor of 0.8. Calculate (a) the line current and (b) the phase current.

We have that:

$$\text{Three-phase power} = \sqrt{3}\,V_L\,I_L\cos\phi\,(\text{W})$$

For (a)

$$I_L = \frac{\text{Power}}{\sqrt{3}\,V_L\cos\phi}\,(\text{A})$$

$$\therefore I_L = \frac{20{,}000\,\text{W}}{1.732 \times 400\,\text{V} \times 0.8}$$

$$I_L = 36.08\,(\text{A})$$

For delta connection

$$I_L = \sqrt{3}\,I_P\,(\text{A})$$

Thus, for (b)

$$I_P = I_L/\sqrt{3}\,(\text{A})$$

$$\therefore I_P = \frac{36.08\,\text{A}}{1.732} = 20.83\,\text{A}$$

Example 3

Three identical loads each having a resistance of 30 Ω and inductive reactance of 40 Ω are connected first in star and then in delta to a 400 V three-phase supply. Calculate the phase currents and line currents for each connection.

For each load

$$Z = \sqrt{R^2 + X_L^2}\ (\Omega)$$

$$\therefore Z = \sqrt{30^2 + 40^2}$$

$$Z = \sqrt{2500} = 50\,\Omega$$

For star connection

$$V_L = \sqrt{3}\,V_P \quad \text{and} \quad I_L = I_P$$

$$V_P = \frac{V_L}{\sqrt{3}}\ (\text{V})$$

$$\therefore V_P = \frac{400\,\text{V}}{1.732} = 230.9\,\text{V}$$

$$I_P = \frac{V_P}{Z_P}\ (\text{A})$$

$$\therefore I_P = \frac{230.9\,\text{V}}{50\,\Omega} = 4.62\,\text{A}$$

$$I_P = I_L$$

Therefore phase and line currents are both equal to 4.62 A.

For delta connection

$$V_L = V_P \quad \text{and} \quad I_L = \sqrt{3}\,I_P$$

$$V_L = V_P = 400\,\text{V}$$

$$I_P = V_P / Z_P\ (\text{A})$$

$$\therefore I = \frac{400\,\text{V}}{50\,\Omega} = 8\,\text{A}$$

$$I_L = \sqrt{3}\,I_P\ (\text{A})$$

$$\therefore I_L = 1.732 \times 8\,\text{A} = 13.86\,\text{A}$$

Transformers

A **transformer** is an electrical machine which is used to change the value of an alternating voltage. They vary in size from miniature units used in electronics to huge power transformers used in power-stations. A transformer will only work when an alternating voltage is connected. It will not normally work from a d.c. supply such as a battery.

A transformer, as shown in Fig. 2.55, consists of two coils, called the primary and secondary coils, or windings, which are insulated from each other and wound on to the same steel or iron core.

COMPANION WEBSITE

Visit the companion website for more on this topic.

Definition

A *transformer* is an electrical machine which is used to change the value of an alternating voltage.

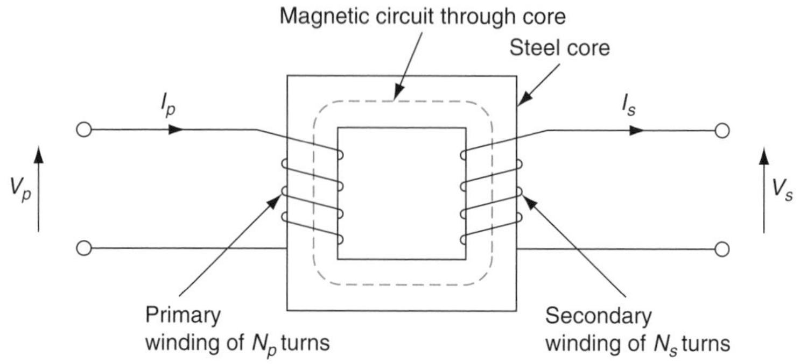

Figure 2.55 A simple transformer.

Figure 2.56 A 3 ph transformer.

An alternating voltage applied to the primary winding produces an alternating current, which sets up an alternating magnetic flux throughout the core. This magnetic flux induces an e.m.f. in the secondary winding, as described by Faraday's law, which says that when a conductor is cut by a magnetic field, an e.m.f. is induced in that conductor. Since both windings are linked by the same magnetic flux, the induced e.m.f. per turn will be the same for both windings. Therefore, the e.m.f. in both windings is proportional to the number of turns. In symbols:

$$\frac{V_p}{N_p} = \frac{V_s}{N_s}$$

where

V_p = the primary voltage
V_s = the secondary voltage
N_p = the number of primary turns
N_s = the number of secondary turns

Moving the terms around, we have a general expression for a transformer:

$$\frac{V_p}{V_s} = \frac{N_p}{N_s}$$

Try this

Maths

Using the general equation for a transformer given above, follow this maths carefully, step by step, in the following example.

Example

A 230 V to 12 V emergency lighting transformer is constructed with 800 primary turns. Calculate the number of secondary turns required. Collecting the information given in the question into a usable form, we have:

$$V_p = 230\text{ V}$$
$$V_s = 12\text{ V}$$
$$N_p = 800$$

From the general equation:

$$\frac{V_p}{V_s} = \frac{N_p}{N_s}$$

the equation for the secondary turn is:

$$N_s = \frac{N_p V_s}{V_p}$$

$$\therefore N_s = \frac{800 \times 12V}{230V} = 42\text{ turns}$$

42 turns are required on the secondary winding of this transformer to give a secondary voltage of 12 V.

Types of transformer

Step down transformers are used to reduce the output voltage, often for safety reasons. Figure 2.57 shows a step down transformer where the primary winding has twice as many turns as the secondary winding. The turns ratio is 2:1 and, therefore, the secondary voltage is halved.

Step up transformers are used to increase the output voltage. The electricity generated in a power-station is stepped up for distribution on the National Grid network. Figure 2.58 shows a step up transformer where the primary winding has only half the number of turns as the secondary winding. The turns ratio is 1:2 and, therefore, the secondary voltage is doubled.

Instrument transformers are used in industry and commerce so that large currents and voltage can be measured by small electrical instruments.

A current transformer (CT) has the large load currents connected to the primary winding of the transformer and the ammeter connected to the secondary winding. The ammeter is calibrated to take account of the turns ratio of the transformer, so that the ammeter displays the actual current being taken by the load when the ammeter is actually only taking a small proportion of the load current.

Definition

Step down transformers are used to reduce the output voltage, often for safety reasons.

Definition

Step up transformers are used to increase the output voltage. The electricity generated in a power-station is stepped up for distribution on the National Grid network.

Key fact

Transformers

- Transformers are very efficient (more than 80%) because they do not have any moving parts.

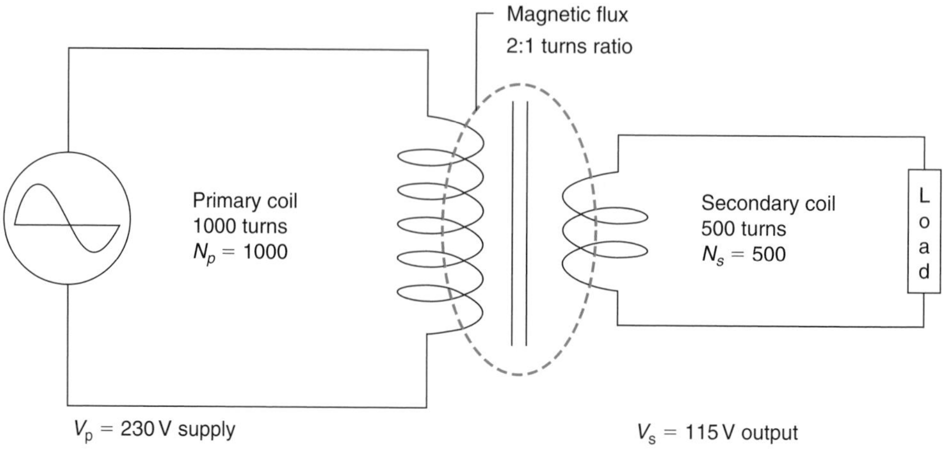

Figure 2.57 A step down transformer.

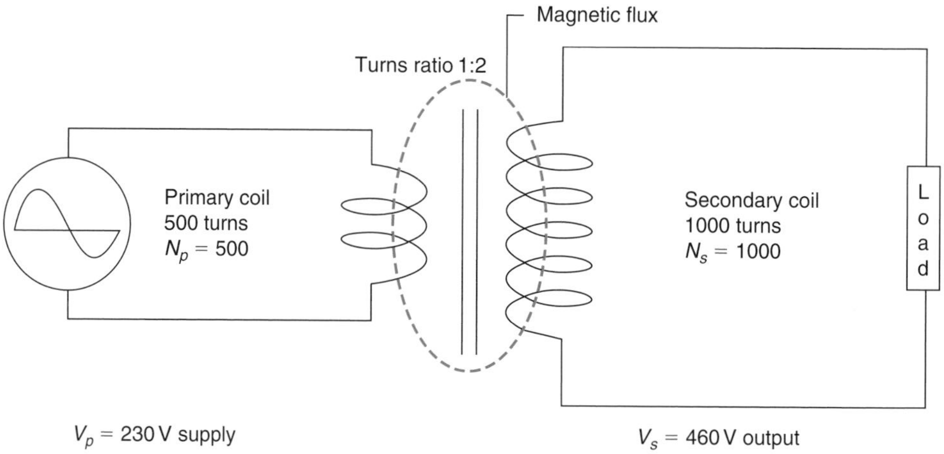

Figure 2.58 A step up transformer.

A voltage transformer (VT) has the main supply voltage connected to the primary winding of the transformer and the voltmeter connected to the secondary winding. The voltmeter is calibrated to take account of the turns ratio of the transformer, so that the voltmeter displays the actual supply voltage.

Isolating transformers – such as separated extra-low voltage (SELV) transformers. If the primary winding and the secondary winding of a double wound transformer have a separate connection to earth, then the output of the transformer is effectively isolated from the input since the only connection between the primary and secondary windings is the magnetic flux in the core.

Discharge lamps

Discharge lamps do not produce light by means of an incandescent filament but by the excitation of a gas or metallic vapour contained within a glass envelope. A voltage applied to two terminals or electrodes sealed into the end of a glass tube containing a gas or metallic vapour will excite the contents and produce light directly. Fluorescent tubes and CFLs operate on this principle.

Definition

Discharge lamps do not produce light by means of an incandescent filament but by the excitation of a gas or metallic vapour contained within a glass envelope.

Fluorescent luminaires

A **luminaire** is equipment which supports an electric lamp and distributes or filters the light created by the lamp. It is essentially the 'light fitting'.

Definition

A *luminaire* is equipment which supports an electric lamp and distributes or filters the light created by the lamp.

A lamp is a device for converting electrical energy into light energy. There are many types of lamps. General lighting service (GLS) lamps and tungsten halogen lamps use a very hot wire filament to create the light and so they also become very hot in use. Fluorescent tubes operate on the 'discharge' principle; that is, the excitation of a gas within a glass tube. They are cooler in operation and very efficient in converting electricity into light. They form the basic principle of most energy-efficient lamps.

Fluorescent lamps are linear arc tubes, internally coated with a fluorescent powder, containing a little low-pressure mercury vapour and argon gas. The lamp construction is shown in Fig. 2.59.

Passing a current through the electrodes of the tube produces a cloud of electrons that ionize the mercury vapour and the argon in the tube, producing invisible ultraviolet light and some blue light. The fluorescent powder on the inside of the glass tube is very sensitive to ultraviolet rays and converts this radiation into visible light.

Fluorescent luminaires require a simple electrical circuit to initiate the ionization of the gas in the tube and a device to control the current once the arc is struck and the lamp is illuminated. Such a circuit is shown in Fig. 2.60.

A typical application for a fluorescent luminaire is in suspended ceiling lighting modules used in many commercial buildings. Energy-efficient lamps use electricity much more efficiently.

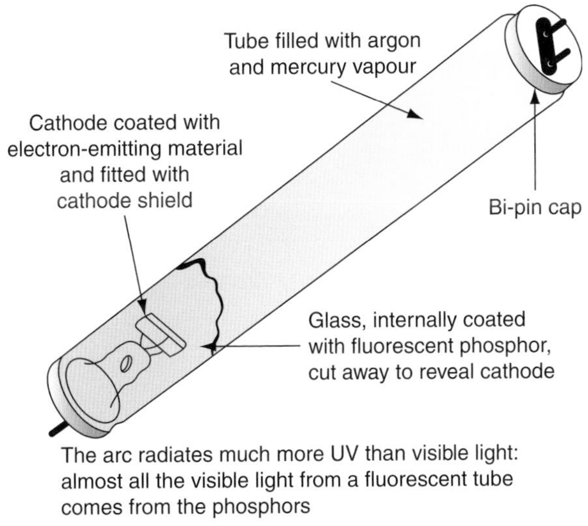

Figure 2.59 Fluorescent lamp construction.

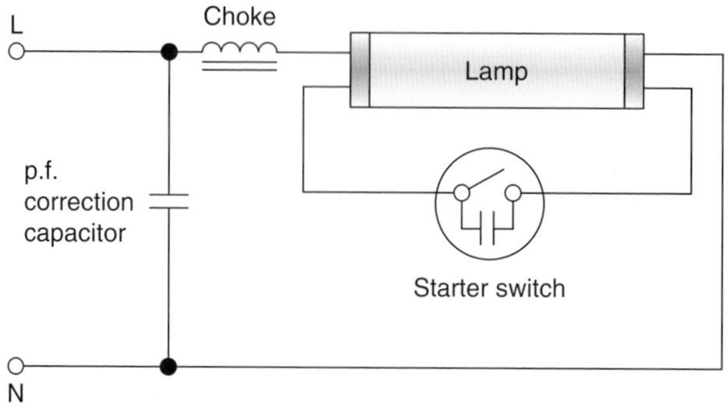

Figure 2.60 Fluorescent lamp circuit arrangement.

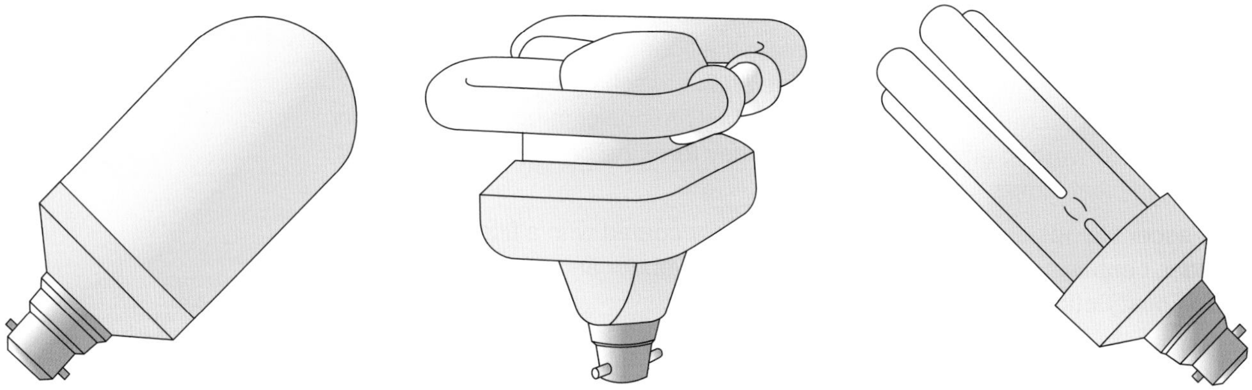

Figure 2.61 Energy-efficient lamps.

Definition

CFLs are miniature fluorescent lamps designed to replace ordinary GLS lamps.

Compact fluorescent lamps

CFLs are miniature fluorescent lamps designed to replace ordinary GLS lamps. They are available in a variety of shapes and sizes so that they can be fitted into existing light fittings. Figure 2.61 shows three typical shapes. The 'stick' type give most of their light output radially while the flat 'double D' type give most of their light output above and below.

Check your understanding

When you have completed the questions, check out the answers at the back of the book.

Note: more than one multiple-choice answer may be correct.

1 The SI unit of mass is the:
   a. kilogram or kg
   b. pound or lb
   c. metre or m
   d. millimetre or mm.

2 The SI unit of length is the:
   a. kilogram or kg
   b. pound or lb
   c. metre or m
   d. millimetre or mm.

3 The SI unit of time is the:
   a. minute or m
   b. second or s
   c. hour or h
   d. day or d.

4 The SI unit of electric current is the:
   a. ohm or Ω
   b. volt or V
   c. watt or W
   d. ampere or A.

5 The SI unit of resistance is the:
   a. ohm or Ω
   b. volt or V
   c. watt or W
   d. ampere or A.

6 The ampere is a measure of:
   a. potential difference
   b. power
   c. force
   d. electric current.

7 The watt is a measure of:
a. potential difference
b. power
c. force
d. electric current.

8 The volt is a measure of:
a. potential difference
b. power
c. force
d. electric current.

9 The newton is a measure of:
a. potential difference
b. power
c. force
d. electric current.

10 Which of the following may be defined as 'a measure of the force which a body exerts on anything which supports it':
a. acceleration
b. force
c. mass
d. weight.

11 Which of the following may be defined as 'a measure of the amount of material in a substance':
a. acceleration
b. force
c. mass
d. weight.

12 Which of the following may be defined as 'may cause a stationary object to move or bring a moving body to rest':
a. acceleration
b. force
c. mass
d. weight.

13 Which of the following may be defined as 'the force applied times the distance moved in the direction of the force':
a. acceleration
b. work done
c. power
d. velocity.

14 Which of the following may be defined as 'the rate of doing work':
a. acceleration
b. work done
c. power
d. velocity.

15 Which of the following may be defined as 'the speed in a given direction':

a. acceleration
b. work done
c. power
d. velocity.

16 Good conductor materials are:

a. copper
b. PVC
c. brass
d. wood.

17 Good insulator materials are:

a. copper
b. PVC
c. brass
d. wood.

18 A good conductor material:

a. has lots of free electrons
b. has no free electrons
c. may be made of copper
d. may be made of plastic.

19 A good insulator material:

a. has lots of free electrons
b. has no free electrons
c. may be made of copper
d. may be made of plastic.

20 In a series circuit:

a. the current is 'common' to all resistors
b. the voltage is 'common' to all resistors
c. $R_T = R_1 + R_2$
d. $\frac{1}{R_T} = \frac{1}{R_1} + \frac{1}{R_2}$.

21 In a parallel circuit:

a. the current is 'common' to all resistors
b. the voltage is 'common' to all resistors
c. $R_T = R_1 + R_2$
d. $\frac{1}{R_T} = \frac{1}{R_1} + \frac{1}{R_2}$.

22 The current taken by a $10\,\Omega$ resistor when connected to a 230 V supply will be:

a. 2.3 A
b. 10 A
c. 23 A
d. 230 A.

23 The resistance of a kettle element which takes 12 A from a 230 A main supply is:

a. 2.88 Ω
b. 5.00 Ω
c. 12.24 Ω
d. 19.16 Ω.

24 A 12 Ω filament lamp was found to be taking a current of 2 A at full brilliance. The voltage across the lamp under these conditions is:

a. 6 V
b. 12 V
c. 24 V
d. 48 V.

25 Current flowing through a solenoid sets up a magnetic flux. If an iron core is added to the solenoid while the current is maintained at a constant value, the magnetic flux will:

a. remain constant
b. totally collapse
c. decrease in strength
d. increase in strength.

26 Resistors of 6 Ω and 3 Ω are connected in series. The combined resistance value will be:

a. 2.0 Ω
b. 3.6 Ω
c. 6.3 Ω
d. 9.0 Ω.

27 Resistors of 6 Ω and 3 Ω are connected in parallel. The combined resistance value will be:

a. 2.0 Ω
b. 3.6 Ω
c. 6.3 Ω
d. 9.0 Ω.

28 Resistors of 60 Ω, 40 Ω and 20 Ω are connected in series. The total resistance value will be:

a. 10.9 Ω
b. 20.0 Ω
c. 60.6 Ω
d. 120 Ω.

29 Resistors of 20 Ω, 40 Ω and 60 Ω are connected in parallel. The total resistance value will be:

a. 10.9 Ω
b. 20.0 Ω
c. 60.0 Ω
d. 120 Ω.

30 Two identical resistors are connected in series across a 24 V battery. The voltage drop across each resistor will be:

a. 2 V
b. 6 V
c. 12 V
d. 24 V.

31 Two identical resistors are connected in parallel across a 24 V battery. The voltage drop across each resistor will be:

a. 2 V
b. 6 V
c. 12 V
d. 24 V.

32 The opposition to current flow in an a.c. resistive circuit is called:

a. resistance
b. inductance
c. reactance
d. impedance.

33 The opposition to current flow in an a.c. capacitive or inductive current is called:

a. resistance
b. inductance
c. reactance
d. impedance.

34 The total opposition to current flow in any a.c. circuit is called:

a. resistance
b. inductance
c. reactance
d. impedance.

35 A straight line having definite length and direction that represents to scale a quantity such as current voltage or impedance is called:

a. a series a.c. circuit
b. capacitive reactance
c. a phasor, as in a phasor diagram
d. the impedance triangle.

36 An a.c. series circuit has an inductive reactance of $4\,\Omega$ and a resistance of $3\,\Omega$. The impedance of this circuit will be:

a. $5\,\Omega$
b. $7\,\Omega$
c. $12\,\Omega$
d. $25\,\Omega$.

37 An a.c. series circuit has a capacitive reactance of 12 Ω and a resistance of 9 Ω. The impedance of this current will be:

a. 3 Ω
b. 15 Ω
c. 20 Ω
d. 108 Ω.

38 The inductive reactance of a 100 mH coil when connected to a 50 Hz supply will be:

a. 5 Ω
b. 20 Ω
c. 31.42 Ω
d. 31.42 kΩ.

39 The capacitive reactance of a 100 μF capacitor when connected to a 50 Hz supply will be:

a. 5 Ω
b. 20 Ω
c. 31.8 Ω
d. 31.8 kΩ.

40 A circuit with *bad* power factor causes:

a. a fall in the supply voltage
b. an increase in the supply voltage
c. more current to be taken from the supply
d. less current to be taken from the supply.

41 One application for a series d.c. motor is:

a. an electric train
b. a microwave oven
c. a central heating pump
d. an electric drill.

42 One application for an a.c. induction motor is:

a. an electric train
b. a microwave oven
c. a central heating pump
d. an electric drill.

43 One application for a shaded pole a.c. motor is:

a. an electric train
b. a microwave oven
c. a central heating pump
d. an electric drill.

44 A step down transformer has 1000 turns on the primary winding and 500 turns on the secondary winding. If the input voltage was 230 V the output voltage will be:

a. 2 V
b. 115 V
c. 200 V
d. 460 V.

45 An electromagnetic switch operated by a solenoid is one definition of:
   a. a transformer
   b. an a.c. motor
   c. a relay
   d. an inductive coil.

46 State the units of resistance and current.

47 Describe, with the aid of a simple diagram, how the atoms and electrons behave in a material said to be a good conductor of electricity.

48 Describe, with the aid of a simple diagram, how the atoms and electrons behave in a material said to be a good insulator.

49 List five materials which are used as good conductors in the electrical industry.

50 List five materials which are used as good insulators in the electrical industry.

51 Sketch a simple circuit of two resistors connected in series across a battery and explain how the current flows in this circuit.

52 Sketch a simple circuit of two resistors connected in parallel across a battery and explain how the current flows in this circuit.

53 Sketch a simple circuit to show how a voltmeter and ammeter would be connected into the circuit to measure total voltage and total current.

54 Sketch the construction of a simple transformer and label the primary and secondary windings. Why is the metal core of the transformer laminated? How do we cool a big power transformer?

55 List five practical applications for a transformer – for example, a shaver socket.

56 Describe the three effects of an electric current.

57 Sketch the magnetic flux patterns:
   a. around a simple bar magnet
   b. a horseshoe magnet
   c. explain the action and state one application for a solenoid.

58 Briefly describe what we mean by 'a turning force' and give five practical examples of this effect.

59 Briefly define what we mean by a 'simple machine' and give five examples.

60 Briefly describe what we mean by 'the efficiency of a machine'.

61 Sketch the construction of a simple alternator and label all the parts.

62 State how an e.m.f. is induced in an alternator. Sketch and name the shape of the generated e.m.f.

63 Calculate or state the average r.m.s. and maximum value of the domestic a.c. mains supply and show these values on a sketch of the mains supply.

64 Use a sketch with notes of explanation to describe ‘good’ and ‘bad’ power factor.

65 State how power factor correction is achieved on:
  a. a fluorescent light fitting
  b. an electric motor.

66 Use a sketch to help you describe the meaning of the words:
  a. inductance
  b. mutual inductance.

67 Use a sketch with notes of explanation to describe how a force is applied to a conductor in a magnetic circuit and how this principle is applied to an electric motor.

68 Use a sketch with notes of explanation to show how a turning force is applied to the rotor and, therefore, the drive shaft of an electric motor.

69 Sketch the magnetic hysteresis loop of a magnetic material suitable for:
  a. a permanent magnet
  b. a transformer.

70 Give three applications for each of the following types of motor:
  a. a d.c. series motor
  b. an a.c. induction motor
  c. an a.c. split-phase motor
  d. an a.c. shaded pole motor.

CHAPTER 3

C&G Unit 203

Electrical installations technology

Unit 203 of the City and Guilds 2365-02 syllabus

Learning outcomes – when you have completed this chapter you should:

- know about electrical industry regulations;
- know the electrical earthing requirements;
- know how electricity is generated and distributed;
- know the types of micro-renewable energies;
- know the types of wiring systems and circuits for different environments;
- know the types of technical information.

COMPANION @ WEBSITE

This chapter has free associated content, including animations and instructional videos, to support your learning.

When you see the logo, visit the companion website for more on this topic www.routledge.com/cw/linsley

Basic Electrical Installation Work. 978-1-138-84884-9

Electrical industry regulations

In Chapter 1 of this book we looked at the regulations which control all work activities. In this chapter, City and Guilds direct us to look at the regulations which control all of our electrical work activities.

If you have already successfully completed Chapter 1 and are confident that you remember this topic, go straight on to the next topic, 'Non-compliance with the Regulations' on page 156.

The most important recent piece of health and safety law was passed by Parliament in 1974 called the Health and Safety at Work Act. This Act gave added powers to the Inspectorate and is the basis of all modern statutory health and safety laws. This law not only increased the employer's liability for safety measures, but also put the responsibility for safety on employees.

Health, safety and welfare legislation has increased the awareness of everyone to the risks involved in the workplace. All statutes within the Acts of Parliament must be obeyed and, therefore, we all need an understanding of the laws as they apply to the electrical industry.

Statutory laws

Definition

Statutory Regulations have been passed by Parliament and have, therefore, become laws.

Acts of Parliament are made up of Statutes. **Statutory Regulations** have been passed by Parliament and have, therefore, become laws. Non-compliance with the laws of this land may lead to prosecution by the Courts and possible imprisonment for offenders.

We shall now look at some of the Statutory Regulations as they apply to the electrical industry.

The Health and Safety at Work Act 1974

Many governments have passed laws aimed at improving safety at work, but the most important recent legislation has been the Health and Safety at Work Act 1974. The purpose of the Act is to provide the legal framework for stimulating and encouraging high standards of health and safety at work; the Act puts the responsibility for safety at work on both workers and managers.

The employer has a duty to care for the health and safety of employees (Section 2 of the Act). To do this the employer must ensure that:

- the working conditions and standard of hygiene are appropriate;
- the plant, tools and equipment are properly maintained;
- the necessary safety equipment – such as personal protective equipment (PPE), dust and fume extractors and machine guards – is available and properly used;
- the workers are trained to use equipment and plant safely.

Employees have a duty to care for their own health and safety and that of others who may be affected by their actions (Section 7 of the Act). To do this they must:

- take reasonable care to avoid injury to themselves or to others as a result of their work activity;

- cooperate with their employer, helping him or her to comply with the requirements of the Act;
- not interfere with or misuse anything provided to protect their health and safety.

Failure to comply with the Health and Safety at Work Act is a criminal offence and any infringement of the law can result in heavy fines, a prison sentence or both.

Enforcement of Health and Safety Regulations

Laws and rules must be enforced if they are to be effective. The system of control under the Health and Safety at Work Act comes from the Health and Safety Executive (HSE) which is charged with enforcing the law. The HSE is divided into a number of specialist inspectorates or sections which operate from local offices throughout the United Kingdom. From the local offices the Inspectors visit individual places of work.

The HSE Inspectors have been given wide-ranging powers to assist them in the enforcement of the law. They can:

1 Enter premises unannounced and carry out investigations, take measurements or photographs.
2 Take statements from individuals.
3 Check the records and documents required by legislation.
4 Give information and advice to an employee or employer about safety in the workplace.
5 Demand the dismantling or destruction of any equipment, material or substance likely to cause immediate serious injury.
6 Issue an improvement notice which will require an employer to put right, within a specified period of time, a minor infringement of the legislation.
7 Issue a prohibition notice which will require an employer to stop immediately any activity likely to result in serious injury, and which will be enforced until the situation is corrected.
8 Prosecute all persons who fail to comply with their safety duties, including employers, employees, designers, manufacturers, suppliers and the self-employed.

Safety documentation

Under the Health and Safety at Work Act, the employer is responsible for ensuring that adequate instruction and information is given to employees to make them safety conscious. Part 1, Section 3 of the Act instructs all employers to prepare a written health and safety policy statement and to bring this to the notice of all employees. Your employer must let you know who your safety representatives are.

To promote adequate health and safety measures the employer must consult with the employees. An employee who identifies a dangerous situation should initially report to his or her site safety representative, who should then bring the dangerous situation to the notice of the management for action which will remove the danger. This may mean changing company policy or procedures, or making modifications to equipment. All actions which relate to health and safety should be documented and recorded as evidence that the company takes its health and safety policy seriously.

The Electricity Safety, Quality and Continuity Regulations 2002 (formerly Electricity Supply Regulations 1989)

The Electricity Safety, Quality and Continuity Regulations 2002 are issued by the Department of Trade and Industry. They are statutory regulations which are enforceable by the laws of the land. They are designed to ensure a proper and safe supply of electrical energy up to the consumer's terminals.

These regulations impose requirements upon the regional electricity companies regarding the installation and use of electric lines and equipment. The regulations are administered by the Engineering Inspectorate of the Electricity Division of the Department of Energy and will not normally concern the electrical contractor except that it is these regulations which lay down the earthing requirement of the electrical supply at the meter position.

The regional electricity companies must declare the supply voltage and maintain its value between prescribed limits or tolerances.

The government agreed on 1 January 1995 that the electricity supplies in the United Kingdom would be harmonized with those of the rest of Europe. Thus the voltages used previously in low-voltage supply systems of 415 V and 240 V have become 400 V for three-phase supplies and 230 V for single-phase supplies. The permitted tolerances to the nominal voltage have also been changed from $\pm 6\%$ to $+10\%$ and -6%. This gives a voltage range of 216–253 V for a nominal voltage of 230 V and 376–440 V for a nominal supply voltage of 400 V.

The next proposed change is for the tolerance levels to be adjusted to $\pm 10\%$ of the declared nominal voltage (IET Regulation, Appendix 2:14).

The frequency is maintained at an average value of 50 Hz over 24 hours so that electric clocks remain accurate.

Regulation 29 gives the area boards the power to refuse to connect a supply to an installation which in their opinion is not constructed, installed and protected to an appropriately high standard. This regulation would only be enforced if the installation did not meet the requirements of the IET Regulations for Electrical Installations.

The Electricity at Work Regulations 1989 (EWR)

This legislation came into force in 1990 and replaced earlier regulations such as the Electricity (Factories Act) Special Regulations 1944. The regulations are made under the Health and Safety at Work Act 1974, and enforced by the Health and Safety Executive. The purpose of the regulations is to 'require precautions to be taken against the risk of death or personal injury from electricity in work activities'.

Section 4 of the EWR tells us that 'all systems must be constructed so as to prevent danger … and be properly maintained. … Every work activity shall be carried out in a manner which does not give rise to danger. … In the case of work of an electrical nature, it is preferable that the conductors be made dead before work commences.'

The EWR do not tell us specifically how to carry out our work activities and ensure compliance, but if proceedings were brought against an individual for breaking the EWR, the only acceptable defence would be 'to prove that all reasonable steps were taken and all diligence exercised to avoid the offence' (Regulation 29).

An electrical contractor could reasonably be expected to have 'exercised all diligence' if the installation was wired according to the IET Wiring Regulations (see below). However, electrical contractors must become more 'legally aware' following the conviction of an electrician for manslaughter at Maidstone Crown Court in 1989. The court accepted that an electrician had caused the death of another man as a result of his shoddy work in wiring up a central heating system. He received a nine-month suspended prison sentence. This case has set an important legal precedent, and in future any tradesman or professional who causes death through negligence or poor workmanship risks prosecution and possible imprisonment.

The Management of Health and Safety at Work Regulations 1999

The Health and Safety at Work Act 1974 places responsibilities on employers to have robust health and safety systems and procedures in the workplace. Directors and managers of any company who employ more than five employees can be held personally responsible for failures to control health and safety.

The Management of Health and Safety at Work Regulations 1999 tell us that employers must systematically examine the workplace, the work activity and the management of safety in the establishment through a process of 'risk assessments'. A record of all significant risk assessment findings must be kept in a safe place and be available to an HSE Inspector if required. Information based on these findings must be communicated to relevant staff and, if changes in work behaviour patterns are recommended in the interests of safety, they must be put in place. The process of risk assessment is considered in detail later in this chapter.

Risks, which may require a formal assessment in the electrical industry, might be:

- working at heights;
- using electrical power tools;
- falling objects;
- working in confined places;
- electrocution and personal injury;
- working with 'live' equipment;
- using hire equipment;
- manual handling – pushing – pulling – lifting;
- site conditions – falling objects – dust – weather – water – accidents and injuries.

And any other risks which are particular to a specific type of workplace or work activity.

The Control of Substances Hazardous to Health Regulations 2002 (COSHH)

The original COSHH Regulations were published in 1988 and came into force in October 1989. They were re-enacted in 1994 with modifications and improvements, and the latest modifications and additions came into force in 2002.

The COSHH Regulations control people's exposure to hazardous substances in the workplace. Regulation 6 requires employers to assess the risks to health from working with hazardous substances, to train employees in techniques which will reduce the risk and provide personal protective equipment (PPE) so that employees will not endanger themselves or others through exposure to hazardous substances. Employees should also know what cleaning, storage and disposal procedures are required and what emergency procedures to follow. The necessary information must be available to anyone using hazardous substances as well as to visiting HSE Inspectors.

Hazardous substances include:

1 Any substance which gives off fumes causing headaches or respiratory irritation.
2 Man-made fibres which might cause skin or eye irritation (e.g. loft insulation).
3 Acids causing skin burns and breathing irritation (e.g. car batteries, which contain dilute sulphuric acid).
4 Solvents causing skin and respiratory irritation (strong solvents are used to cement together PVC conduit fittings and tube).
5 Fumes and gases causing asphyxiation (burning PVC gives off toxic fumes).
6 Cement and wood dust causing breathing problems and eye irritation.
7 Exposure to asbestos – although the supply and use of the most hazardous asbestos material is now prohibited, huge amounts were installed between 1950 and 1980 in the construction industry and much of it is still in place today. In their latest amendments, the COSHH Regulations focus on giving advice and guidance to builders and contractors on the safe use and control of asbestos products. These can be found in Guidance Notes EH 71 or visit www.hse.uk/hiddenkiller.

Where PPE is provided by an employer, employees have a duty to use it to safeguard themselves.

Provision and Use of Work Equipment Regulations 1998

These regulations tidy up a number of existing requirements already in place under other regulations such as the Health and Safety at Work Act 1974, the Factories Act 1961 and the Offices, Shops and Railway Premises Act 1963.

The Provision and Use of Work Equipment Regulations 1998 place a general duty on employers to ensure minimum requirements of plant and equipment. If an employer has purchased good-quality plant and equipment which is well maintained, there is little else to do. Some older equipment may require modifications to bring it in line with modern standards of dust extraction, fume extraction or noise, but no assessments are required by the regulations other than those generally required by the Management Regulations 1999 discussed previously.

The Construction (Health, Safety and Welfare) Regulations 1996

An electrical contractor is a part of the construction team, usually as a subcontractor, and therefore the regulations particularly aimed at the construction industry also influence the daily work procedures and environment of an electrician. The most important recent piece of legislation is the Construction Regulations.

The temporary nature of construction sites makes them one of the most dangerous places to work. These regulations are made under the Health and Safety at Work Act 1974 and are designed specifically to promote safety at work in the construction industry. Construction work is defined as any building or civil engineering work, including construction, assembly, alterations, conversions, repairs, upkeep, maintenance or dismantling of a structure.

The general provision sets out minimum standards to promote a good level of safety on-site. Schedules specify the requirements for guardrails, working platforms, ladders, emergency procedures, lighting and welfare facilities. Welfare facilities set out minimum provisions for site accommodation: washing facilities, sanitary conveniences and protective clothing. There is now a duty for all those working on construction sites to wear head protection, and this includes electricians working on-site as subcontractors.

The Construction (Design and Management) Regulations 1994

The Construction (Design and Management) Regulations (CDM) are aimed at improving the overall management of health, safety and welfare throughout all stages of the construction project.

The person requesting that construction work commence, the client, must first of all appoint a '**duty holder**', someone who has a duty of care for health, safety and welfare matters on-site. This person will be called a 'planning supervisor'. The planning supervisor must produce a 'pre-tender' health and safety plan and coordinate and manage this plan during the early stages of construction.

> **Definition**
>
> *'Duty holder'*, someone who has a duty of care for health, safety and welfare matters on-site. This phrase recognizes the level of responsibility which electricians are expected to take on as part of their job in order to control electrical safety in the work environment.

The client must also appoint a principal contractor who is then required to develop the health and safety plan made by the planning supervisor, and keep it up to date during the construction process to completion.

The degree of detail in the health and safety plan should be in proportion to the size of the construction project and recognize the health and safety risks involved on that particular project. Small projects will require simple, straightforward plans; large projects, or those involving significant risk, will require more detail. The CDM Regulations will apply to most large construction projects but they do not apply to the following:

- construction work, other than demolition work, that does not last longer than 30 days and does not involve more than four people;
- construction work carried out inside commercial buildings such as shops and offices, which does not interrupt the normal activities carried out on those premises;
- construction work carried out for a domestic client;
- the maintenance and removal of pipes or lagging which form a part of a heating or water system within the building.

The Management of Health and Safety at Work Regulations 1999

The Health and Safety at Work Act 1974 places responsibilities on employers to have robust health and safety systems and procedures in the workplace. Directors and managers of any company who employ more than five people can be held personally responsible for failures to control health and safety.

The Management of Health and Safety at Work Regulations 1999 tell us that employers must systematically examine the workplace, the work activity and the management of safety in the establishment through a process of 'risk assessments'. A record of all significant risk assessment findings must be kept in a safe place and be available to an HSE Inspector if required. Information based on these findings must be communicated to relevant staff and if changes in work behaviour patterns are recommended in the interests of safety, then they must be put in place. The process of risk assessment is considered in detail in Chapter 1 of this book.

Noise at work

Figure 3.1 Ear protectors protect workers from noise at work.

Noise is accepted as a problem in most industries and surprisingly there has been very little control legislation. The Health and Safety Executive have published a 'Code of Practice' and 'Guidance Notes' HSG 56 for reducing the exposure of employed persons to noise. A continuous exposure limit of below 85 dB for an eight-hour working day is recommended by the Code.

Noise may be defined as any disagreeable or undesirable sound or sounds, generally of a random nature, which do not have clearly defined frequencies. The usual basis for measuring noise or sound level is the decibel scale. Whether noise of a particular level is harmful or not also depends on the length of exposure to it. This is the basis of the widely accepted limit of 85 dB of continuous exposure to noise for eight hours per day.

A peak sound pressure of above 200 pascals or about 120 dB is considered unacceptable and 130 dB is the threshold of pain for humans. If a person has to shout to be understood at 2 m, the background noise is about 85 dB. If the distance is only 1 m, the noise level is about 90 dB. Continuous noise at work causes deafness, makes people irritable, affects concentration, causes fatigue and accident proneness, and may mask sounds which need to be heard in order to work efficiently and safely.

It may be possible to engineer out some of the noise, for example, by placing a generator in a separate sound-proofed building. Alternatively, it may be possible to provide job rotation, to rearrange work locations or provide acoustic refuges.

Where individuals must be subjected to some noise at work, it may be reduced by ear protectors. These may be disposable ear plugs, reusable ear plugs or ear muffs. The chosen ear protector must be suited to the user and suitable for the type of noise, and individual personnel should be trained in its correct use.

Noise at Work Regulations 1989

The Noise at Work Regulations, unlike the previous vague or limited provisions, apply to all workplaces and require employers to carry out assessments of the noise levels within their premises and to take appropriate action where necessary. The 1989 Regulations came into force on 1 January 1990 implementing in the United Kingdom the EC Directive 86/188/EEC 'The Protection of Workers from Noise'.

Three action levels are defined by the regulations:

1 The first action level is a daily personal noise exposure of 85 dB, expressed as 85 dB(A).
2 The second action level is a daily personal noise exposure of 90 dB(A).
3 The third defined level is a peak action level of 140 dB(A) or 200 Pa of pressure which is likely to be linked to the use of cartridge-operated tools, shooting guns or similar loud explosive noises. This action level is likely to be most important where workers are subjected to a small number of loud impulses during an otherwise quiet day.

Safety first

The Noise at Work Regulations are intended to reduce hearing damage caused by loud noise.

The Noise at Work Regulations are intended to reduce hearing damage caused by loud noise. So, what is a loud noise? If you cannot hear what someone is saying when they are 2 m away from you or if they have to shout to make themselves heard, then the noise level is probably above 85 dB and should be measured by a competent person.

At the first action level an employee must be provided with ear protection (ear muffs or ear plugs) on request. At the second action level the employer must reduce, so far as is reasonably practicable, other than by providing ear protection, the exposure to noise of that employee.

Hearing damage is cumulative, it builds up, leading eventually to a loss of hearing ability. Young people, in particular, should get into the routine of avoiding noise exposure before their hearing is permanently damaged. The damage can also take the form of permanent tinnitus (ringing noise in the ears) and an inability to distinguish words of similar sound such as bit and tip.

Vibration is also associated with noise. Direct vibration through vibrating floors or from vibrating tools can lead to damage to the bones of the feet or hands. A condition known as 'vibration white finger' is caused by an impaired blood supply to the fingers, associated with vibrating hand tools.

Employers and employees should not rely too heavily on ear protectors. In practice, they reduce noise exposure far less than is often claimed, because they may be uncomfortable or inconvenient to wear. To be effective, ear protectors need to be worn all the time when in noisy places. If left off for even a short time, the best protectors cannot reduce noise exposure effectively.

Protection against noise is best achieved by controlling it at source. Wearing ear protection must be a last resort. Employers should:

- design machinery and processes to reduce noise and vibration (mounting machines on shock-absorbing materials can dampen out vibration);
- when buying new equipment, where possible, choose quiet machines. Ask the supplier to specify noise levels at the operator's working position;
- enclose noisy machines in sound-absorbing panels;
- fit silencers on exhaust systems;
- install motor drives in a separate room away from the operator;
- inform workers of the noise hazard and get them to wear ear protection;
- reduce a worker's exposure to noise by job rotation or provide a noise refuge.

New regulations introduced in 2006 reduce the first action level to 80 dB(A) and the second level to 85 dB(A) with a peak action level of 98 dB(A) or 140 Pa of pressure. Every employer must make a 'noise' assessment and provide workers with information about the risks to hearing if the noise level approaches the first action level. They must do all that is reasonably practicable to control the noise exposure of their employees and clearly mark ear protection zones. Employees must wear personal ear protection while in such a zone.

Personal Protective Equipment (PPE) at Work Regulations 1998

Definition

PPE is defined as all equipment designed to be worn, or held, to protect against a risk to health and safety.

PPE is defined as all equipment designed to be worn, or held, to protect against a risk to health and safety. This includes most types of protective clothing, and equipment such as eye, foot and head protection, safety harnesses, life-jackets and high-visibility clothing.

Under the Health and Safety at Work Act, employers must provide free of charge any PPE and employees must make full and proper use of it. Safety signs are useful reminders of the type of PPE to be used in a particular area. The vulnerable parts of the body which may need protection are the head, eyes, ears, lungs, torso, hands and feet and, in addition, protection from falls may need to be considered. Objects falling from a height present the major hazard against which head protection is provided. Other hazards include striking the head against projections and hair becoming entangled in machinery. Typical methods of protection include helmets, light-duty scalp protectors called 'bump caps' and hairnets.

Safety first

PPE

Always wear or use the PPE (personal protective equipment) provided by your employer for your safety.

Figure 3.2 Employers must provide Personal Protective Equipment.

Safety first

Safety signs

- Always follow the instructions given in the safety signs where you are working.
- It will help to keep you safe.

The eyes are very vulnerable to liquid splashes, flying particles and light emissions such as ultraviolet light, electric arcs and lasers. Types of eye protectors include safety spectacles, safety goggles and face shields. Screen-based workstations are being used increasingly in industrial and commercial locations by all types of personnel. Working with VDUs (visual display units) can cause eye strain and fatigue.

Some people may prove to be allergic to quite innocent products such as flour dust in the food industry or wood dust in the construction industry. The main effect of inhaling dust is a measurable impairment of lung function. This can be avoided by wearing an appropriate mask, respirator or breathing apparatus as recommended by the company's health and safety policy and indicated by local safety signs.

A worker's body may need protection against heat or cold, bad weather, chemical or metal splash, impact or penetration and contaminated dust. Alternatively, there may be a risk of the worker's own clothes causing contamination of the product, as in the food industry. Appropriate clothing will be recommended in the company's health and safety policy. Ordinary working clothes and clothing provided for food hygiene purposes are not included in the Personal Protective Equipment at Work Regulations.

Hands and feet may need protection from abrasion, temperature extremes, cuts and punctures, impact or skin infection. Gloves or gauntlets provide protection from most industrial processes, but should not be worn when operating machinery because they may become entangled in it. Care in selecting the appropriate protective device is required; for example, barrier creams provide only a limited protection against infection.

Try this

PPE

- Make a list of any PPE which you have used at work.
- What was this PPE protecting you from?

Boots or shoes with in-built toe-caps can give protection against impact or falling objects and, when fitted with a mild steel sole plate, can also provide protection from sharp objects penetrating through the sole. Special slip-resistant soles can also be provided for employees working in wet areas.

Environmental laws and regulations

The environment describes the world in which we live, work and play. It relates to our neighbourhood and surroundings and the situation in which we find ourselves.

> **Definition**
>
> *The environment* describes the world in which we live, work and play. It relates to our neighbourhood and surroundings and the situation in which we find ourselves.

Environmental laws protect the environment in which we live by setting standards for the control of pollution to land, air and water.

If a wrong is identified in the area in which we now think of as 'environmental' it can be of two kinds.

1 An offence in common law which means damage to property, nuisance or negligence leading to a claim for damages.
2 A statutory offence against one of the laws dealing with the protection of the environment. These offences are nearly always 'crimes' and punished by fines or imprisonment rather than by compensating any individual.

The legislation dealing with the environment has evolved for each part – air, water, land noise and radioactive substances. Where organizations' activities impact upon the environmental laws they are increasingly adopting environmental management systems which comply with ISO 14001. Let us now look at some of the regulations and try to see the present picture at the beginning of the new millennium.

Environmental Protection Act 1990

In the context of environmental law, the Environmental Protection Act 1990 was a major piece of legislation. The main sections of the Act are:

Part 1	Integrated pollution control by HM Inspectorate of Pollution, and air pollution control by Local Authorities
Part 2	Wastes on land
Part 3	Statutory nuisances and clean air
Part 4	Litter
Part 5	Radioactive Substances Act 1960
Part 6	Genetically modified organisms
Part 7	Nature conservation
Part 8	Miscellaneous, including contaminated land.

The Royal Commission of 1976 identified that a reduction of pollutant to one medium, air, water or land, then led to an increase of pollutant in another. It therefore stressed the need to take an integrated approach to pollution control. The processes subject to an integrated pollution control are:

- air emissions;
- processes which give rise to significant quantities of special waste, that is, waste defined in law in terms of its toxicity or flammability;
- processes giving rise to emissions to sewers or 'Red List' substances. These are 23 substances including mercury, cadmium and many pesticides which are subject to discharge consent to the satisfaction of the Environment Agency.

Where a process is under investigation the Inspectorate is empowered to set conditions to ensure that the best practicable environmental option (BPEO) is employed to control pollution. This is the cornerstone of the Environmental Protection Act.

Pollution Prevention and Control Regulations 2000

The system of Pollution Prevention and Control replaced that of Integrated Pollution Control established by the Environmental Protection Act 1990, thus bringing environmental law into the new millennium and implementing the European Directive (EC/96/61) on integrated pollution prevention and control. The new system was fully implemented in 2007.

Pollution Prevention and Control is a regime for controlling pollution from certain industrial activities. This regime introduces the concept of Best Available Technique (BAT) for reducing and preventing pollution to an acceptable level.

Industrial activities are graded according to their potential to pollute the environment:

- A(1) installations are regulated by the Environment Agency;
- A(2) installations are regulated by the Local Authorities;
- Part B installations are also regulated by the Local Authorities.

All three systems require the operators of certain industrial installations to obtain a permit to operate. Once an operator has submitted a permit application, the regulator then decides whether to issue a permit. If one is issued it will include conditions aimed at reducing and preventing pollution to acceptable levels. A(1) installations are generally perceived as having the greatest potential to pollute the environment. A(2) installations and Part B installations would have the least potential to pollute.

The industries affected by these regulations are those dealing with petrol vapour recovery, incineration of waste, mercury emissions from crematoria, animal rendering, non-ferrous foundry processes, surface treating of metals and plastic materials by powder coating, galvanizing of metals and the manufacture of certain specified composite wood-based boards.

Clean Air Act 1993

We are all entitled to breathe clean air but until quite recently the only method of heating houses and workshops was by burning coal, wood or peat in open fires. The smoke from these fires created air pollution and the atmosphere in large towns and cities was of poor quality. On many occasions in the 1950s the burning of coal in London was banned as the city was grinding to a halt because of the combined effect of smoke and fog, called smog. Smog was a very dense fog in which you could barely see more than a metre in front of you and which created serious breathing difficulties. In the new millennium we are no longer dependent upon coal and wood to heat our buildings; smokeless coal has been created and the gaseous products of combustion are now diluted and dispersed by new chimney design regulations. Using well-engineered combustion equipment together with the efficient arrestment of small particles in commercial chimneys of sufficient height, air pollution has been much reduced. This is what the Clean Air Act set out to achieve and it has been largely successful.

The **Clean Air Act** applies to all small and medium-sized companies operating furnaces, boilers or incinerators. Compliance with the Act does not require an application for authorization and so companies must make sure that they do not commit an offence. In general the emission of dark smoke from any chimney is unacceptable. The emission of dark smoke from any industrial premises is also unacceptable. This might be caused by, for example, the burning of old tyres or old cable.

Definition

The *Clean Air Act* applies to all small and medium-sized companies operating furnaces, boilers or incinerators.

In England, Scotland and Wales it is not necessary for the Local Authority to have witnessed the emission of dark smoke before taking legal action. Simply the evidence of burned materials, which potentially give rise to dark smoke when burned, is sufficient. In this way the law aims to stop people creating dark smoke under the cover of darkness.

A **public nuisance** is 'an act unwarranted by law or an omission to discharge a legal duty which materially affects the life, health, property, morals or reasonable comfort or convenience of Her Majesty's subjects'. This is a criminal offence and Local Authorities can prosecute, defend or appear in proceedings that affect the inhabitants of their area.

Definition

A *public nuisance* is 'an act unwarranted by law or an omission to discharge a legal duty which materially affects the life, health, property, morals or reasonable comfort or convenience of Her Majesty's subjects'.

Controlled Waste Regulations 1998

Under these regulations we have a 'Duty of Care to handle, recover and dispose of all waste responsibly'. This means that all waste must be handled, recovered and disposed of by individuals or businesses that are authorized to do so under a system of signed Waste Transfer Notes.

The Environmental Protection (Duty of Care) Regulations 1991 state that as a business you have a duty to ensure that any waste you produce is handled safely and in accordance with the law. This is the Duty of Care and applies to anyone who produces, keeps, carries, treats or disposes of waste from business or industry.

You are responsible for the waste that you produce, even after you have passed it on to another party such as a skip hire company, a scrap metal merchant, recycling company or local council. The Duty of Care has no time limit and extends until the waste has either been finally and properly disposed of or fully recovered.

So what does this mean for your company?

- Make sure that waste is only transferred to an authorized company.
- Make sure that waste being transferred is accompanied by the appropriate paperwork showing what was taken, where it was to be taken and by whom.
- Segregate the different types of waste that your work creates.
- Label waste skips and waste containers so that it is clear to everyone what type of waste goes into what skip.

Definition

Special waste is covered by the Special Waste Regulations 1996 and is waste that is potentially hazardous or dangerous and which may, therefore, require special precautions during handling, storage, treatment or disposal. Examples of special waste are asbestos, lead-acid batteries, used engine oil, solvent-based paint, solvents, chemical waste and pesticides.

- Minimize the waste that you produce and do not leave waste behind for someone else to clear away. Remember there is no time limit on your Duty of Care for waste.

Occupiers of domestic properties are exempt from the Duty of Care for the household waste that they produce. However, they do have a Duty of Care for the waste produced by, for example, a tradesperson working at a domestic property.

Special waste is covered by the Special Waste Regulations 1996 and is waste that is potentially hazardous or dangerous and which may, therefore, require special precautions during handling, storage, treatment or disposal. Examples of special waste are asbestos, lead-acid batteries, used engine oil, solvent-based paint, solvents, chemical waste and pesticides. The disposal of special waste must be carried out by a competent person, with special equipment and a licence.

Hazardous Waste Regulations 2005

New Hazardous Waste Regulations were introduced in July 2005 and under these regulations electric discharge lamps and tubes such as fluorescent, sodium, metal halide and mercury vapour are classified as hazardous waste. While each lamp only contains a very small amount of mercury, vast numbers are used and disposed of each year, resulting in a significant environmental threat. The environmentally responsible way to dispose of lamps and tubes is to recycle them and this process is now available through electrical wholesalers, as described in Chapter 1.

Electrical companies produce relatively small amounts of waste and even smaller amounts of special waste. Most companies buy in the expertise of specialist waste companies these days and build these costs into the contract.

Packaging (Essential Requirements) Regulations 2003

Definition

The new *Packaging Regulations* were introduced on 25 August 2003 bringing the UK into harmony with Europe. The regulations deal with the essential requirements of packaging for the storage and transportation of goods.

The new **Packaging Regulations** were introduced on 25 August 2003 bringing the UK into harmony with Europe. The regulations deal with the essential requirements of packaging for the storage and transportation of goods. There are two essential elements to the regulations:

1. The packaging shall be designed and manufactured so that the volume and weight is to the minimum amount required to maintain the necessary level of safety for the packaged product.
2. The packaging shall be designed and manufactured in such a way that the packaging used is either reusable or re-recyclable.

The regulations are enforced by the Weights and Measures Authority in Great Britain, the Department of Enterprise Trade and Investment in Northern Ireland and the Procurator-fiscal or Lord Advocate in Scotland.

Waste Electrical and Electronic Equipment EU Directive 2007

The Waste Electrical and Electronic Equipment (WEEE) Regulations will ensure that Britain complies with its EU obligation to recycle waste from electrical products. The regulations came into effect in July 2007 and from that date any company which makes, distributes or trades in electrical or electronic goods such as household appliances, sports equipment and even torches and toothbrushes has to make arrangements for recycling these goods at the end of their useful life. Batteries will be covered separately by yet another forthcoming EU directive.

Some sectors are better prepared for the new regulations than others. Mobile phone operators, O2, Orange, Virgin and Vodafone, along with retailers such as Currys and Dixons, have already joined together to recycle their mobile phones collectively. In Holland the price of a new car now includes a charge for the recycling costs.

Further information is available on the DTI and DEFRA website under WEEE.

Manual Handling Operations Regulations 1992

Manual handling is lifting, transporting or supporting loads by hand or by bodily force. The load might be any heavy object: a printer, a VDU, a box of tools or a step-ladder. Whatever the heavy object is, it must be moved thoughtfully and carefully, using appropriate lifting techniques if personal pain and injury are to be avoided. *Many people hurt their back, arms and feet, and over one-third of all three-day reported injuries submitted to the HSE each year are the result of manual handling*.

When lifting heavy loads, correct lifting procedures must be adopted to avoid back injuries. Figure 3.3 demonstrates the technique. Do not lift objects from the floor with the back bent and the legs straight as this causes excessive stress on the spine. Always lift with the back straight and the legs bent so that the powerful leg muscles do the lifting work. Bend at the hips and knees to get down to the level of the object being lifted, positioning the body as close to the object as possible. Grasp the object firmly and, keeping the back straight and the head erect, use the leg muscles to raise in a smooth movement. Carry the load close to the body. When putting the object down, keep the back straight and bend at the hips and knees, reversing the lifting procedure. A bad lifting technique will result in sprains, strains and pains. *There have been too many injuries over the years resulting from bad manual handling techniques. The problem has become so serious that the HSE has introduced new legislation* under the Health and Safety at Work Act 1974, the Manual Handling Operations Regulations 1992. Publications such as *Getting to Grips with Manual Handling* can be obtained from HSE Books; the address and Infoline are given in Appendix B.

Definition

Manual handling is lifting, transporting or supporting loads by hand or by bodily force.

Safety first

Lifting

- bend your legs;
- keep your back straight;
- use the leg muscles to raise the weight in a smooth movement.

COMPANION WEBSITE — Visit the companion website for more on this topic.

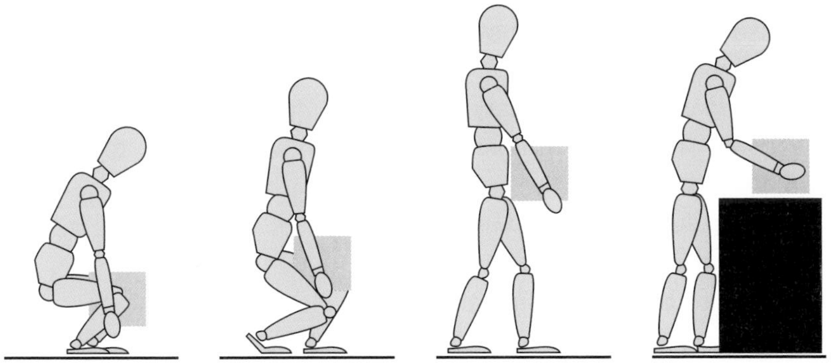

Figure 3.3 Correct manual lifting and carrying procedure.

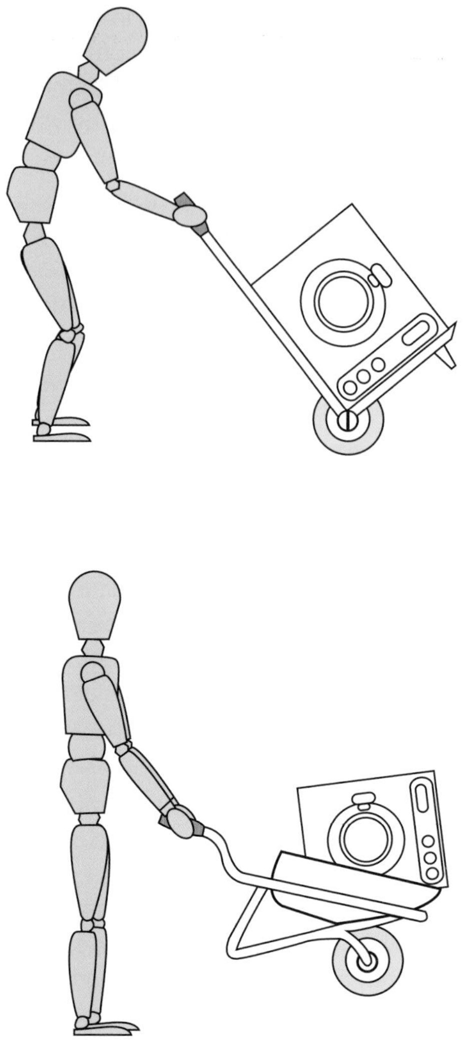

Figure 3.4 Always use a mechanical aid to transport a heavy load when available.

Where a job involves considerable manual handling, employers must now train employees in the correct lifting procedures and provide the appropriate equipment necessary to promote the safe manual handling of loads.

Consider some 'good practice' when lifting loads.

- Do not lift the load manually if it is more appropriate to use a mechanical aid. Only lift or carry what you can easily manage.
- Always use a trolley, wheelbarrow or truck such as that shown in Fig. 3.4 when these are available.
- Plan ahead to avoid unnecessary or repeated movement of loads.
- Take account of the centre of gravity of the load when lifting – the weight acts through the centre of gravity.
- Never leave a suspended load unsupervised.
- Always lift and lower loads gently.
- Clear obstacles out of the lifting area.
- Use the manual lifting techniques described above and avoid sudden or jerky movements.
- Use gloves when manual handling to avoid injury from rough or sharp edges.

- Take special care when moving loads wrapped in grease or bubble-wrap.
- Never move a load over other people or walk under a suspended load.

Non-statutory regulations

Statutory laws and regulations are written in a legal framework. Some don't actually tell us how to comply with the laws at an everyday level.

Non-statutory regulations and codes of practice interpret the statutory regulations telling us how we can comply with the law.

They have been written for every specific section of industry, commerce and situation, to enable everyone to comply with or obey the written laws.

When the Electricity at Work Regulations (EWR) tell us to 'ensure that all systems are constructed so as to prevent danger' they do not tell us how to actually do this in a specific situation. However, the IET Regulations tell us precisely how to carry out our electrical work safely in order to meet the statutory requirements of the EWR. In Part 1 of the IET Regulations, at 114, it states, 'the Regulations are non-statutory. They may, however, be used in a court of law in evidence to claim compliance with a statutory requirement.' If your electrical work meets the requirements of the IET Regulations, you will also meet the requirements of EWR.

Over the years, non-statutory regulations and codes of practice have built upon previous good practice and responded to changes by bringing out new editions of the various regulations and codes of practice to meet the changing needs of industry and commerce.

We will now look at one non-statutory regulation, what is sometimes called 'the electrician's bible', the most important set of regulations for anyone working in the electrical industry, the BS 7671: 2008 Requirements for Electrical Installations, IET Wiring Regulations 17th Edition.

Definition

Statutory laws and regulations are written in a legal framework.

Definition

Non-statutory regulations and codes of practice interpret the statutory regulations telling us how we can comply with the law.

The IET Wiring Regulations 17th edition requirements for electrical installations to BS 7671: 2008

The Institution of Engineering and Technology Requirements for Electrical Installations (the IET Regulations) are non-statutory regulations. They relate principally to the design, selection, erection, inspection and testing of electrical installations, whether permanent or temporary, in and about buildings generally and to agricultural and horticultural premises, construction sites and caravans and their sites. Paragraph 7 of the introduction to the EWR says: 'the IET Wiring Regulations is a code of practice which is widely recognized and accepted in the United Kingdom and compliance with them is likely to achieve compliance with all relevant aspects of the Electricity at Work Regulations.' The IET Wiring Regulations are the national standard in the United Kingdom and apply to installations operating at a voltage up to 1000 V a.c. They do not apply to electrical installations in mines and quarries, where special regulations apply because of the adverse conditions experienced there.

The current edition of the IET Wiring Regulations is the 17th edition 2008 (Amendment 3 : 2015). The main reason for incorporating the IET Wiring Regulations into British Standard BS 7671: 2008 was to create harmonization with European Standards.

The IET Regulations take account of the technical intent of the CENELEC European Standards, which in turn are based on the IEC International Standards.

The purpose in harmonizing British and European Standards is to help develop a single European market economy so that there are no trade barriers to electrical goods and services across the European Economic Area.

To assist electricians in their understanding of the regulations a number of guidance notes have been published. The guidance notes which I will frequently make reference to in this book are those contained in the *On Site Guide.* Eight other guidance notes booklets are also currently available. These are:

- *Selection and Erection*;
- *Isolation and Switching*;
- *Inspection and Testing*;
- *Protection against Fire*;
- *Protection against Electric Shock*;
- *Protection against Overcurrent*;
- *Special Locations*;
- *Earthing and Bonding.*

These guidance notes are intended to be read in conjunction with the regulations.

The IET Wiring Regulations are the electrician's bible and provide the authoritative framework of information for anyone working in the electrical industry.

Codes of practice

Codes of practice are, by definition, recommendations drawn up by a regulatory authority describing what is regarded as good practice in a particular situation.

In our own electrical industry, we have the *On Site Guide* and the *Electricians Guide to the Building Regulations*, both published by the IET.

In addition to these two excellent publications, we have the UNITE Union toolbox guide, and the NICEIC (National Inspection Council for Electrical Installation Contractors) produces many publications, guides and information cards.

Laws protecting people

In Chapter 1 and this chapter we have looked at some of the major pieces of legislation that affect our working environment and some of the main pieces of environmental law. Let us now look at some of the laws and regulations that protect and affect us as individuals, and our human rights and responsibilities.

Employment Rights Act 1996

If you work for a company you are an employee and you will have a number of legal rights under the Employment Rights Act 1996.

As a trainee in the electrical industry you are probably employed by a company and, therefore, are an employee. There are strict guidelines regarding those who are employed and those who are self-employed. Indicators of being employed are listed below:

- you work wholly or mainly for one company and work is centred upon the premises of the company;

- you do not risk your own money;
- you have no business organization such as a storage facility or stock in trade;
- you do not employ anyone;
- you work a set number of hours in a given period and are paid by the hour and receive a weekly or monthly wage or salary;
- someone else has the right to control what you do at work even if such control is rarely practised.

Indicators of being self-employed are as follows:

- you supply the materials, plant and equipment necessary to do the job;
- you give a price for doing a job and will bear the consequences if your price is too low or something goes wrong;
- you have the right to hire other people who will answer to you and are paid by you to do a job;
- you may be paid an agreed amount for a job regardless of how long it takes or be paid according to some formula, for example, a fee to 'first fix' a row of houses;
- within an overall deadline, you have the right to decide how and when the work will be done.

The titles 'employed' or 'self-employed' are not defined by statute but have emerged through cases coming before the courts. The above points will help in deciding the precise nature of the working relationship.

Home working is a growing trend which prompts the question as to whether home workers are employed or self-employed. As in any circumstance, it will depend upon the specific conditions of employment, and the points mentioned above may help to decide the question.

The Inland Revenue look with concern at those people who claim to be self-employed but do all or most of their work for one company. There is a free leaflet available from local Inland Revenue Offices, IR 56 – titled '*Employed or Self-employed*' – which will give further guidance if required.

If you are an employee you have a special relationship in law with your employer which entitles you to the following benefits:

- a written statement of the particulars of your employment. It is clearly in the interests of both parties to understand at the outset of their relationship the terms and conditions of employment. The legal relationship between employer and employee is one of contract. Both parties are bound by the agreed terms but the contract need not necessarily be in writing, although contracts of apprenticeship must be in writing;
- the date your employment started;
- the continuity of service, that is, whether employment with a previous employer is to count as part of an employee's continuous service. Continuous service is normally with one employer but there are exceptions, for example, if a business is transferred or taken over, or there is a change of partners or trustees. This is important because many employees' rights depend on the need to show that he or she has worked for the 'appropriate period' and this is known as '**continuous service**';
- the job title;
- the normal place from which you work;
- a brief description of your work;
- the hours to be worked;

- holiday entitlement and holiday pay;
- sick pay entitlement;
- pension scheme arrangements;
- the length of notice which an employee is obliged to give and is entitled to receive to terminate his or her contract of employment;
- where the employment is not intended to be permanent, the period for which it is expected to continue and the date when it is to end;
- disciplinary and grievance procedures;
- the rate of pay and frequency, weekly or monthly;
- an itemized pay statement showing:
  - the gross amount of the wage or salary;
  - the amounts of any deductions and the purpose for which they have been made. This will normally be tax and National Insurance contributions, but may also include payments to professional bodies or trade unions;
  - the net amount of salary being paid.

An **employer** has responsibilities to all employees. Even if the responsibilities are not written down in the contract of employment, they are implied by law. Case histories speak of a relationship of trust, confidence and respect. These responsibilities include:

- the obligation to pay an employee for work done;
- the obligation to treat an employee fairly;
- the obligation to take reasonable care of an employee's health and safety;
- an obligation to provide equal treatment both for men and women.

An **employee** also has responsibilities to his or her employer. These include:

- carrying out the tasks for which you are employed with all reasonable skill and care;
- conducting yourself in such a way as would best serve your employer's interests;
- carrying out all reasonable orders.

An employee is not expected to carry out any order that is plainly illegal or unreasonable. 'Illegal' is quite easy to define – anything which is against the law, for example, driving a vehicle for which you do not hold a licence or falsifying documents or accounts. 'Unreasonable' is more difficult to define; what is reasonable to one person may be quite unreasonable to another person.

Finally, employees are under a general duty not to disclose confidential information relating to their employer's affairs that they might obtain in the course of their work. Employees are also under a general duty not to assist a competitor of their employer. This is one aspect of the employee's duty to ensure that the relationship between employer and employee is one of trust. Even when an employee has left an employer, confidential information is not to be disclosed.

Health and Safety (First Aid) Regulations 1981

People can suffer an injury or become ill while at work. It does not matter whether the injury or illness is caused by the work they do or not; what is important is that they are able to receive immediate attention by a competent person or that an ambulance is called in serious cases. First aid at work covers the arrangements an employer must make to ensure that this happens. It can save lives and prevent a minor incident from becoming a major one.

The Health and Safety (First Aid) Regulations 1981 require employers to provide '**adequate**' and '**appropriate**' equipment, facilities and personnel to enable first aid to be given to employees if they are injured or become ill at work. What is adequate and appropriate will depend upon the type of work being carried out by the employer. The minimum provision is a suitably stocked first aid box and a competent person to take charge of first aid arrangements.

Employers must consider:

- How many people are employed and, therefore, how many first aid boxes will be required?
- What is the pattern of working hours, shift work, night work, and is a 'first aider' available for everyone at all times?
- How many trained 'first aiders' will be required?
- Where will first aid boxes be made available?
- Do employees travel frequently or work alone?
- Will it be necessary to issue personal first aid boxes if employees travel or work away from the company's main premises?
- How hazardous is the work being done – what are the risks?
- Are different employees at different levels of risk?
- What has been the accident or sickness record of staff in the past?

Although there is no legal responsibility for employers to make provision for non-employees, the HSE strongly recommends that they are included in any first aid provision.

We looked at first aid provision at work in Chapter 1 of this book.

Data Protection Act 1998

The right to privacy is a fundamental human right and one that many of us take for granted. Most of us, for instance, would not want our medical records freely circulated, and many people are sensitive about revealing their age, religious beliefs, family circumstances or academic qualifications. In the United Kingdom, even the use of name and address files for mail shots is often felt to be an invasion of privacy.

With the advent of large computerized databases it is now possible for sensitive personal information to be stored without the individual's knowledge and accessed by, say, a prospective employer, credit card company or insurance company in order to assess somebody's suitability for employment, credit or insurance.

The Data Protection Act 1984 grew out of public concern about personal privacy in the face of rapidly developing computer technology.

The Act covers 'personal data' which is 'automatically processed'. It works in two ways, giving individuals certain rights while requiring those who record and use personal information on computer to be open about that use and to follow proper practices.

The Data Protection Act 1998 was passed in order to implement a European Data Protection Directive. This Directive sets a standard for data protection throughout all the countries of the European Union, and the new Act was brought into force in March 2000. The Act gives the following useful definitions:

- *Data subjects*: the individuals to whom the personal data relate – we are all data subjects.

- *Data users*: those who control the contents and use a collection of personal data. They can be any type of company or organization, large or small, within the public or private sector.
- *Personal data*: information about living, identifiable individuals. Personal data does not have to be particularly sensitive information and can be as little as a name and address.
- *Automatically processed*: processed by computer or other technology such as document image-processing systems. The Act does not currently cover information which is held on manual records, for example, in ordinary paper files.

Registered data users must comply with the eight data protection principles of good information handling practice contained in the Act. Broadly these state that data must be:

1. Obtained and processed fairly and lawfully.
2. Held for the lawful purposes described in the data users' register entry.
3. Used for the purposes and disclosed only to those people described in the register entry.
4. Adequate, relevant and not excessive in relation to the purposes for which they are held.
5. Accurate and, where necessary, kept up to date.
6. Held no longer than is necessary for the registered purpose.
7. Accessible to the individual concerned who, where appropriate, has the right to have information about themselves corrected or erased.
8. Surrounded by proper security.

Exemptions from the Act

- The Act does not apply to payroll, pensions and accounts data, nor to names and addresses held for distribution purposes.
- Registration may not be necessary if the data are for personal, family, household or recreational use.
- Data subjects do not have a right to access data if the sole aim of collecting it is for statistical or research purposes.
- Data can be disclosed to the data subject's agent (e.g. lawyer or accountant), to persons working for the data user, and in response to urgent need to prevent injury or damage to health.

In addition, there are exemptions for special categories, including data held:

- in connection with national security;
- for prevention of crime;
- for the collection of tax or duty.

The rights of data subjects

The Data Protection Act allows individuals to have access to information held about themselves on computer and where appropriate to have it corrected or deleted.

As an individual you are entitled, on making a written request to a data user, to be supplied with a copy of any personal data held about yourself. The data user may charge a fee of up to £10 for each register entry for supplying this information but in some cases it is supplied free.

Usually the request must be responded to within 40 days. If not, you are entitled to complain to the Registrar or apply to the courts for correction or deletion of the data.

Apart from the right to complain to the Registrar, data subjects also have a range of rights which they may exercise in the civil courts. These are:

- right to compensation for unauthorized disclosure of data;
- right to compensation for inaccurate data;
- right of access to data and to apply for rectification or erasure where data are inaccurate;
- right to compensation for unauthorized access, loss or destruction of data.

For more information see www.gov.uk/data – protection.

Equal Opportunity Act 2010

The objectives of the Equal Opportunity Act are to encourage the identification and elimination of discrimination, sexual harassment and victimization, and to promote and facilitate the progressive realization of equality. Under this Act there is a duty to eliminate discrimination.

Equal opportunity is about giving and getting a fair go. It does not assume everyone is the same and it does not mean treating everyone the same.

Discrimination is the unfavourable treatment of a person in an area of public life, for example, at work.

Prejudice and discrimination

It is because we are all different to each other that life is so interesting and varied. Our culture is about the way of life that we have, the customs, ideas and experiences that we share and the things that we find acceptable and unacceptable. Different groups of people have different cultures. When people have a certain attitude towards you, or the group of people to which you belong, or a belief about you that is based upon lack of knowledge, understanding or myth, this is prejudice.

When prejudice takes form or action it becomes discrimination and this often results in unfair treatment of people. Regardless of our age, ability, sex, religion, race or sexuality we should all be treated equally and with respect. If we are treated differently because of our differences, we are being discriminated against.

If you are being discriminated against or you see it happening to someone else, you do not have to put up with it. Stay calm and do not retaliate but report it to someone, whoever is the most appropriate person: your supervisor, trainer or manager. If you are a member of a trade union you may be able to get help from them if it is an employment-related matter.

There are three areas covered by legislation at the moment: these are race, sex and disability. In the next few years the law will change to make it unlawful to discriminate in the training or workplace on the grounds of sexual orientation, religious belief and age.

The Race Relations Act 1976 and Amendment Act 2000

The 1976 Race Relations Act (RRA) made employers liable for acts of racial discrimination committed by their employees in the course of their employment. However, police officers are office holders, not employees, and, therefore, chief officers of the police were not liable under the 1976 Act for acts of racial discrimination. The Commission for Racial Equality proposed that the Act be extended to include all public services and the amendment came into force in 2000.

It is illegal to discriminate against someone because of their race, colour, nationality, citizenship or ethnic origin.

Institutional racism is when the policies or practises of an organization or institution result in its failure to provide an appropriate service to people because of their colour, culture or ethnic origin. It may mean that the organization or institution does or does not do something, or that someone is treated less favourably. This includes public services as well as educational institutions.

There are some exceptions in the RRA. It does not apply to certain jobs where people from a certain ethnic or racial background are required for authenticity. These are known as 'genuine occupational qualifications' and might apply to actors and restaurants.

The Commission for Racial Equality website can be found at www.cre.gov.uk.

Sex Discrimination Act 1975

'Sexism' takes place every time a person, usually a woman, is discriminated against because of their sex. The Sex Discrimination Act of 1975 makes it unlawful to discriminate against people on sexual grounds in areas relating to recruitment, promotion or training. Job advertisements must not discriminate in their language but they can make it clear that they are looking for people of a particular sex. If, though, a person of either sex applies, they must be treated equally and fairly.

There are some exceptions in the Sex Discrimination Act (SDA) known as '**genuine occupational qualifications**' that might apply to artists, models, actors and some parts of the priesthood in the Church. Some exceptions can also apply when appointing people to occupations where 'decency' is required, for example, changing-room attendants in swimming pools, gymnasiums, etc., and women are not allowed to work underground.

Sex discrimination is when someone is treated less favourably because of sex or marital status. It includes sexual harassment and unfavourable treatment because a woman is pregnant. Employers fear a high level of absenteeism, often unjustified, from a mother who is trying to juggle the conflicting demands of work and motherhood. This is known as 'direct sex discrimination'.

'Indirect sex discrimination' occurs when a condition of the job is applied to both sexes but excludes or disadvantages a larger proportion of one sex and is not justifiable. For example, an unnecessary height requirement of 180 cm (5' 10") would discriminate against women because fewer women would be able to meet this requirement.

The Equal Opportunities Commission has published a Code of Practice that gives guidance on best practice in the promotion of equality of opportunity in employment. Further information can be found on the SDA website at www.eoc.org.uk.

Disability Discrimination Act 1995

There are more than 8.5 million disabled people in the United Kingdom. The Disability Discrimination Act (DDA) makes it unlawful to discriminate against a disabled person in the areas of employment, access to goods and services, and buying or renting land or property.

Why would anyone want to discriminate against a disabled person if they have the skills, qualifications and abilities to do the job? The 2012 Paralympics have shown that many disabled people have extraordinary abilities, drive, motivation and determination. These extraordinary abilities can be of value to individual companies, industry and commerce.

It is now unlawful for employers of more than 15 people to discriminate against employees or job applicants on the grounds of disability. Reasonable adjustments must be made for people with disabilities and employers must ensure that discrimination does not occur in the workplace.

Under Part 111 of the DDA, from 1 October 2004 service providers have had to take reasonable steps to remove, alter or provide reasonable means of avoiding physical features that make it impossible or unreasonably difficult for disabled people to use their services. The duty requires service providers to make '**reasonable**' adjustments to their premises so that disabled people can use the service and are not restricted by physical barriers. If this is not possible, the service should be provided by means of a reasonable alternative such as bringing goods to the disabled person or helping the person to find items.

All organizations which provide goods, facilities or services to the public are covered by the DDA including shops, offices, public houses, leisure facilities, libraries, museums, banks, cinemas, churches and many more; in fact there are few exemptions.

Some service providers will need to incur significant capital expenditure in order to comply with the DDA. What is '**reasonable**' will depend upon the state and condition of the service provider's premises. A subjective standard will apply when determining what is reasonable under the circumstances at a given location. Whether or not an adjustment is reasonable will ultimately be a question of fact for the courts.

Further information can be found on the DDA website at www.disability.gov.uk.

The Human Rights Act 1998

The Human Rights Act (HRA) 1998 came into force on 2 October 2000 bringing the European Convention on Human Rights into UK law. It means that if you think your human rights have been violated you can take action through the British court system, rather than taking it to the European Court of Human Rights. The Act makes it unlawful for a '**public authority**' to act in a way that goes against any of the rights laid down in the Convention unless an Act of Parliament meant that it could not have acted differently. The basic human rights in the Human Rights Act are:

- the right to life;
- the right to a fair trial;

- the right to respect for your private and family life;
- the right to marry;
- the right to liberty and security;
- prohibition of torture;
- prohibition of slavery and forced labour;
- prohibition of discrimination;
- prohibition of the abuse of rights;
- freedom of thought, conscience and religion;
- freedom of expression;
- freedom of assembly and association;
- no punishment without law.

If you feel that your human rights have been violated, you should seek advice from a solicitor. Rights under the Act can only be used against a public authority such as the police or a local authority. They cannot be used against a private company. For more information see www.humanrights.gov.uk.

Non-compliance with the Regulations

In the case of any electrical business the purpose of the Regulations is to 'require precautions to be taken against the risk of death or personal injury from electricity in work activities'.

Section 4 of the EWR tells us that 'all systems must be constructed so as to prevent danger … and be properly maintained. … Every work activity shall be carried out in a manner which does not give rise to danger. … In the case of work of an electrical nature, it is preferable that the conductors be made dead before work commences.'

The Regulations do not tell us specifically how to carry out our work activities and ensure compliance, but if proceedings were brought against an individual for breaking the Regulations, the only acceptable defence would be 'to prove that all reasonable steps were taken and all diligence exercised to avoid the offence' (Regulation 29 of the Electricity at Work Regulations).

An electrical contractor could reasonably be expected to have 'exercised all diligence' if the installation was wired according to the IET Wiring Regulations. However, electrical contractors must become more 'legally aware' following the conviction of an electrician for manslaughter at Maidstone Crown Court in 1989. The court accepted that an electrician had caused the death of another man as a result of his shoddy work in wiring up a central heating system. He received a nine-month suspended prison sentence. This case has set an important legal precedent, and in future any tradesman or professional who causes death through negligence or poor workmanship risks prosecution and possible imprisonment.

Just think how disastrous that would be for the individual and the company they worked for. If a loss of life or serious injury occurs at work, the HSE Inspectors will investigate the circumstances of the accident. The HSE Inspectors have the power to stop work continuing if they believe there is significant risk or danger in the workplace under a 'prohibition notice'.

Under an 'improvement notice', the HSE can demand that safety precautions be improved in the workplace. See Chapter 1, Enforcement of Health and Safety Regulations.

At a personal level, a work injury might stop you playing your favourite sport, either temporarily or permanently. You would probably suffer a loss of earnings if you were sick as a result of an injury, and it may mean that you were no longer able to work as an electrician if the injury was more serious.

The company involved in a legal dispute for non-compliance with the regulations will incur significant legal costs and possible loss of reputation. In this situation, clients or customers may lose confidence with the company in dispute and find other service providers. The resulting loss of income may then lead to redundancies in the company.

Non-compliance with the regulations has serious implications for individuals and for an employer.

Electricity supplies and earthing systems

COMPANION WEBSITE Visit the companion website for more on this topic.

Electricity is generated in modern power-stations at 25 kV and fed through transformers to the consumer over a complex network of cables known as the National Grid system. This is a network of cables, mostly at a very high voltage, suspended from transmission towers, linking together the 175 power-stations and millions of consumers. There are approximately 5,000 miles of high-voltage transmission lines in England and Wales, running mostly through the countryside.

Man-made structures erected in rural areas often give rise to concern, but every effort is made to route the overhead lines away from areas where they might spoil a fine view. There is full consultation with Local Authorities and interested parties as to the route which lines will take. Farmers are paid a small fee for having transmission line towers on their land. Over the years many different tower designs and colours have been tried, but for the conditions in the United Kingdom, galvanized steel lattice towers are considered the least conspicuous and most efficient.

For those who consider transmission towers unsightly, the obvious suggestion might be to run all cables underground. In areas of exceptional beauty this is done, but underground cables are about 10 times more expensive than the equivalent overhead lines. The cost of running the largest lines underground is about £18 million per kilometre compared with about £1.8 million per kilometre overhead. On long transmission lines the losses can be high, but by raising the operating voltage and therefore reducing the current for a given power, the I^2R losses are reduced, the cable diameter is reduced and the overall efficiency of transmission is increased. In order to standardize equipment, standard voltages are used. These are:

- 400 and 275 kV for the super-grid;
- 132 kV for the original grid;
- 66 and 33 kV for secondary transmission;
- 11 kV for high-voltage distribution;
- 400 V for commercial consumer supplies;
- 230 V for domestic consumer supplies.

A diagrammatic representation showing the distribution of power from the power-station to the consumer is given in Fig. 3.11.

All local distribution in the United Kingdom is by underground cables from substations placed close to the load centre and supplied at 11 kV. Transformers in these local substations reduce the voltage to 400 V, and three-phase and neutral distributor cables connect this supply to consumers. Connecting to one-phase and neutral of a three-phase 400 V supply gives a 230 V single-phase supply suitable for domestic consumers.

When single-phase loads are supplied from a three-phase supply, the load should be 'balanced' across the phases. That is, the load should be equally distributed across the three phases so that each phase carries approximately the same current. This prevents any one phase from being overloaded.

Safe electrical installations

The provision of a safe electrical system is fundamental to the whole concept of using electricity in and around buildings safely. The electrical installation as a whole must be protected against overload and short-circuit damage, and the people using the installation must be protected against electric shock. An installation that meets the requirements of the IET Wiring Regulations Requirements for Electrical Installations will be so protected. The method most universally used in the United Kingdom to provide for the safe use of electrical energy is Basic Protection and Fault Protection, as described in Chapter 41 of the IET Regulations. So, let us look at these essential safety elements.

The consumer's mains equipment is normally fixed close to the point at which the supply cable enters the building. To meet the requirements of the IET Regulations it must provide:

- protection against electric shock (Chapter 41);
- protection against overcurrent (Chapter 43);
- isolation and switching (Chapter 53).

Protection against electric shock, both 'basic protection' and 'fault protection', is provided by insulating and placing live parts out of reach in suitable enclosures, earthing and bonding metalwork and providing fuses or circuit-breakers so that the supply is automatically disconnected under fault conditions.

To provide **overcurrent protection** it is necessary to provide a device which will disconnect the supply automatically before the overload current can cause a rise in temperature which would damage the installation. A fuse or MCB would meet this requirement.

An **isolator** is a mechanical device which is operated manually and is provided so that the whole of the installation, one circuit or one piece of equipment may be cut off from the live supply. In addition, a means of switching off for maintenance or emergency switching must be provided. A switch may provide the means of isolation, but an isolator differs from a switch in that it is intended to be opened when the circuit concerned is not carrying current. Its purpose is to ensure the safety of those working on the circuit by making dead those parts which are live in normal service. One device may provide both isolation and switching provided that the characteristics of the device meet the regulations for both functions. The switching of electrically operated equipment in normal service is referred to as **functional switching**.

Circuits are controlled by switchgear which is assembled so that the circuit may be operated safely under normal conditions, isolated automatically under fault conditions, or isolated manually for safe maintenance. These requirements are met by good workmanship carried out by **electrically skilled persons** and the installation of approved British Standard materials such as switches, isolators, fuses or circuit-breakers (IET Regulation 131.1.1). The switchgear must also be non-combustible or housed in a non-combustible cabinet or enclosure (IET Regulation 421.1.200). The equipment belonging to the supply authority is sealed to prevent unauthorized entry, because if connection were made to

Definition

To provide *overcurrent protection* it is necessary to provide a device which will disconnect the supply automatically before the overload current can cause a rise in temperature which would damage the installation. A fuse or MCB would meet this requirement.

Definition

An *isolator* is a mechanical device which is operated manually and is provided so that the whole of the installation, one circuit or one piece of equipment may be cut off from the live supply.

Definition

The switching of electrically operated equipment in normal service is referred to as *functional switching*.

Definition

An 'electrically skilled person' is someone who has the necessary technical skills, training and expertise to safely carry out a particular activity.

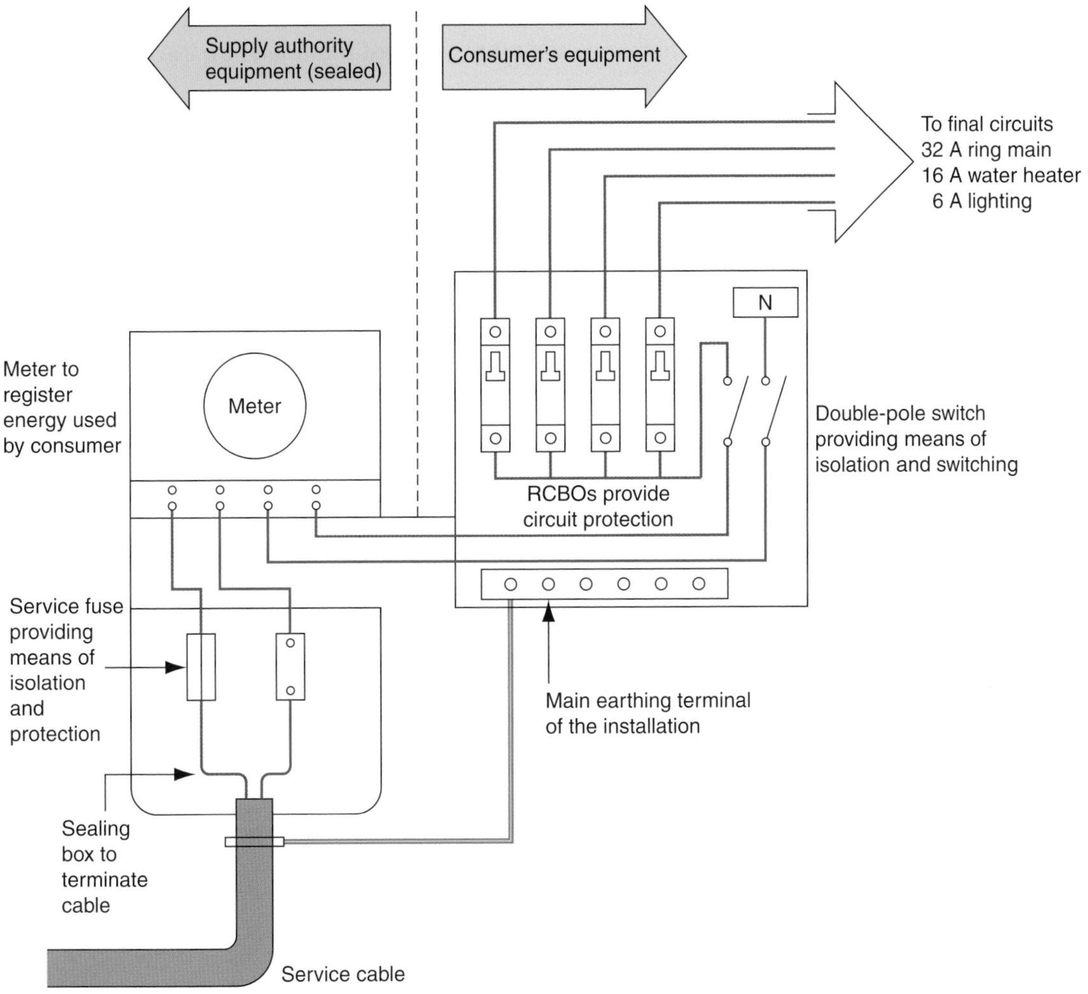

Figure 3.5 Simplified diagram of connections and equipment at a domestic service position.

the supply before the meter, the energy used by the consumer would not be recorded on the meter. Figure 3.5 shows the connections and equipment at a domestic service position.

Protective electrical bonding to earth

The purpose of the bonding regulations is to keep all the exposed metalwork of an installation at the same earth potential as the metalwork of the electrical installation, so that no currents can flow and cause an electric shock. For a current to flow there must be a difference of potential between two points, but if the points are joined together there can be no potential difference. This bonding or linking together of the exposed metal parts of an installation is known as 'protective bonding' and gives protection against electric shock.

Let us now define some of the important new words as they apply to electrical installations.

Earthing is the connection of the exposed conductive parts of an electrical installation to the main protective earthing terminal of the installation.

Bonding is the linking together of the exposed or extraneous metal parts of an electrical installation for the purpose of safety.

Definition

Earthing is the connection of the exposed conductive parts of an electrical installation to the main protective earthing terminal of the installation.

Exposed conductive parts are the metalwork of the electrical installation: the conduit, trunking, metal boxes and equipment that make up the electrical installation.

Extraneous conductive parts are the other metal parts which do not form a part of the electrical installation: the structural steelwork of the building, gas, water and central heating pipes and radiators.

Basic protection is protection against electric shock under fault-free conditions and is provided by insulating live parts in accordance with section 416 of the IET Regulations.

Fault protection is protection against electric shock under single fault conditions and is provided by protective bonding and automatic disconnection of the supply (by a fuse or MCB) in accordance with IET Regulations 411.3 to 6.

Protection from electric shock is provided by basic protection and fault protection.

Protective bonding is bonding for the purpose of safety.

Definition

Bonding is the linking together of the exposed or extraneous metal parts of an electrical installation for the purpose of safety.

Definition

Exposed conductive parts are the metalwork of the electrical installation: the conduit, trunking, metal boxes and equipment that make up the electrical installation.

Definition

Extraneous conductive parts are the other metal parts which do not form a part of the electrical installation: the structural steelwork of the building, gas, water and central heating pipes and radiators.

Try this

Memory Aid

- Writing out important definitions helps you to remember them.

Main protective bonding

Where earthed electrical equipment may come into contact with the metalwork of other services, they too must be effectively connected to the main protective earthing terminal of the installation (IET Regulations Chapter 544).

Other services are described as:

- main water pipes;
- main gas pipes;
- other service pipes and ducting;
- central heating and air-conditioning systems;
- exposed metal parts of the building structure;
- lightning protective conductors.

Protective bonding should be made to gas and water services at their point of entry into the building, as shown in Fig. 3.6, using insulated bonding conductors of not less than half the cross-section of the incoming main earthing conductor. The minimum permitted size is 6 mm^2 but the cross-section need not exceed 25 mm^2 (IET Regulation 544.1.1). The bonding clamp must be fitted on the consumer's side of the gas meter between the outlet union, before any branch pipework but within 600 mm of the meter (IET Regulation 544.1.2).

A permanent label must also be fixed in a visible position at or near the point of connection of the bonding conductor with the words 'Safety Electrical Connection – Do Not Remove' (IET Regulation 514.13.1). Supplementary bonding is described later in this book.

Definition

Basic protection is protection against electric shock under fault-free conditions and is provided by insulating live parts in accordance with section 416 of the IET Regulations.

Definition

Fault protection is protection against electric shock under single fault conditions and is provided by protective bonding and automatic disconnection of the supply (by a fuse or MCB) in accordance with IET Regulations 411.3 to 6.

Definition

Protective bonding is bonding for the purpose of safety.

Electric shock and overload protection

Electric shock is normally caused either by touching a conductive part that is normally live, or by touching an exposed conductive part made live by a fault. The

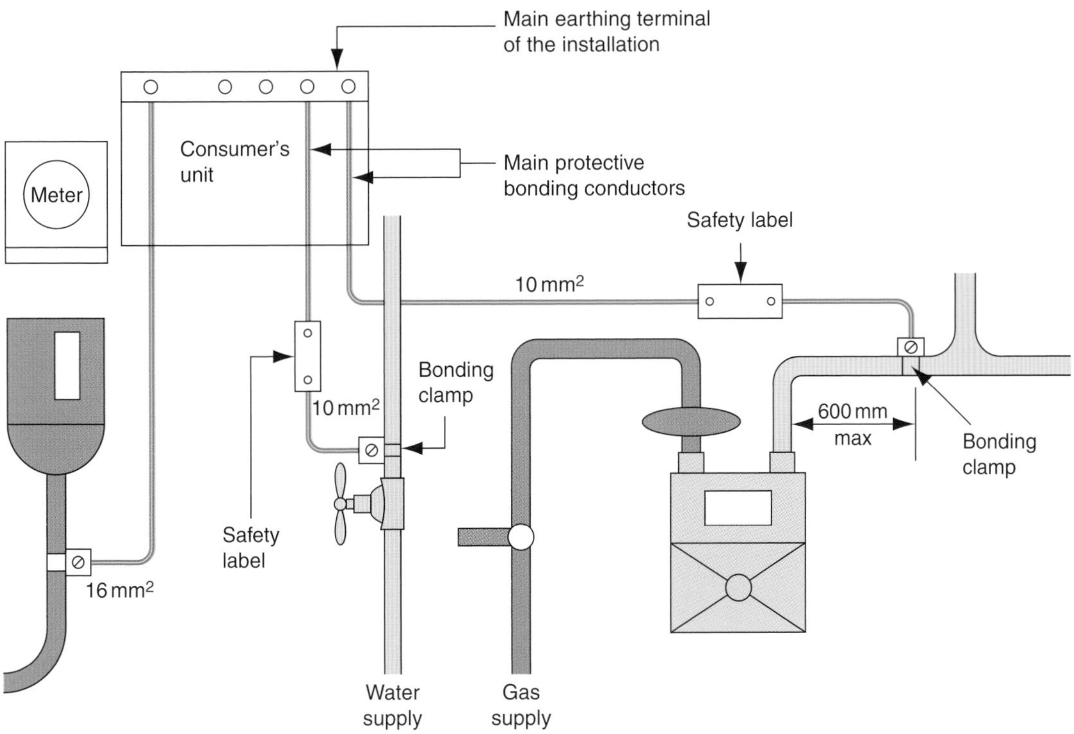

Figure 3.6 Main protective bonding of gas and water supplies.

touch voltage curve in Fig. 3.8 shows that a person in contact with 230V must be released from this danger in 40ms if harmful effects are to be avoided. Similarly, a person in contact with 400V must be released in 15ms to avoid being harmed.

In general, protection against touching live parts is achieved by insulating live parts and is called 'basic protection'. Protection against touching something made live as a result of a fault, and called 'fault protection' (IET Regulation 131.2.2), is achieved by protective bonding and automatic disconnection of the supply in the event of a fault occurring. Separated extra low-voltage supplies (SELV) provide protection against both 'basic' and 'fault' protection.

Part 4 of the IET Regulations deals with the application of protective measures for safety and Chapter 53 with the regulations for switching devices or switchgear required for protection, isolation and switching of a consumer's installation.

The consumer's main switchgear must be readily accessible to the consumer and be able to:

- isolate the complete installation from the supply;
- protect against overcurrent;
- cut off the current in the event of a serious fault occurring.

Switchgear assemblies including consumer unit enclosures must be:

- manufactured from non-combustible or not readily combustible material; or be
- enclosed in a cabinet or enclosure that is constructed of non-combustible or not readily combustible material. (IET Amendment 3 Regulation 421.1.200.)

This new regulation has been introduced to protect against fire that can result from the overheating of connections within consumer units. Overheating may occur because connections have not been properly made or because they have become loose.

Figure 3.7 Be careful when working with live wires!

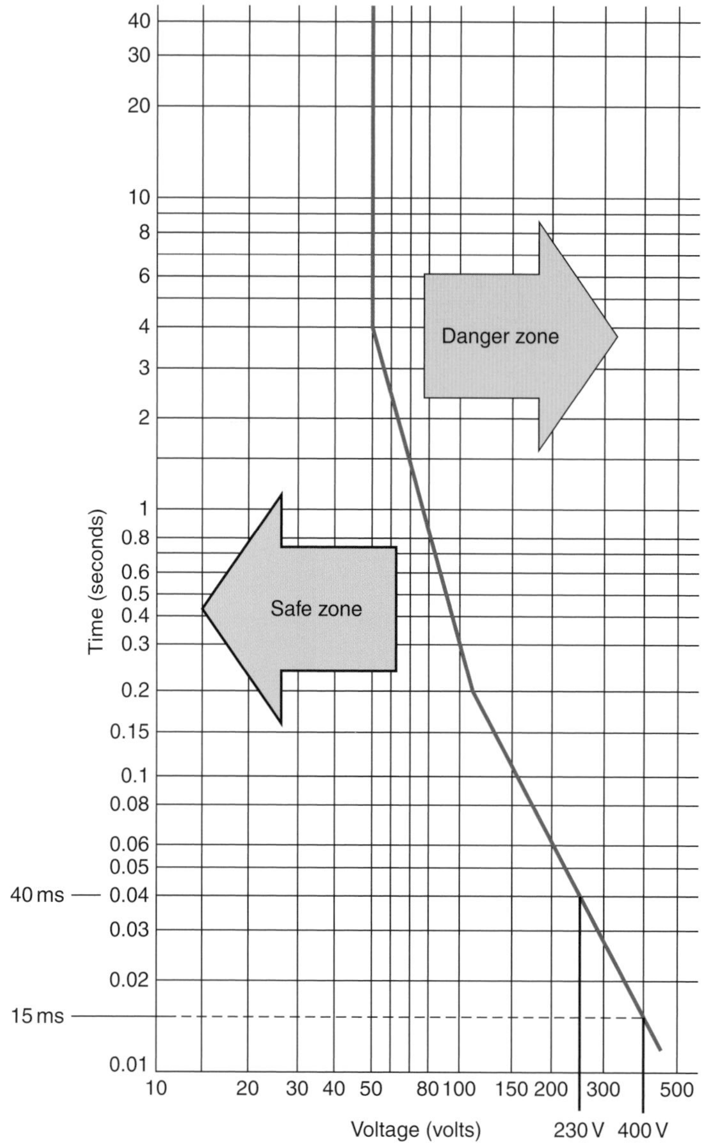

Figure 3.8 Touch voltage curve.

Visit the companion website for more on this topic.

Protection against overcurrent

Excessive current may flow in a circuit as a result of an overload or a short-circuit. An **overload** or **overcurrent** is defined as a current which exceeds the rated value in an otherwise healthy circuit. A short-circuit is an overcurrent resulting from a fault of negligible impedance between live conductors having a difference in potential under normal operating conditions. Overload currents usually occur in a circuit because it is abused by the consumer or because it has been badly designed or modified by the installer. Short-circuits usually occur as a result of an accident which could not have been predicted before the event.

Definition

An *overload* or *overcurrent* is defined as a current which exceeds the rated value in an otherwise healthy circuit. A short-circuit is an overcurrent resulting from a fault of negligible impedance.

An overload may result in currents of two or three times the rated current flowing in the circuit, while short-circuit currents may be hundreds of times greater than the rated current. In both cases, the basic requirement for protection is that the circuit should be interrupted before the fault causes a temperature rise which might damage the insulation, terminations, joints or the surroundings of the conductors. If the device used for overload protection is also capable of breaking a prospective short-circuit current safely, then one device may be used to give protection from both faults (IET Regulation 432.1). Devices which offer protection from overcurrent are:

- semi-enclosed fuses manufactured to BS 3036;
- cartridge fuses manufactured to BS 88–3: 2010;
- high-breaking capacity fuses (HBC fuses) manufactured to BS 88–2: 2010;
- MCBs manufactured to BS EN 60898.

We will look at overcurrent protection, fuses and MCBs in more detail later in this chapter.

Key fact

Fuses

How does a fuse work?

- under fault conditions excessive current flows;
- the fuse element gets hot;
- the fuse element melts;
- disconnecting the circuit it protected.

Generation, transmission and distribution of electricity

Generation

Figure 2.7 (page 76) shows a simple a.c. generator or alternator producing an a.c. waveform. We generate electricity in large, modern power-stations using the same basic principle of operation. However, in place of a single loop of wire, the power-station alternator has a three-phase winding and powerful electromagnets. The generated voltage is three identical sinusoidal waveforms each separated by 120°, as shown in Fig. 3.9. The prime mover is not, of course, a simple crank handle, but a steam turbine. Hot water is heated until it becomes superheated steam, which drives the vanes of a steam turbine which is connected to the alternator. The heat required to produce the steam may come from burning coal or oil or from a nuclear reactor. Whatever the primary source of energy, it is only being used to drive a turbine which is connected to an alternator, to generate electricity.

Safety first

Bonding clamps:

- must be of an approved type;
- must be fitted to clean pipework;
- must be tight and secure;
- must have a visible label.

IET Regulation 514.13.1

Transmission

Electricity is generated in the power-station alternator at 25 kV. This electrical energy is fed into a transformer to be stepped up to a very high voltage for transmission on the National Grid network at 400 kV, 275 kV or 132 kV. These very

Key fact

Information

The individual conductors on the National Grid network are only about 2.5 cm in diameter because they operate at a very high voltage.

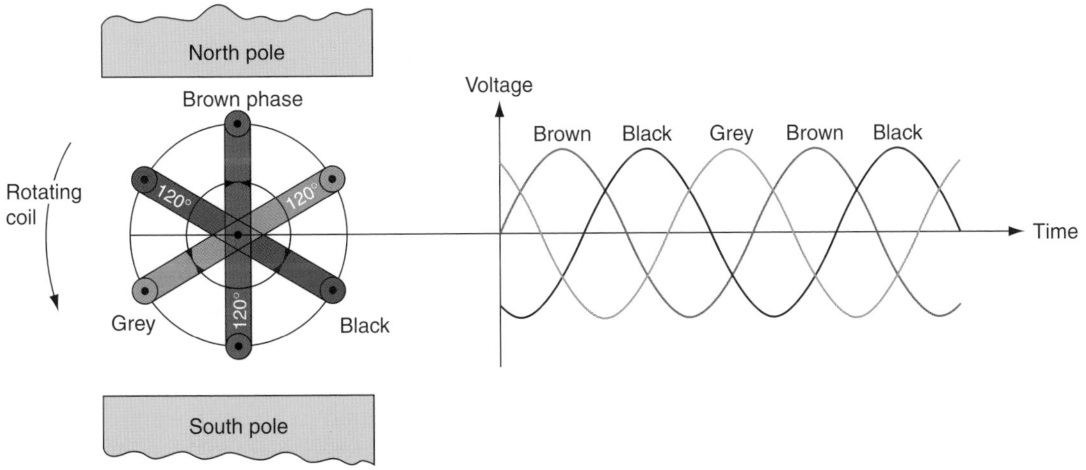

Figure 3.9 Generation of a three-phase voltage.

Figure 3.10 Transmission line steel pylon.

high voltages are necessary because, for a given power, the current is greatly reduced, which means smaller grid conductors and the transmission losses are reduced.

The National Grid network was established in 1934 and consists of over 5,000 miles of overhead aluminium conductors suspended from steel pylons which link together all the power-stations. Environmentalists say that these steel towers are ugly, but this method is about 10 times cheaper than the equivalent underground cable at these high voltages. Figure 3.10 shows a transmission line steel pylon.

Electricity is taken from the National Grid by appropriately located substations which eventually transform the voltage down to 11 kV at a local substation. At the local substation the neutral conductor is formed for single-phase domestic supplies and three-phase supplies to shops, offices and garages. These supplies are usually underground radial supplies from the local substation, but in rural areas we still see transformers and overhead lines suspended on wooden poles. Figures 3.11 and 3.12 give an overview of the system from power-station to consumer.

Distribution to the consumer

The electricity leaves the local substation and arrives at the consumer's mains intake position. The final connections are usually by simple underground radial feeders at 400 V/230 V. Underground cable distribution is preferred within a city, town or village because people find the overhead distribution, which we see in rural and remote areas, unsightly. Further, at these lower distribution voltages, the cost of underground cables is not prohibitive. The 400 V/230 V is derived from the 11 kV/400 V substation transformer by connecting the secondary winding-in star, as shown in Fig. 3.13. The star point is earthed to an earth electrode sunk into the ground below the substation and from this point is taken the fourth conductor and the neutral. Loads connected between phases are fed at 400 V and those fed between one phase and neutral at 230 V. A three-phase

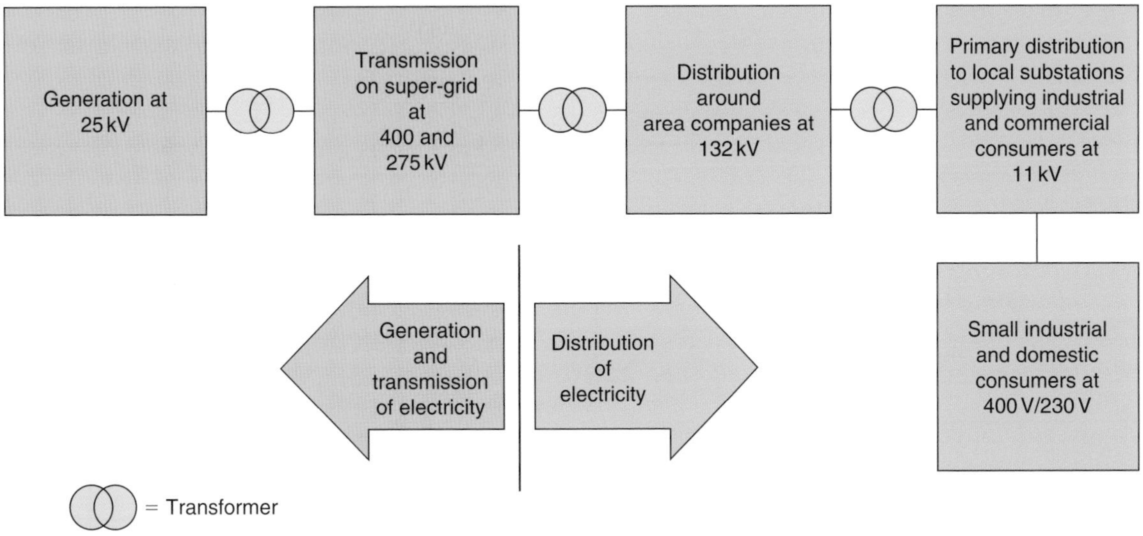

Figure 3.11 Generation, transmission and distribution of electrical energy.

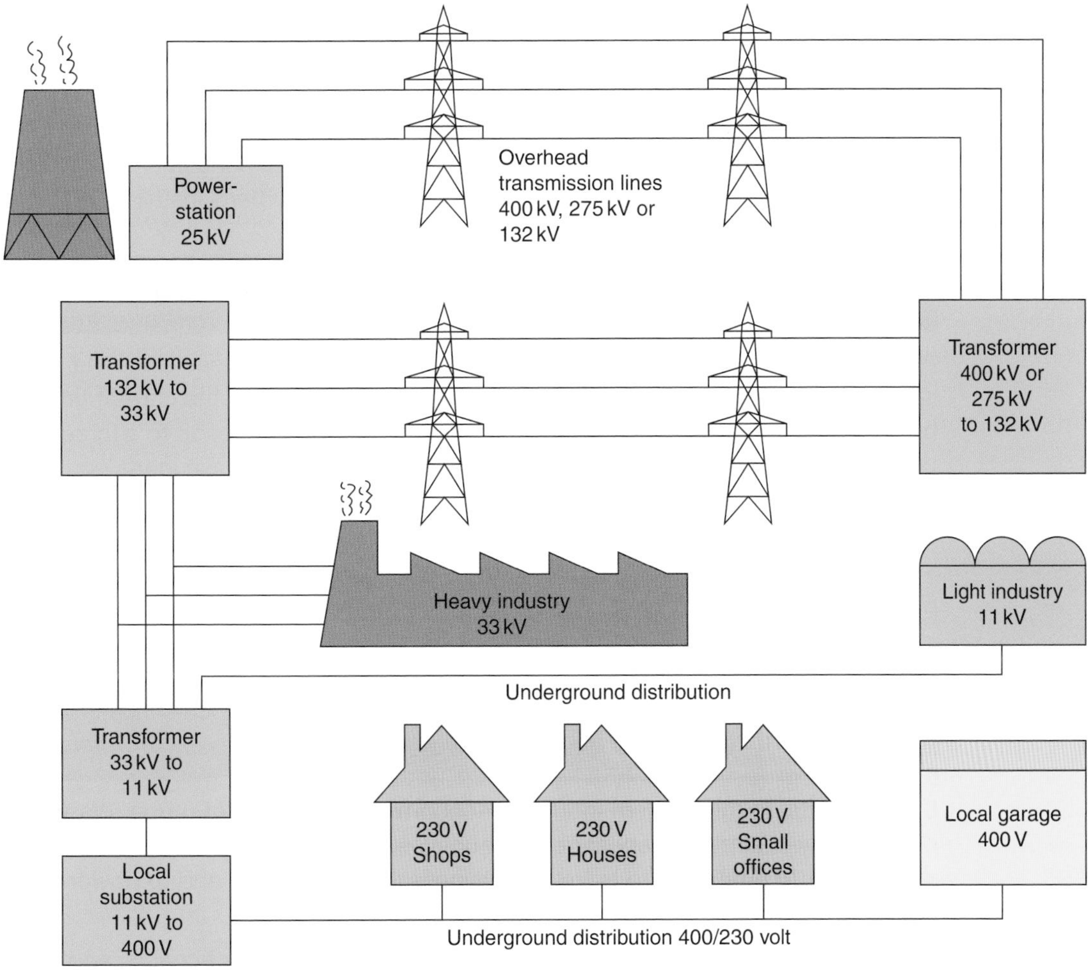

Figure 3.12 Simplified diagram of the distribution of electricity from power station to consumer.

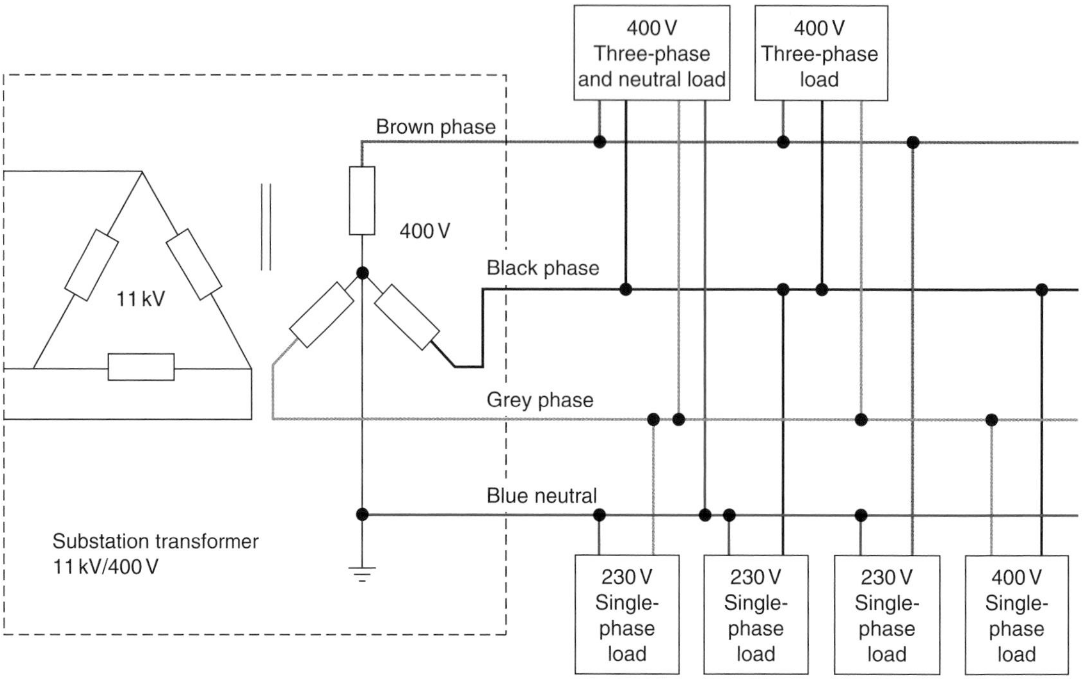

Figure 3.13 Three-phase four-wire distribution.

400 V supply is used for supplying small industrial and commercial loads such as garages, schools and blocks of flats. A single-phase 230 V supply is usually provided for individual domestic consumers.

At the mains intake position, the supplier will provide a sealed HBC fuse (BS 88–2: 2010) and a sealed energy meter to measure the consumer's electricity consumption. It is after this point that we reach the consumer's installation.

Balancing single-phase loads

A three-phase load such as a motor has equally balanced phases since the resistance of each phase winding will be the same. Therefore, the current taken by each phase will be equal. When connecting single-phase loads to a three-phase supply, care should be taken to distribute the single-phase loads equally across the three phases so that each phase carries approximately the same current. Equally distributing the single-phase loads across the three-phase supply is known as 'balancing' the load. A lighting load of 18 luminaires would be 'balanced' if six luminaires were connected to each of the three phases.

Environmental technology systems and renewable energy

Energy is vital to the modern industrial economy in the UK and Europe. We also need energy in almost every aspect of our lives, to heat and light our homes and offices, to enable us to travel on business or for pleasure, and to power our business and industrial machines.

In the past the UK has benefited from its fossil fuel resources of coal oil and gas but respectable scientific sources indicate that the fossil fuel era is drawing to a close. Popular estimates suggest that gas and oil will reach peak production in the year 2060 with British coal reserves lasting only a little longer. Therefore we must look to different ways of generating electricity so that:

- the remaining fossil fuel is conserved;
- our CO_2 emissions are reduced to avoid the consequences of climate change;
- we ensure that our energy supplies are secure, and not dependent upon supplies from other countries.

Following the introduction of the Climate Change Act in 2008 the UK and other Member States agreed an EU-wide target of 20% renewable energy by the year 2020 and 60% by 2050. Meeting these targets will mean basing much of the new energy infrastructure around renewable energy, particularly offshore wind power.

The 'Energy Hierarchy' states that organizations and individuals should address energy issues in the following order so as to achieve the agreed targets.

1 Reduce the need for energy – reducing energy demand is cost saving, reduces greenhouse gas emissions and contributes to the security of supply. Reducing the energy loss from buildings by better insulation and switching off equipment when not in use is one way of achieving this target.
2 Use energy more efficiently – use energy-efficient lamps and 'A'-rated equipment.
3 Use renewable energy – renewable energy refers to the energy that occurs naturally and repeatedly in the environment. This energy may come from wind, waves or water, the sun, or heat from the ground or air.
4 Any continuing use of fossil fuels should use clean and efficient technology. Power-stations generating electricity from coal and oil (fossil fuel) release a lot of CO_2 in the generating process. New-build power-stations must now be fitted with carbon capture filters to reduce the bad environmental effects.

Funding for environmental technology systems

Renewable energy is no less reliable than energy generated from more traditional sources. Using renewable energy does not mean that you have to change your lifestyle or your domestic appliances. There has never been a better time to consider generating energy from renewable technology than now because grants and funding are available to help individuals and companies.

Key fact

Renewable energy is no less reliable than energy generated from more traditional sources.

The Low Carbon Building Programme implemented by the Department of Energy and Climate Change (DECC) provides grants towards the installation of renewable technologies and is available to householders, public non-profit-making organizations and commercial organizations across the UK.

Key fact

The government's 'feed in tariff' pays a tax-free sum which is guaranteed for 25 years.

The government's 'feed in tariff' pays a tax-free sum which is guaranteed for 25 years. It is called 'Clean energy cash back' and has been introduced to promote the uptake of small-scale renewable and low carbon electricity generation technologies. The customer receives a generation tariff from the electricity supplier, whether or not any electricity generated is exported to the National Grid, and an additional export tariff when electricity is transported to the electricity grid through a smart meter.

From April 2010, clean energy generators were paid 41.3p for each kWh of electricity generated. Surplus energy fed back into the National Grid earned an extra 3p per unit. However, the scheme has been very popular and the feedback tariffs were soon reduced to 21p. From 1 August 2012 the fee is only 16p for each kWh. If you add to this the electricity bill savings, a normal householder could still make some saving. Savings vary according to energy use and type of system used. The Energy Saving Trust at www.energysavingtrust.org.uk (tel. 01752 823600) and www.britishgas.co.uk and Ofgem at www.ofgem.gov.uk/ provide an on-line calculator to determine the cost, size of system and CO_2 savings for PV systems.

Microgeneration technologies

Microgeneration is defined in The Energy Act 2004 Section 82 as the generation of heat energy up to 45 kW and electricity up to 50 kW.

Today, microgeneration systems generate relatively small amounts of energy at the site of a domestic or commercial building. However, it is estimated that by 2050, 30 to 40 per cent of the UK's electricity demand could be met by installing microgeneration equipment to all types of buildings.

In the USA, the EU and the UK buildings consume more than 70% of the nations' electricity and contribute almost 40% of the polluting CO_2 greenhouse gases. Any reductions which can be made to these figures will be good for the planet, and hence the great interest today in the microgeneration systems. Micro-generation technologies include small wind turbines, solar photovoltaic (PV) systems, small-scale hydro and micro CHP (combined heat and power) systems. Microgenerators that produce electricity may be used as stand-alone systems, or may be run in parallel with the low-voltage distribution network, that is, the a.c. mains supply.

Key fact

Microgeneration technologies include small wind turbines, solar photovoltaic (PV) systems, small-scale hydro and micro CHP (combined heat and power) systems.

The April 2008 amendments to the Town and Country Planning Act 1990 now allow for the installation of microgeneration systems within the boundary of domestic premises without obtaining planning permission. However, size limitations have been set to reduce the impact upon neighbours. For example, solar panels attached to a building must not protrude more than 200 mm from the roof slope, and stand-alone panels must be no higher than four metres above ground level and no nearer than five metres from the property boundary. See the Electrical Safety Council site for advice on connecting microgeneration systems at www.esc.org.uk/bestpracticeguides.html.

Smart electricity meters

Key fact

Smart electricity meters are designed to be used in conjunction with micro-generators.

Smart electricity meters are designed to be used in conjunction with micro-generators. Electricity generated by the consumer's microgenerator can be sold back to the energy supplier using the 'smart' two-way meter.

The Department for Energy and Climate Change are planning to introduce smart meters into consumers' homes from 2012 and this is expected to run through until 2020 with the aim being to help consumers reduce their energy bills.

When introducing the proposal Ed Miliband, the Energy and Climate Change Secretary, said, 'the meters which most of us have in our homes were designed for a different age, before climate change. Now we need to get smarter with our energy. This is a big project affecting 26 million homes and several million businesses. The project will lead to extra work for electrical contractors through installing the meters on behalf of the utility companies and implementing more energy efficient devices once customers can see how much energy they are using.'

Already available is the Real Time Display (RTD) wireless monitor which enables consumers to see exactly how many units of electricity they are using through an easy-to-read portable display unit. By seeing the immediate impact in pence per hour of replacing existing lamps with low-energy ones or switching off unnecessary devices throughout the home or office, consumers are naturally

Figure 3.14 An example of a smart electricity meter.

motivated to consider saving energy. RTD monitors use a clip-on sensor on the meter tails and include desk-top software for PC and USB links.

Let us now look at some of these microgeneration technologies.

Solar photovoltaic (PV) power supply systems

A **solar photovoltaic** (PV) system is a collection of PV cells known as a PV string, that forms a PV array and collectively are called a PV generator which turns sunlight directly into electricity. PV systems may be 'stand-alone' power supplies or be designed to operate in parallel with the public low-voltage distribution network, that is, the a.c. mains supply.

> **Definition**
>
> A *solar photovoltaic* (PV) system is a collection of PV cells known as a PV string, that forms a PV array and collectively are called a PV generator which turns sunlight directly into electricity.

Stand-alone PV systems are typically a small PV panel of maybe 300 mm by 300 mm tilted to face the southern sky, where it receives the maximum amount of sunlight. They typically generate 12 to 15 volts and are used to charge battery supplies on boats, weather stations, road signs and any situation where electronic equipment is used in remote areas away from a reliable electrical supply.

The developing nations are beginning to see stand-alone PV systems as the way forward for electrification of rural areas beyond the National Grid rather than continuing with expensive diesel generators and polluting kerosine lamps.

The cost of PV generators is falling. The period 2009 to 2010 saw PV cells fall by 30% and, with new 'thin-film' cells being developed, the cost is expected to continue downwards. In the rural areas of the developing nations they see PV systems linked to batteries bringing information technology, radio and television to community schools. This will give knowledge and information to the next generation which will help these countries to develop a better economy, a better way of life and to have a voice in the developed world.

Stand-alone systems are not connected to the electricity supply system and are therefore exempt from much of BS 7671, the IET Regulations. However, Regulation 134.1.1 'good workmanship by competent persons and proper materials shall be used in all electrical installations' will apply to any work done by an electrician who must also pay careful attention to the manufacturer's installation instructions.

Solar photovoltaic (PV) systems designed to operate in parallel with the public low-voltage distribution network are the type of microgenerator used on commercial and domestic buildings. The PV cells operate in exactly the same way as the stand-alone system described above, but will cover a much greater area. The PV cells are available in square panels which are clipped together and laid over the existing roof tiles as shown in Fig. 3.15, or the PV cells may be manufactured to look just like the existing roof tiles which are integrated into the existing roof.

A solar PV system for a domestic three-bedroomed house will require approximately 15 to 20 square metres generating 2 to 3 kilowatts of power and the cost at the time of going to press of the PV cells alone will be in the region of £10,000 to £12,000, although grants are available. On the positive side, a PV system for a three-bedroomed house will save around 1,200 kg of CO_2 per year.

These bigger microgeneration systems are designed to be connected to the power supply system and the installation must therefore comply with Section 712 of BS 7671: 2008. Section 712 contains the requirements for protective measures comprising automatic disconnection of the supply wiring systems, isolation, switching and control, earthing arrangements and labelling. In addition, the installation must meet the requirements of the Electricity Safety Quality and Continuity Regulations 2006. This is a mandatory requirement. However, where the output does not exceed 16A per line, they are exempt from some of the requirements providing that:

- the equipment is type tested and approved by a recognized body;
- the installation complies with the requirements of BS 7671, the IET Regulations;
- the PV equipment must disconnect from the distributor's network in the event of a network fault;
- the distributor must be advised of the installation before or at the time of commissioning.

Figure 3.15 PV system in a domestic situation.

Installations of less than 16A per phase but up to 5 kilowatt peak (kWp) will also be required to meet the requirements of the Energy Network Association's Engineering Recommendation G83/1 for small-scale embedded generators in parallel with public low-voltage distribution networks. Installations generating more than 16A must meet the requirements of G59/1 which requires approval from the distributor before any work commences.

Solar thermal hot water heating

Solar thermal hot water heating systems are recognized as a reliable way to use the energy of the sun to heat water. The technology is straightforward and solar thermal panels for a three-bedroomed house cost at the time of going to press between £3,000 and £6,000 for a 3 to 6 m^2 panel and they will save about 260 kg of CO_2 annually.

Definition

Solar thermal hot water heating systems are recognized as a reliable way to use the energy of the sun to heat water.

The solar panel comprises a series of tubes containing water that is pumped around the panel and a heat exchanger in the domestic water cylinder as shown in Fig. 3.16. Solar energy heats up the domestic hot water. A solar panel of about 4 m^2 will deliver about 1,000 kWh per year which is about half the annual water demand of a domestic dwelling. However, most of the heat energy is generated during the summer and so it is necessary to supplement the solar system with a boiler in the winter months. Figure 3.17 shows a photo of an installed solar hot water panel.

If you travel to Germany, you will see a lot of photovoltaic and solar thermal panels on the roofs there. In the UK, planning requirements for solar thermal and PV installations have already been made much easier. A website detailing planning requirements for solar and wind can be found at www.planningportal.gov.uk/uploads/hhghouseguide.html.

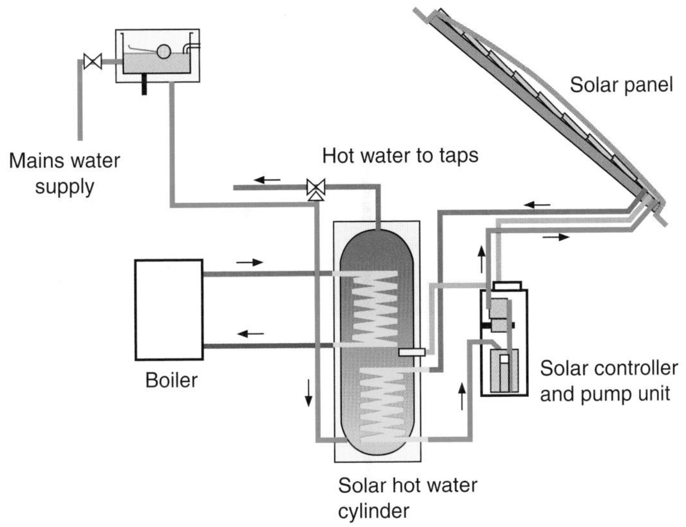

Figure 3.16 Solar-powered hot water system.

Wind energy generation

Modern large-scale wind machines are very different from the traditional windmill of the last century which gave no more power than a small car engine. Very large structures are needed to extract worthwhile amounts of energy from the wind.

Figure 3.17 Solar hot water panel.

Figure 3.18 Wind farm energy generators.

Modern large-scale wind generators are taller than electricity pylons, with a three-blade aeroplane-type propeller to catch the wind and turn the generator. If a wind turbine was laid down on the ground, it would be longer and wider than a football pitch. They are usually sited together in groups in what has become known as 'wind energy farms', as shown in Fig. 3.18.

Each modern grid-connected wind turbine generates about 600 kW of electricity. A wind energy farm of 20 generators will therefore generate 12 MW, a useful contribution to the National Grid, using a naturally occurring, renewable, non-polluting source of energy. The Department of Energy and Climate Change considers wind energy to be the most promising of the renewable energy sources.

In 2010 there were 253 wind energy farms in operation in the UK with 12 operating offshore. The 3,000 turbines on these farms have the capacity to generate 4,600 MW of electricity, enough for 2.5 million homes. There are a further 500 wind energy farms planned or in construction. However, because of the unpredictable nature of the wind, and inefficiencies in the generation process, the amount of power produced will be less than the installed capacity.

The Countryside Commission, the government's adviser on land use, has calculated that to achieve a target of generating 10% of the total electricity supply by wind power will require 40,000 generators of the present size. At the time of writing, 2012, we are generating only about 2.5% of the total electricity supply from wind power and all hopes are pinned on large offshore wind farms to achieve the government and EU targets for 2020.

Wind power is an endless renewable source of energy, is safer than nuclear power and provides none of the polluting emissions associated with fossil fuel. If there was such a thing as a morally pure form of energy, then wind energy would be it. However, wind farms are, by necessity, sited in some of the most beautiful landscapes in the UK. Building wind energy farms in these areas of outstanding natural beauty has outraged conservationists. Prince Charles has reluctantly joined the debate, saying that he was in favour of renewable energy sources but he believed that 'wind farms are an horrendous blot on the landscape'. He believes that if they are to be built at all they should be constructed well out to sea.

Definition

Wind power is an endless renewable source of energy, is safer than nuclear power and provides none of the polluting emissions associated with fossil fuel.

The next generation of wind farms will mostly be built offshore, and this is the place to build them, where there is more space and more wind, but the proposed size of these turbines creates considerable engineering problems. From the sea bed foundations to the top of the turbine blade will measure up to a staggering 250 metres, three times the height of the Statue of Liberty. Each offshore turbine, generating between 5 and 7 MW, will weigh between 200 and 300 tonnes. When you put large wind forces onto that structure you will create a very big cantilever effect which creates a major engineering challenge.

The world's largest offshore wind energy farm built so far was opened in September 2010. The 100-turbine 'Thanet' wind farm just off the Kent coast will generate enough power to supply 200,000 homes. The Thanet project cost £780 million to build. The turbines are up to 380 feet high and cover an area as large as 4,000 football pitches. The Thanet project will not retain its title as the world's largest wind farm for long because the 'Greater Gabbard' wind farm, off the north-east coast and having 140 turbines, is due to open in late 2014. The Thanet project brings Britain's total wind energy capacity above 5,000 megawatts for the first time.

The Department of Energy and Climate Change has calculated that 10,000 wind turbines could provide the energy equivalent of 8 million tonnes of coal per year and reduce CO_2 emissions. While this is a worthwhile saving of fossil fuel, opponents point out the obvious disadvantages of wind machines; among them the need to maintain the energy supply during periods of calm means that wind machines can only ever supplement a more conventional electricity supply.

Key fact

The Department of Energy and Climate Change has calculated that 10,000 wind turbines could provide the energy equivalent of 8 million tonnes of coal per year and reduce CO_2 emissions.

Small wind microgenerators can be used to make a useful contribution to a domestic property or a commercial building. They can be stand-alone and about the size of a tall street lamp. A 12 m-high turbine costs about £24,000 and, with

Figure 3.19 Small wind generator on a domestic property.

a good wind, will generate 10,000 kWh per year. However, if you live in a village, town or city you are unlikely to obtain the Local Authority Building and Planning permissions to install a wind generator because your neighbours will object.

Small wind generators of the type shown in Fig. 3.19 typically generate between 1.5A and 15A in wind speeds of 10 mph to 40 mph.

Heat pumps

Definition

In applications where heat must be upgraded to a higher temperature so that it can be usefully employed, a *heat pump* must be used.

In applications where heat must be upgraded to a higher temperature so that it can be usefully employed, a **heat pump** must be used. Energy from a low-temperature source such as the earth, the air, a lake or river is absorbed by a gas or liquid contained in pipes, which is then mechanically compressed by an electric pump to produce a temperature increase. The high-temperature energy is then transferred to a heat exchanger so that it might do useful work, such as providing heat to a building. For every 1 kWh of electricity used to power the heat pump compressor, approximately 3 to 4 kWh of heating are produced.

How a heat pump works

1. A large quantity of low-grade energy is absorbed from the environment and transferred to the refrigerant inside the heat pump (called the evaporator). This causes the refrigerant temperature to rise, causing it to change from liquid to a gas.
2. The refrigerant is then compressed, using an electrically driven compressor, reducing its volume but causing its temperature to rise significantly.
3. A heat exchanger (condenser) extracts the heat from the refrigerant to heat the domestic hot water or heating system.
4. After giving up its heat energy, the refrigerant turns back into a liquid, and, after passing through an expansion valve, is once more ready to absorb energy from the environment and the cycle is repeated, as shown in Fig. 3.20.

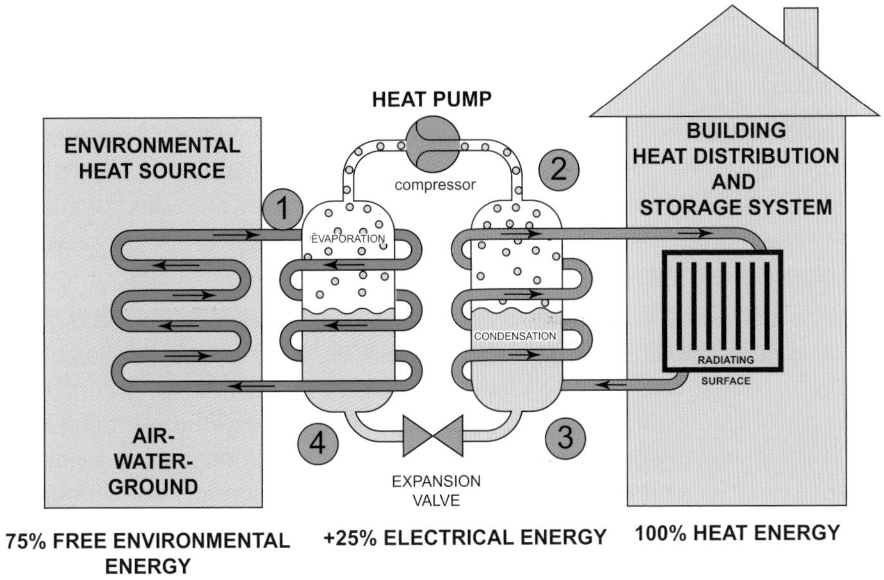

Figure 3.20 Heat pump working principle.

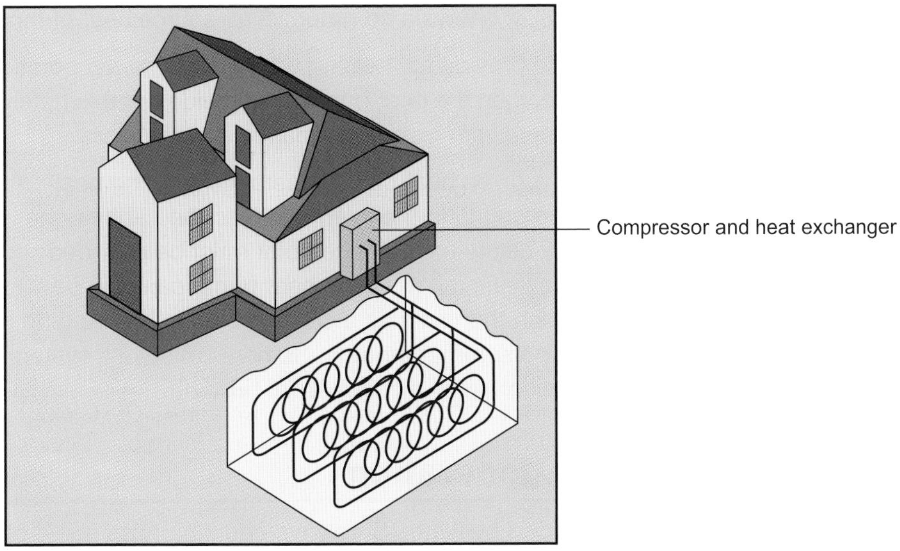

Figure 3.21 Ground source heating system.

A refrigerator works on this principle. Heat is taken out of the food cabinet, compressed and passed on to the heat exchanger or radiator at the back of the fridge. This warm air then radiates by air convection currents into the room. Thus the heat from inside the cabinet is moved into the room, leaving the sealed refrigerator cabinet cold.

Heat pumps (ground source)

Ground source heat pumps extract heat from the ground by circulating a fluid through polythene pipes buried in the ground in trenches or in vertical boreholes, as shown in Fig. 3.21. The fluid in the pipes extracts heat from the ground and a heat exchanger within the pump extracts heat from the fluid. These systems are most effectively used to provide underfloor radiant heating or water heating.

Calculations show that the length of pipe buried at a depth of 1.5 m required to produce 1.2 kW of heat will vary between 150 m in dry soil and 50 m in wet soil.

> **Definition**
>
> *Ground source heat pumps* extract heat from the ground by circulating a fluid through polythene pipes buried in the ground in trenches or in vertical boreholes.

The average heat output can be taken as 28 watts per metre of pipe. A rule of thumb guideline is that the surface area required for the ground heat exchanger should be about 2.5 times the area of the building to be heated.

This type of installation is only suitable for a new-build project and the ground heat exchanger will require considerable excavation and installation. The installer must seek Local Authority Building Control permissions before work commences.

Heat pumps (air source)

The performance and economics of heat pumps are largely determined by the temperature of the heat source and so we seek to use a high-temperature source. The heat sources used by heat pumps may be soil, the air, ground or surface water. Unfortunately all these sources follow the external temperature, being lower in winter when demand is highest. Normal atmosphere is an ideal heat source in that it can supply an almost unlimited amount of heat although unfortunately at varying temperatures, but relatively mild winter temperatures in the UK mean excellent levels of efficiency and performance throughout the year. For every 1 kWh of electricity used to power the heat pump compressor, between 3 and 4 kWh of heating energy is produced. They also have the advantage over ground source heat pumps of lower installation costs because they do not require any groundwork. Figure 3.22 shows a commercial air source heat pump.

Figure 3.22 Air source heat pump unit.

If the air heat pump is designed to provide full heating with an outside temperature of two to four degrees centigrade, then the heat pump will provide approximately 80% of the total heating requirement with high performance and efficiency.

The point at which the output of a given heat pump meets the building heat demand is known as the 'balance point'. In the example described above, the 20% shortfall of heating capacity below the balance point must be provided by some supplementary heat. However, an air-to-air heat pump can also be operated in the reverse cycle which then acts as a cooling device, discharging cold air into the building during the summer months. So here we have a system which could be used for air-conditioning in a commercial building.

Hydroelectric power generation

The UK is a small island surrounded by water. Surely we could harness some of the energy contained in tides, waves, lakes and rivers? Many different schemes have been considered in the past 20 years and a dozen or more experimental schemes are being tested now.

Water power makes a useful contribution to the energy needs of Scotland but the possibility of building similar hydroelectric schemes in England is unlikely chiefly due to the topographical nature of the country.

The Severn Estuary has a tidal range of 15 m, the largest in Europe, and is a reasonable shape for building a dam across the estuary. This would allow the basin to fill with water as the tide rose, and then allow the impounded water to flow out through electricity-generating turbines as the tide falls. However, such a tidal barrier might have disastrous ecological consequences for the many wildfowl and wading bird species by the submerging of the mudflats which now provide winter shelter for these birds. Therefore, the value of the power which might be produced must be balanced against the possible ecological consequences.

France has successfully operated a 240 MW tidal power-station at Rance in Brittany for the past 25 years.

Marine Current Turbines Ltd are carrying out research and development on submerged turbines which will rotate by exploiting the principle of flowing water in general and tidal streams in particular. The general principle is that an 11 m diameter water turbine is lowered into the sea down a steel column drilled in the sea bed. The tidal movement of the water then rotates the turbine and generates electricity.

The prototype machines were submerged in the sea off Lynmouth in Devon. In May 2008 they installed the world's first tidal turbine in the Strangford Narrows in Northern Ireland where it is now grid connected and generating 1.2 MW.

All the above technologies are geared to providing hydroelectric power connected to the national grid, but other micro-hydro schemes are at the planning and development stage.

Figure 3.23 Tidal flow water turbines.

Micro-hydro generation

The use of small hydropower (SHP) or micro-hydropower has grown over recent decades led by continuous technical developments, brought about partly in the UK by the 2010 coalition government's 'feed in tariff' where green electricity producers are paid a premium to produce electricity from renewable sources.

The normal perception of hydropower is of huge dams, but there is a much bigger use of hydropower in smaller installations. Asia, and especially China, is set to become a leader in hydroelectric generation. Australia and New Zealand are focusing on small hydro plants. Canada, a country with a long tradition of hydropower, is developing small hydropower as a replacement for expensive diesel generation in remote off-grid communities.

Small hydropower schemes generate electricity by converting the power available in rivers, canals and streams. The object of a hydropower scheme is to convert the potential energy of a mass of water flowing in a stream with a certain fall, called the head, into electrical energy at the lower end of the stream where the powerhouse is located. The power generated is proportional to the flow, called the discharge, and to the head of water available. The fundamental asset of hydropower is that it is a clean and a renewable energy source and the fuel used, water, is not consumed in the electricity-generating process.

In the Derbyshire Peak District along the fast-flowing River Goyt there were once 16 textile mills driven by water wheels. The last textile mill closed in the year 2000 but the Old Torr Mill has been saved. Where once the water wheel stood is now a gigantic 12-tonne steel screw, 2.4 metres in diameter. The water now drives the Reverse Archimedian Screw, affectionately called 'Archie', to produce enough electricity for 70 homes. The electricity-generating project is owned by the residents of New Mills in a sharing co-operative in which surplus electricity is sold back to the grid. Archie will produce 250 MWh of electricity each year and the installation cost was £300,000 in 2008. See Figure 3.24.

In July 2010 The Lake District National Park Authority granted permission for a water turbine to be built on a 350 m-long stretch of the fast-flowing River Kent. Approximately 1225 MWh of electricity will be generated each year, enough to power 250 homes. The project will cost £1.6 M funded by grants and loans and completion is anticipated in late 2011 or early 2012. The Kentmere Hydro Project is a community project and the power generated will be fed back to the National Grid. The Trust will receive money from the government's clean energy cash-back scheme which it will use to pay back the construction loans, leaving a surplus which will be used to support local projects in this small Lakeland community.

Key fact

Green electricity producers are paid a premium to produce electricity from renewable sources.

Key fact

The normal perception of hydropower is of huge dams, but there is a much bigger use of hydropower in smaller installations.

Definition

Small hydropower schemes generate electricity by converting the power available in rivers, canals and streams.

Figure 3.24 An example of an Archimedian Screw at the River Dart country park, Devon.Credit: WRE Limited.

Key fact

The type of turbine chosen for any hydro scheme will depend upon the discharge rate of the water and the head of water available.

The type of turbine chosen for any hydro scheme will depend upon the discharge rate of the water and the head of water available. A Pelton Wheel is a water turbine in which specially shaped buckets attached to the periphery of the wheel are struck by a jet of water. The kinetic energy of the water turns the wheel which is coupled to the generator.

Axial turbines comprise a large cylinder in which a propeller-type water turbine is fixed at its centre. The water moving through the cylinder causes the turbine blade to rotate and generate electricity.

A Francis Turbine or Kaplan Turbine is also an axial turbine but the pitch of the blades can be varied according to the load and discharge rate of the water.

Small water turbines will reach a mechanical efficiency at the coupling of 90%.

Up and down the country, riverside communities must be looking at the relics of our industrial past and wondering if they might provide a modern solution for clean, green, electrical energy. However, despite the many successes and obvious potential, there are many barriers to using waterways for electricity generation in the European countries. It is very difficult in this country to obtain permission from the Waterways Commission to extract water from rivers, even though, once the water has passed through the turbine, it is put back into the river. Environmental pressure groups are opposed to micro-hydro generation because of its perceived local environmental impact on the river ecosystem and the disturbance to fishing. Therefore, once again, the value of the power produced would have to be balanced against the possible consequences.

Combined heat and power (CHP)

Definition

CHP is the simultaneous generation of usable heat and power in a single process. That is, heat is produced as a by-product of the power-generation process.

CHP is the simultaneous generation of usable heat and power in a single process. That is, heat is produced as a by-product of the power-generation process. A chemical manufacturing company close to where I live has a small power-station which meets some of their electricity requirements using the smart meter principle. Their 100 MW turbine is driven by high-pressure steam. When the steam condenses after giving up its energy to the turbine, there remains a lot of very hot water which is then piped around the offices and some production plant buildings for space heating. Combining heat and power in this way can increase the overall efficiency of the fuel used because it is performing two operations.

CHP can also use the heat from incinerating refuse to heat a nearby hospital, school or block of flats.

Biomass heating

Definition

Biomass is derived from plant materials and animal waste. It can be used to generate heat and to produce fuel for transportation.

Biomass is derived from plant materials and animal waste. It can be used to generate heat and to produce fuel for transportation. The biomass material may be straw and crop residues, crops grown specially for energy production, or rape seed oil and waste from a range of sources including food production. The nature of the fuel will determine the way that energy can best be recovered from it.

There is a great deal of scientific research being carried out at the moment into **'biomass renewables'**; that is, energy from crops. This area of research is at an early stage, but is expected to flourish in the next decade. The first renewable energy plant, which is to be located at Teesport on the River Tees in the north-east of England, has received approval from the Department for Energy and Climate Change for building to commence.

The facility will be one of the largest biomass plants to be built in the world and is scheduled to enter commercial operation in late 2012. Young trees will be grown as a crop to produce wood chips. The plant will use 2.5 million tonnes of wood chips each year to produce 300 MW of electrical energy. The plant will operate 24 hours a day, all year round to meet some of the National Grid base load.

Definition

'Biomass renewables'; that is, energy from crops.

Water conservation

Conservation is the preservation of something important, especially of the natural environment. Available stored water is a scarce resource in England and Wales where there are only 1,400 cubic metres per person per year; very little compared with France, which has 3,100 cubic metres per person per year, Italy which has 2,900 and Spain 2,800. About half of the water used by an average home is used to shower, bathe and wash the laundry, and another third is used to flush the toilet.

Definition

Conservation is the preservation of something important, especially of the natural environment.

At a time when most domestic and commercial properties have water meters installed it saves money to harvest and reuse water.

The City and Guilds has asked us to look at two methods of water conservation: rainwater harvesting and grey water recycling.

Rainwater harvesting

Rainwater harvesting is the collection and storage of rainwater for future use. Rainwater has in the past been used for drinking, water for livestock and water for irrigation. It is now also being used to provide water for car cleaning and garden irrigation in domestic and commercial buildings.

Definition

Rainwater harvesting is the collection and storage of rainwater for future use.

Many gardeners already harvest rainwater for garden use by collecting run-off from a roof or greenhouse and storing it in a water butt or water barrel. However, a 200-litre water butt does not give much drought protection in spite of garden plants preferring rainwater to fluoridated tap water. To make a useful contribution the rainwater storage tank should be between 2,000- and 7,000-litre capacity. The rainwater-collecting surfaces will be the roof of the building and any hard paved surfaces such as patios. Downpipes and drainage pipes then route the water to the storage tank situated, perhaps, under the lawn. An electric pump lifts the water from the storage tank to the point of use, possibly a dedicated outdoor tap. The water is then distributed through a hose-pipe or sprinkler system to the garden in the normal way.

With a little extra investment, rainwater can be filtered and used inside the house to supply washing machines and WCs. Installing domestic pipes and interior plumbing can be added to existing homes although it is more straightforward in a new-build home.

With the move towards more sustainable homes UK architects are becoming more likely to specify rainwater harvesting in their design to support alternatives to a mains water supply. In Germany, rainwater harvesting systems are now installed as standard in all new commercial buildings.

Grey water recycling

Grey water is tap water which has already performed one operation and is then made available to be used again instead of flushing it down the drain. Grey water recycling offers a way of getting double the use out of the world's most precious resource.

Definition

Grey water is tap water which has already performed one operation and is then made available to be used again instead of flushing it down the drain.

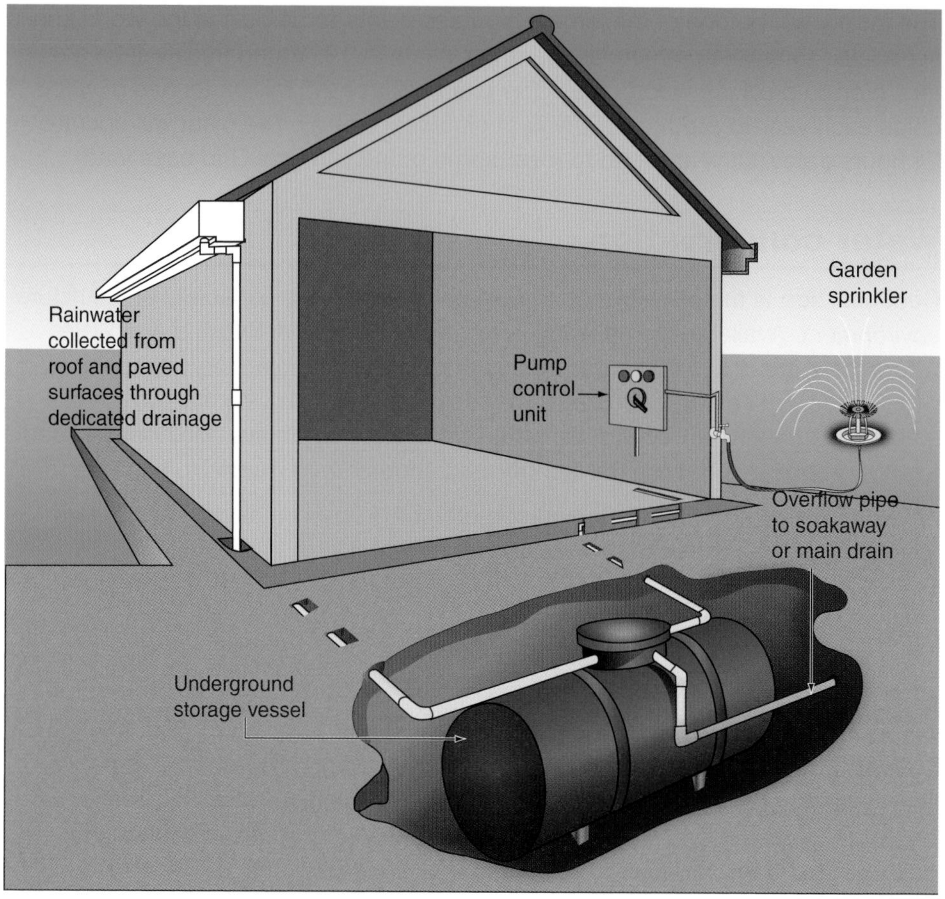

Figure 3.25 Underground rainwater storage vessel.

There are many products on the market such as the BRAC system which takes in water used in the shower, bath and laundry, cleans it by filtering and then reuses it for toilet flushing. It is only a matter of routing the grey waste water drain-pipe from the bath, shower and laundry to the filter unit and then plumbing the sanitized grey water to the toilet tank.

These systems are easy to install, particularly in a new-build property. It is only a matter of re-routing the drain-pipes. Another option for your grey water is to route it into the rainwater storage tank for further use in the garden.

Protecting electrical equipment, circuits and people

We know from the earlier chapters in this book that using electricity is one of the causes of accidents in the workplace. Using electricity is a hazard because it has the 'potential' and the possibility to cause harm. Therefore, the provision of protective devices in an electrical installation is fundamental to the whole concept of the safe use of electricity in buildings. The electrical installation as a whole must be protected against overload or short-circuit and the people using the building must be protected against the risk of shock, fire or other risks arising from their own misuse of the installation or from a fault. The installation and maintenance of adequate and appropriate protective measures is a vital part of the safe use of electrical energy. I want to look at protection against an electric shock by both basic and fault protection, at protection by protective bonding and automatic disconnection of the supply, and protection against excess current.

Let us first define some of the words we will be using. Chapter 54 of the IET Regulations describes the earthing arrangements for an electrical installation. It gives the following definitions:

Earth – the conductive mass of the earth whose electrical potential is taken as zero.

Earthing – the act of connecting the exposed conductive parts of an installation to the main earthing terminal of the installation.

Bonding conductor – a protective conductor providing bonding.

Bonding – the linking together of the exposed or extraneous metal parts of an electrical installation.

Circuit protective conductor (CPC) – a protective conductor connecting exposed conductive parts of equipment to the main earthing terminal of the installation. This is the green and yellow insulated conductor in twin and earth cable.

Exposed conductive parts – the metalwork of an electrical appliance or the trunking and conduit of an electrical system which can be touched because they are not normally live, but which may become live under fault conditions.

Extraneous conductive parts – the structural steelwork of a building and other service pipes such as gas, water, radiators and sinks. They do not form a part of the electrical installation but may introduce a potential, generally earth potential, to the electrical installation.

Shock protection – protection from electric shock is provided by basic protection and fault protection.

Basic protection – is provided by the insulation of live parts in accordance with Section 416 of the IET Regulations.

Fault protection – is provided by protective bonding and automatic disconnection of the supply (by a fuse or miniature circuit-breaker, MCB) in accordance with IET Regulations 411.3 to 6.

Protective bonding – this is bonding for the purpose of safety, as shown in Fig. 3.6.

Try this

Definitions

To help you to remember these definitions, try writing them out in your own words. Identify the essential words.

Basic protection and fault protection

The human body's movements are controlled by the nervous system. Very tiny electrical signals travel between the central nervous system and the muscles, stimulating operation of the muscles, which enable us to walk, talk and run, and remember that the heart is also a muscle.

If the body becomes part of a more powerful external circuit, such as the electrical mains, and current flows through it, the body's normal electrical operations are disrupted. The shock current causes unnatural operation of

Definition

Earth – the conductive mass of the earth whose electrical potential is taken as zero.

Earthing – the act of connecting the exposed conductive parts of an installation to the main protective earthing terminal of the installation.

Bonding conductor – a protective conductor providing bonding.

Bonding – the linking together of the exposed or extraneous metal parts of an electrical installation.

Circuit protective conductor (CPC) – a protective conductor connecting exposed conductive parts of equipment to the main earthing terminal.

Exposed conductive parts – the metalwork of an electrical appliance or the trunking and conduit of an electrical system which can be touched because they are not normally live, but which may become live under fault conditions.

Extraneous conductive parts – the structural steelwork of a building and other service pipes such as gas, water, radiators and sinks.

Shock protection – protection from electric shock is provided by basic protection and fault protection.

Basic protection – is provided by the insulation of live parts in accordance with Section 416 of the IET Regulations.

Fault protection – is provided by protective bonding and automatic disconnection of the supply (by a fuse or miniature circuit-breaker, MCB) in accordance with IET Regulations 411.3 to 6.

Protective bonding – this is bonding for the purpose of safety, as shown in Fig. 3.6.

the muscles and the result may be that the person is unable to release the live conductor causing the shock, or the person may be thrown across the room. The current which flows through the body is determined by the resistance of the human body and the surface resistance of the skin on the hands and feet.

This leads to the consideration of exceptional precautions where people with wet skin or wet surfaces are involved, and the need for special consideration in bathroom installations.

Two types of contact will result in a person receiving an electric shock. *Direct contact* with live parts involves touching a terminal or line conductor that is actually live. The regulations call this ***basic protection***. *Indirect contact* results from contact with an exposed conductive part such as the metal structure of a piece of equipment that has *become live as a result of a fault*. The regulations call this ***fault protection***.

In installations operating at normal mains voltage, the primary method of *protection against direct contact* is by insulation. All live parts are enclosed in insulating material such as rubber or plastic, which prevents contact with those parts. The insulating material must, of course, be suitable for the circumstances in which they will be used and the stresses to which they will be subjected. The IET Regulations call this *basic protection* (IET Regulation 131.2.1).

Other methods of *basic protection* include the provision of barriers or enclosures which can only be opened by the use of a tool, or when the supply is first disconnected. Protection can also be provided by fixed obstacles such as a guardrail around an open switchboard or by placing live parts out of reach as with overhead lines.

Fault protection

Protection against indirect contact, called fault protection (IET Regulation 131.2.2), is achieved by connecting exposed conductive parts of equipment to the main earthing terminal of the installation.

In Chapter 13 of the IET Regulations we are told that where the metalwork of electrical equipment may become charged with electricity in such a manner as to cause danger, that metalwork will be connected with earth so as to discharge the electrical energy without danger. The application of protective bonding conductors is one of the important principles for safety (IET Regulation 131.2.2).

There are five methods of protection against contact with metalwork which has become unintentionally live, that is, indirect contact with exposed conductive parts recognized by the IET Regulations. These are:

1. Protective bonding coupled with automatic disconnection of the supply.
2. The use of Class II (double-insulated) equipment.
3. The provision of a non-conducting location.
4. The use of earth-free bonding.
5. Electrical separation.

Methods 3 and 4 are limited to special situations under the effective supervision of trained personnel.

Method 5, electrical separation, is little used but does find an application in the domestic electric shaver supply unit which incorporates an isolating transformer.

Method 2, the use of Class II insulated equipment, is limited to single pieces of equipment such as tools used on construction sites, because it relies upon effective supervision to ensure that no metallic equipment or extraneous earthed metalwork enters the area of the installation.

The method which is most universally used in the United Kingdom is, therefore, Method 1 – protective bonding coupled with automatic disconnection of the supply.

This method relies upon all exposed metalwork being electrically connected together to an effective earth connection. Not only must all the metalwork associated with the electrical installation be so connected, that is, conduits, trunking, metal switches and the metalwork of electrical appliances, but Regulation 411.3.1.2 tells us to connect the extraneous metalwork of water service pipes, gas and other service pipes and ducting, central heating and air-conditioning systems, exposed metallic structural parts of the building and lightning protective systems to the protective earthing terminal. In this way the possibility of a voltage appearing between two exposed metal parts is removed. Protective bonding is shown in Fig. 3.6 and 3.34 to 3.36.

The second element of this protection method is the provision of a means of automatic disconnection of the supply in the event of a fault occurring that causes the exposed metalwork to become live.

IET Regulation 411.3.2 tells us that for final circuits not exceeding 32 A the maximum disconnection time shall not exceed 0.4 s.

The achievement of these disconnection times is dependent upon the type of protective device used, fuse or circuit-breaker, the circuit conductors to the fault and the provision of adequate protective bonding. The resistance, or we call it the impedance, of the earth fault loop must be less than the values given in Appendix 2 of the *On Site Guide* and Tables 41.2 to 41.4 of the IET Regulations. (Table 3.2 shows the maximum value of the earth fault loop impedance for circuits protected by MCBs to BS EN 60898.) We will look at this again later in this chapter under the heading 'Earth fault loop impedance Z_S'. Chapter 54 of the IET Regulations gives details of the earthing arrangements to be incorporated into the supply system to meet these regulations and these are described later in this chapter under the heading 'Electricity supply systems'.

Residual current protection by RCD

The IET Regulations recognize the particular problems created when electrical equipment such as lawnmowers, hedge-trimmers, drills and lights are used outside buildings. In these circumstances the availability of an adequate earth return path is a matter of chance. The regulations, therefore, require that any socket outlet with a rated current not exceeding 20 A, for use by ordinary people and intended for general use, shall have the additional protection of a residual current device (RCD) which has a rated operating current of not more than 30 milliamperes (mA) (IET Regulation 411.3.3).

COMPANION WEBSITE Visit the companion website for more on this topic.

Try this

Definitions

- What do the regulations mean by 'ordinary people'?
- Look at the definitions earlier in this book and write it in the margin here.

An RCD is a type of circuit-breaker that continuously compares the current in the line and neutral conductors of the circuit. The currents in a healthy circuit will be equal, but in a circuit that develops a fault, some current will flow to earth and the line and neutral currents will no longer balance. The RCD detects the imbalance and disconnects the circuit. Figure 3.33 shows an RCD's construction.

Isolation and switching

Part 4 of the IET Regulations deals with the application of protective measures for safety and Chapter 53 with the regulations for switching devices or switchgear required for protection, isolation and switching of a consumer's installation.

The consumer's main switchgear must be readily accessible to the consumer and be able to:

- isolate the complete installation from the supply;
- protect against overcurrent;
- cut off the current in the event of a serious fault occurring.

The regulations identify four separate types of switching: switching for isolation, switching for mechanical maintenance, emergency switching and functional switching.

Definition

Isolation is defined as cutting off the electrical supply to a circuit or item of equipment in order to ensure the safety of those working on the equipment by making dead those parts which are live in normal service.

Isolation is defined as cutting off the electrical supply to a circuit or item of equipment in order to ensure the safety of those working on the equipment by making dead those parts which are live in normal service.

The purpose of isolation switching is to enable electrical work to be carried out safely on an isolated circuit or piece of equipment. Isolation is intended for use by electrically skilled or supervised persons.

Visit the companion website for more on this topic.

An isolator is a mechanical device which is operated manually and used to open or close a circuit off load. An isolator switch must be provided close to the supply point so that all equipment can be made safe for maintenance. Isolators for motor circuits must isolate the motor and the control equipment, and isolators for discharge lighting luminaires must be an integral part of the luminaire so that it is isolated when the cover is removed or be provided with effective local isolation (IET Regulation 537.2.1.6). Devices which are suitable for isolation are isolation switches, fuse links, circuit-breakers, plugs and socket outlets. They must isolate all live supply conductors and provision must be made to secure the isolation (IET Regulation 537.2.2.4).

Isolation at the consumer's service position can be achieved by a double-pole switch which opens or closes all conductors simultaneously. On three-phase supplies the switch need only break the live conductors with a solid link in the neutral, provided that the neutral link cannot be removed before opening the switch.

Definition

The switching for mechanical maintenance requirements is similar to those for isolation except that the control switch must be capable of switching the full load current of the circuit or piece of equipment.

The switching for mechanical maintenance requirements is similar to those for isolation except that the control switch must be capable of switching the full load current of the circuit or piece of equipment.

The purpose of switching for mechanical maintenance is to enable non-electrical work to be carried out safely on the switched circuit or equipment.

Mechanical maintenance switching is intended for use by skilled but non-electrical persons. Switches for mechanical maintenance must be manually operated, not have exposed live parts when the appliance is opened, must be connected in the main electrical circuit and have a reliable on/off indication or visible contact gap (IET Regulation 537.3.2.2). Devices which are suitable for switching off for mechanical maintenance are switches, circuit-breakers, plug and socket outlets.

Emergency switching involves the rapid disconnection of the electrical supply by a single action to remove or prevent danger.

The purpose of emergency switching is to cut off the electrical energy *rapidly* to remove an unexpected hazard.

Emergency switching is for use by anyone. The device used for emergency switching must be immediately accessible and identifiable, and be capable of cutting off the full load current.

Electrical machines must be provided with a means of emergency switching, and a person operating an electrically driven machine must have access to an emergency switch so that the machine can be stopped in an emergency. The remote stop/start arrangement could meet this requirement for an electrically driven machine (IET Regulation 537.4.2.2). Devices which are suitable for emergency switching are switches, circuit-breakers and contactors. Where contactors are operated by remote control they should *open* when the coil is de-energized, that is, fail safe. Push-buttons used for emergency switching must be coloured red and latch in the stop or off position. They should be installed where danger may arise and be clearly identified as emergency switches. Plugs and socket outlets cannot be considered appropriate for emergency disconnection of supplies.

Functional switching involves the switching on or off, or varying the supply, of electrically operated equipment in normal service.

The purpose of functional switching is to provide control of electrical circuits and equipment in normal service.

Functional switching is for the user of the electrical installation or equipment. The device must be capable of interrupting the total steady current of the circuit or appliance. When the device controls a discharge lighting circuit it must have a current rating capable of switching an inductive load. The regulations acknowledge the growth in the number of electronic dimmer switches being used for the control and functional switching of lighting circuits. The functional switch must be capable of performing the most demanding duty it may be called upon to perform (IET Regulations 537.5.2.1 and 2).

Definition

Emergency switching involves the rapid disconnection of the electrical supply by a single action to remove or prevent danger.

Definition

Functional switching involves the switching on or off, or varying the supply, of electrically operated equipment in normal service.

Overcurrent protection

The consumer's mains equipment must provide protection against overcurrent, that is, a current exceeding the rated value (IET Regulation 430.3). Fuses provide overcurrent protection when situated in the live conductors; they must not be connected in the neutral conductor. Circuit-breakers may be used in place of fuses, in which case the circuit-breaker may also provide the means of isolation, although a further means of isolation is usually provided so that maintenance can be carried out on the circuit-breakers themselves.

When selecting a protective device we must give consideration to the following factors:

- the prospective fault current;
- the circuit load characteristics;

Safety first

Isolation

Isolation and switching for protection is an important safety measure in the electrical industry.

- the current-carrying capacity of the cable;
- the disconnection time requirements for the circuit.

The essential requirements for a device designed to protect against overcurrent are:

- it must operate automatically under fault conditions;
- have a current rating matched to the circuit design current;
- have a disconnection time which is within the design parameters;
- have an adequate fault breaking capacity;
- be suitably located and identified.

We will look at these requirements below.

Definition

An *overload current* can be defined as a current which exceeds the rated value in an otherwise healthy circuit.

A *short-circuit* is an overcurrent resulting from a fault of negligible impedance connected between conductors.

An overcurrent may be an overload current, or a short-circuit current. An **overload current** can be defined as a current which exceeds the rated value in an otherwise healthy circuit. Overload currents usually occur because the circuit is abused or because it has been badly designed or modified. A **short-circuit** is an overcurrent resulting from a fault of negligible impedance connected between conductors. Short-circuits usually occur as a result of an accident which could not have been predicted before the event.

An overload may result in currents of two or three times the rated current flowing in the circuit. Short-circuit currents may be hundreds of times greater than the rated current. In both cases the basic requirements for protection are that the fault currents should be interrupted quickly and the circuit isolated safely before the fault current causes a temperature rise or mechanical effects which might damage the insulation, connections, joints and terminations of the circuit conductors or their surroundings (IET Regulations 131).

The selected protective device should have a current rating which is not less than the full load current of the circuit but which does not exceed the cable current rating. The cable is then fully protected against both overload and short-circuit faults (IET Regulation 435.1). Devices which provide overcurrent protection are:

- High breaking capacity (HBC) fuses to BS 88-2: 2010. These are for industrial applications having a maximum fault capacity of 80 kA.
- Cartridge fuses to BS 88-3: 2010. These are used for a.c. circuits on industrial and domestic installations having a fault capacity of about 30 kA.
- Cartridge fuses to BS 1362. These are used in 13 A plug tops and have a maximum fault capacity of about 6 kA.
- Semi-enclosed fuses to BS 3036. These were previously called rewirable fuses and are used mainly on domestic installations having a maximum fault capacity of about 4 kA.
- MCBs to BS EN 60898. These are miniature circuit-breakers (MCBs) which may be used as an alternative to fuses for some installations. The British Standard includes ratings up to 100 A and maximum fault capacities of 9 kA. They are graded according to their instantaneous tripping currents – that is, the current at which they will trip within 100 ms. This is less than the time taken to blink an eye.
- The 3rd Amendment to the 17th Edition of the IET Regulations tells us at Note 5 of Table 53.4 that circuit protective devices and RCDs are not intended for frequent load switching. However, infrequent switching of MCBs is permissible for the purpose of isolation or emergency switching.

Definition

By definition a *fuse* is the weakest link in the circuit. Under fault conditions it will melt when an overcurrent flows, protecting the circuit conductors from damage.

By definition a **fuse** is the weakest link in the circuit. Under fault conditions it will melt when an overcurrent flows, protecting the circuit conductors from damage.

Semi-enclosed fuses (BS 3036)

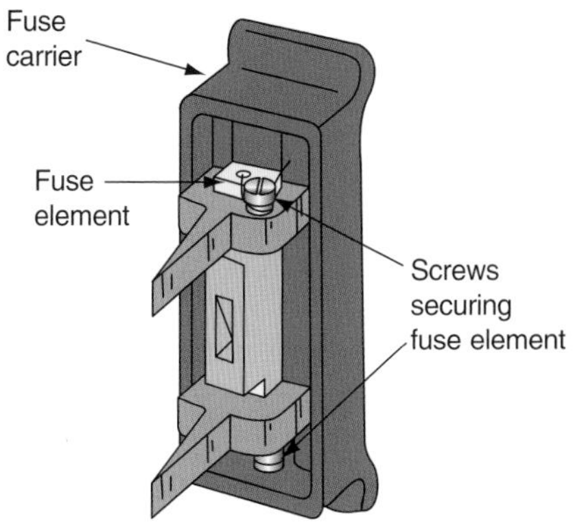

Figure 3.26 A semi-enclosed fuse.

The semi-enclosed fuse consists of a fuse wire, called the fuse element, secured between two screw terminals in a fuse carrier. The fuse element is connected in series with the load and the thickness of the element is sufficient to carry the normal rated circuit current. When a fault occurs an overcurrent flows and the fuse element becomes hot and melts or 'blows'.

This type of fuse is illustrated in Fig. 3.26. The fuse element should consist of a single strand of plain or tinned copper wire having a diameter appropriate to the current rating of the fuse. *This type of fuse was very popular in domestic installations, but less so these days because of its disadvantages.*

Advantages of semi-enclosed fuses

- They are very cheap compared with other protective devices both to install and to replace.
- There are no mechanical moving parts.
- It is easy to identify a 'blown' fuse.

Disadvantages of semi-enclosed fuses

- The fuse element may be replaced with wire of the wrong size either deliberately or by accident.
- The fuse element weakens with age due to oxidization, which may result in a failure under normal operating conditions.
- The circuit cannot be restored quickly since the fuse element requires screw fixing.
- They have low breaking capacity since, in the event of a severe fault, the fault current may vaporize the fuse element and continue to flow in the form of an arc across the fuse terminals.
- They are not guaranteed to operate until up to twice the rated current is flowing.
- There is a danger from scattering hot metal if the fuse carrier is inserted into the base when the circuit is faulty.

Cartridge fuses [(BS 88-3: 2012 (previously BS 1361))]

The cartridge fuse breaks a faulty circuit in the same way as a semi-enclosed fuse, but its construction eliminates some of the disadvantages experienced with an open-fuse element. The fuse element is encased in a glass or ceramic tube and secured to end-caps which are firmly attached to the body of the fuse so that they do not blow off when the fuse operates. Cartridge fuse construction is illustrated in Fig. 3.27. With larger size cartridge fuses, lugs or tags are sometimes brazed on the end-caps to fix the fuse cartridge mechanically to the carrier. They may also be filled with quartz sand to absorb and extinguish the energy of the arc when the cartridge is brought into operation.

Advantages of cartridge fuses

- They have no mechanical moving parts.
- The declared rating is accurate.
- The element does not weaken with age.

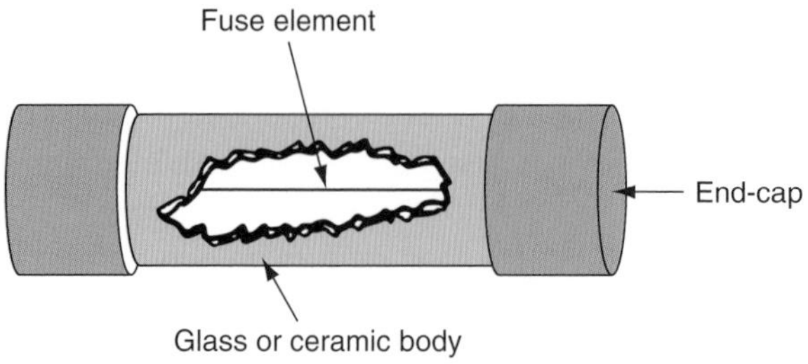

Figure 3.27 Cartridge fuse.

Figure 3.28 MCBs – B Breaker, fits Wylex standard consumer unit (courtesy of Wylex).

- They have small physical size and no external arcing which permits their use in plug tops and small fuse carriers.
- Their operation is more rapid than semi-enclosed fuses. Operating time is inversely proportional to the fault current, so the bigger the fault current the quicker the fuse operates.
- They are easy to replace.
- Small current sizes clip in place.
- Larger current sizes have bolt-hole fixings.

Disadvantages of cartridge fuses

- They are more expensive to replace than fuse elements that can be rewired.
- They can be replaced with an incorrect cartridge.
- The cartridge may be shorted out by wire or silver foil in extreme cases of bad practice.
- It is not possible to see if the fuse element is broken.

Miniature circuit-breakers (BS EN 60898)

The disadvantage of all fuses is that when they have operated they must be replaced. An MCB overcomes this problem since it is an automatic switch which opens in the event of an excessive current flowing in the circuit and can be closed when the circuit returns to normal.

An MCB of the type shown in Fig. 3.28 incorporates a thermal and magnetic tripping device. The load current flows through the thermal and the electromagnetic devices in normal operation but under overcurrent conditions they activate and trip the MCB.

The circuit can be restored when the fault is removed by pressing the ON toggle. This latches the various mechanisms within the MCB and 'makes' the switch contact. The toggle switch can also be used to disconnect the circuit for maintenance or isolation, or to test the MCB for satisfactory operation.

Advantages of MCBs

- They have factory-set operating characteristics.
- Tripping characteristics and therefore circuit protection is set by the installer.
- The circuit protection is difficult to interfere with.
- The circuit is provided with discrimination.
- A faulty circuit may be quickly identified.
- A faulty circuit may be easily and quickly restored.
- The supply may be safely restored by an unskilled operator.

Disadvantages of MCBs

- They are relatively expensive but look at the advantages to see why they are so popular these days.
- They contain mechanical moving parts and therefore require regular testing to ensure satisfactory operation under fault conditions.

Characteristics of MCBs

MCB Type B to BS EN 60898 will trip instantly at between three and five times its rated current and is also suitable for domestic and commercial installations.

MCB Type C to BS EN 60898 will trip instantly at between five and ten times its rated current. It is more suitable for highly inductive commercial and industrial loads.

MCB Type D to BS EN 60898 will trip instantly at between 10 and 25 times its rated current. It is suitable for welding and X-ray machines where large inrush currents may occur.

RCBO

A residual current operated circuit-breaker with integral overcurrent protection (RCBO) provides protection against overload and/or short-circuit. RCBOs give the combined protection of an MCB and an RCD in one device.

Installing overcurrent protective devices

The general principle to be followed is that a protective device must be placed at a point where a reduction occurs in the current carrying capacity of the circuit conductors (IET Regulations 433.2 and 434.2). A reduction may occur because of a change in the size or type of conductor, or because of a change in the method of installation or a change in environmental conditions. The only exceptions to this rule are where an overload protective device opening a circuit might cause a greater danger than the overload itself – for example, a circuit feeding an overhead electromagnet in a scrapyard.

Fault protection

The overcurrent protection device protecting circuits not exceeding 32 A shall have a disconnection time not exceeding 0.4 s (IET Regulation 411.3.2.2).

The IET Regulations permit us to assume that where an overload protective device is also intended to provide short-circuit protection, and has a rated breaking capacity greater than the prospective short-circuit current at the point of its installation, the conductors on the load side of the protective device are considered to be adequately protected against short-circuit currents without further proof. This is because the cable rating and the overload rating of the device are compatible. However, if this condition is not met or if there is some doubt, it must be verified that fault currents will be interrupted quickly before they can cause a dangerously high temperature rise in the circuit conductors. IET Regulation 434.5.2 provides an equation for calculating the maximum operating time of the protective device to prevent the permitted conductor temperature rise from being exceeded as follows:

$$t = \frac{k^2 S^2}{I^2} \text{ (s)}$$

where

t = duration time in seconds

S = cross-sectional area of conductor in square millimetres

I = short-circuit r.m.s. current in amperes

k = a constant dependent upon the conductor metal and type of insulation (see Table 43 A of the IET Regulations).

Example

A 10 mm PVC sheathed mineral insulated (MI) copper cable is short-circuited when connected to a 400 V supply. The impedance of the short-circuit path is 0.1 Ω. Calculate the maximum permissible disconnection time and show that a 50 A Type B MCB to BS EN 60898 will meet this requirement.

$$I = \frac{V}{Z} \text{ (A)} \qquad I = \frac{400\,\text{V}}{0.1\,\Omega} = 4000\,\text{A}$$

$$\therefore \text{ Fault current} = 4000\,\text{A}$$

For PVC sheathed MI copper cables, Table 43.1 gives a value for k of 115. So,

$$t = \frac{k^2 S^2}{I^2} \text{ (s)}$$

$$\therefore t = \frac{115^2 \times 10^2\,\text{mm}^2}{4000\,\text{A}} = 82.66 \times 10^{-3}\,\text{s}$$

The maximum time that a 4000 A fault current can be applied to this 10 mm^2 cable without dangerously raising the conductor temperature is 82.66 ms. Therefore, the protective device must disconnect the supply to the cable in less than 82.66 ms under short-circuit conditions. Manufacturers' information and Appendix 3 of the IET Regulations give the operating times of protective devices at various short-circuit currents in the form of graphs. Let us come back to this problem in a few moments.

Time/current characteristics of protective devices

Disconnection times for various overcurrent devices are given in the form of a logarithmic graph. This means that each successive graduation of the axis represents a 10-times change over the previous graduation.

These logarithmic scales are shown in the graphs of Figs 3.29 and 3.30. From Fig. 3.29 it can be seen that the particular protective device represented by this characteristic will take 8 s to disconnect a fault current of 50 A and 0.08 s to clear a fault current of 1000 A.

Let us now go back to the problem and see if the Type B MCB will disconnect the supply in less than 82.66 ms.

Figure 3.30(a) shows the time/current characteristics for a Type B MCB to BS EN 60898. This graph shows that a fault current of 4000 A will trip the protective device in 20 ms. Since this is quicker than 82.66 ms, the 50 A Type B MCB is suitable and will clear the fault current before the temperature of the cable is raised to a dangerous level.

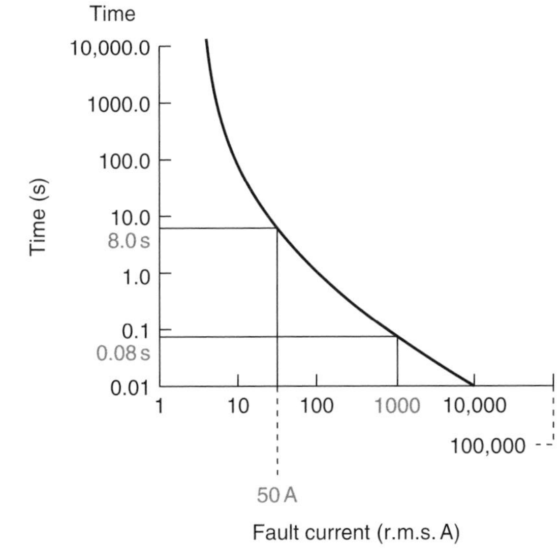

Figure 3.29 Time/current characteristic of an overcurrent protective device.

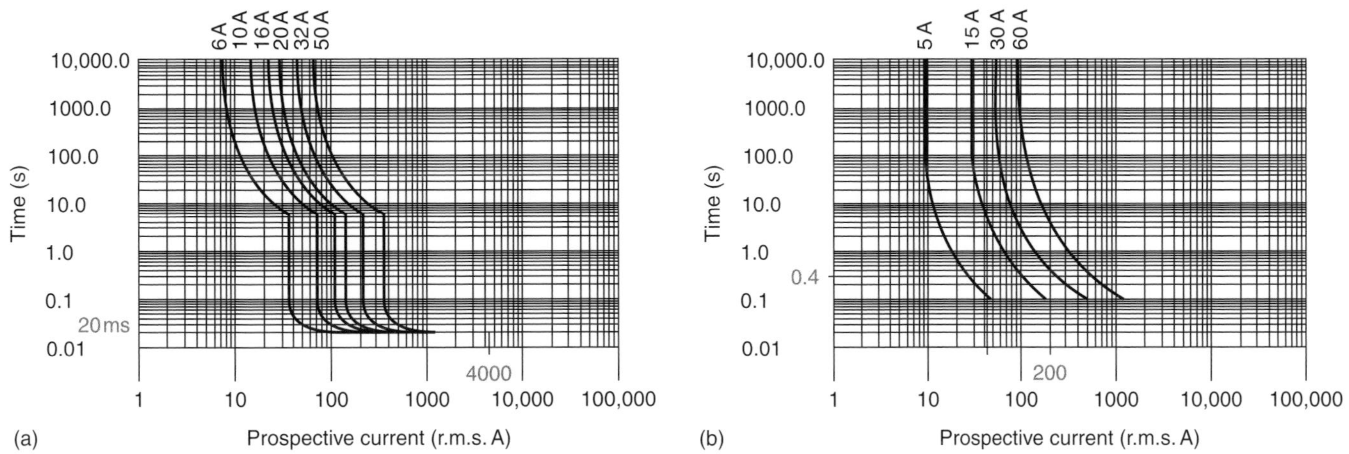

Figure 3.30 Time/current characteristics of (a) a Type B MCB to BS EN 60898, and (b) semi-enclosed fuse to BS 3036.

Appendix 3 of the IET Regulations gives the time/current characteristics and specific values of prospective short-circuit current for a number of protective devices.

These indicate the value of fault current which will cause the protective device to operate in the times indicated by IET Regulation 411.

Figures 3.1, 3.2 and 3.3 in Appendix 3 of the IET Regulations deal with fuses, and Figs 3.4, 3.5 and 3.6 with MCBs.

It can be seen that the prospective fault current required to trip an MCB in the required time is a multiple of the current rating of the device. The multiple depends upon the characteristics of the particular devices. Thus:

- Type B MCB to BS EN 60898 has a multiple of 5
- Type C MCB to BS EN 60898 has a multiple of 10
- Type D MCB to BS EN 60898 has a multiple of 20.

Example

A 6 A Type B MCB to BS EN 60898 which is used to protect a domestic lighting circuit will trip within 0.4 s when 6 A times a multiple of 5, that is, 30 A, flows under fault conditions.

Therefore if the earth fault loop impedance is low enough to allow at least 30 A to flow in the circuit under fault conditions, the protective device will operate within the time required by IET Regulation 411.

The characteristics shown in Appendix 3 of the IET Regulations give the specific values of prospective short-circuit current for all standard sizes of protective device.

Effective coordination of protective devices

In the event of a fault occurring on an electrical installation only the protective device nearest to the fault should operate, leaving other healthy circuits unaffected. A circuit designed in this way would be considered to have effective discrimination or selectivity (IET Regulation 536). Effective discrimination or selectivity can be achieved by coordinated protection since the speed of operation of the protective device increases as the rating decreases. This can be seen in Fig. 3.30(b). A fault current of 200 A will cause a 15 A semi-enclosed fuse to operate in about 0.1 s, a 30 A semi-enclosed fuse in about 0.4 s and a 60 A semi-enclosed fuse in about 5.0 s. If a circuit is arranged as shown in Fig. 3.31 and a fault occurs on the appliance, effective discrimination or selectivity will be achieved because the 16 A fuse will operate more quickly than the other protective devices if they were, for example, all semi-enclosed type fuses with the characteristics shown in Fig. 3.30(b).

Security of supply, and therefore effective discrimination or selectivity, is an important consideration for an electrical designer and is also a requirement of IET Regulation 536.

Visit the companion website for more on this topic.

Earth fault loop impedance Z_S

In order that an overcurrent protective device can operate successfully it must meet the required disconnection times of IET Regulation 411.3.2.2, that is, final

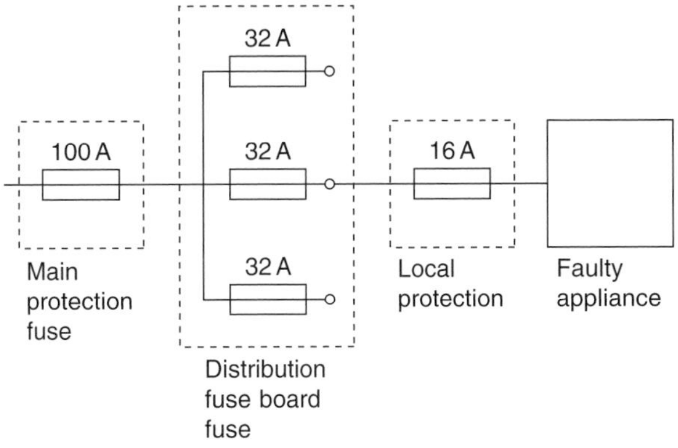

Figure 3.31 Effective discrimination or selectivity achieved by coordination of the protective devices.

circuits not exceeding 32 A shall have a disconnection time not exceeding 0.4 s. To achieve this, the earth fault loop impedance value measured in ohms must be less than those values given in Appendix B of the *On Site Guide* and Tables 41.2 and 41.3 of the IET Regulations. The value of the earth fault loop impedance may be verified by means of an earth fault loop impedance test as described in Chapter 4 of this book. The formula is:

$$Z_S = Z_E + (R_1 + R_2)\ (\Omega)$$

Here Z_E is the impedance of the supply side of the earth fault loop. The actual value will depend upon many factors: the type of supply, the ground conditions, the distance from the transformer, etc. The value can be obtained from the area electricity companies, but typical values are 0.35 Ω for TN-C-S (protective multiple earthing, PME) supplies and 0.8 Ω for TN-S (cable sheath earth) supplies. Also in the above formula, R_1 is the resistance of the line conductor and R_2 is the resistance of the earth conductor. The complete earth fault loop path is shown in Fig. 3.32.

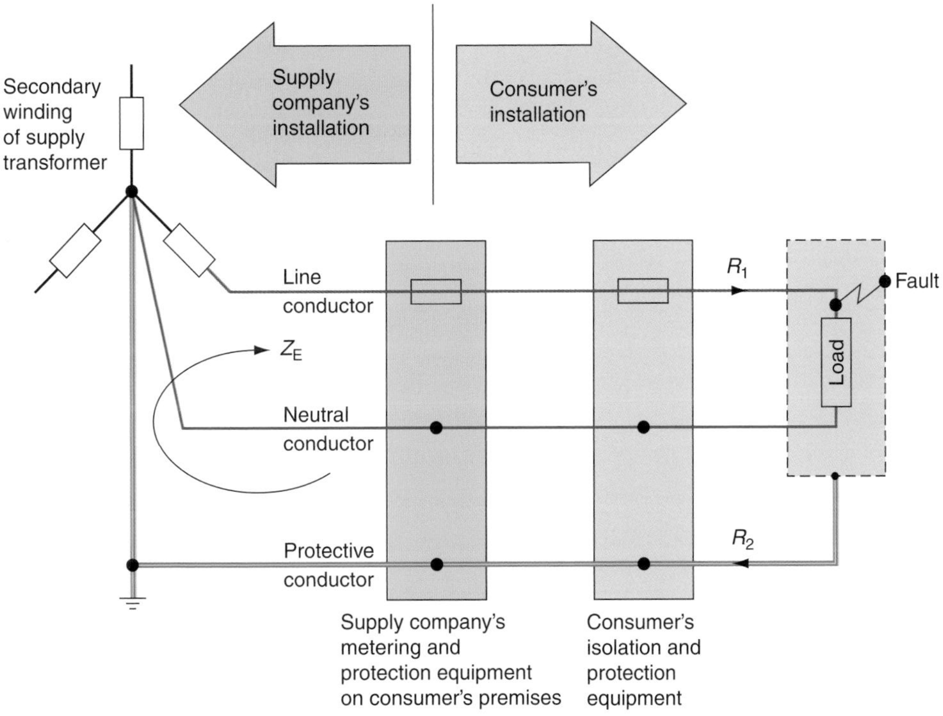

Figure 3.32 Earth fault loop path for a TN-S system.

Example

A 20 A radial socket outlet circuit is wired in 2.5 mm² PVC cable incorporating a 1.5 mm² CPC. The cable length is 30 m installed in an ambient temperature of 20°C and the consumer's protection is by 20 A MCB Type B to BS EN 60898. The earth fault loop impedance of the supply is 0.5 Ω. Calculate the total earth fault loop impedance Z_S, and establish that the value is less than the maximum value permissible for this type of circuit.
We have:

$$Z_S = Z_E + (R_1 + R_2)\ (\Omega)$$
$$Z_E = 0.5\,\Omega \text{ (value given in the question)}$$

From the value given in Table I1 of the *On Site Guide* and reproduced in Table 3.1 a 2.5 mm phase conductor with a 1.5 mm protective conductor has an $(R_1 + R_2)$ value of $19.51 \times 10^{-3}\,\Omega/\text{m}$.

$$\text{For 30m cable } (R_1 + R_2) = 19.51 \times 10^{-3}\,\Omega/\text{m} \times 30\,\text{m} = 0.585\,\Omega$$

However, under fault conditions, the temperature and therefore the cable resistance will increase. To take account of this, we must multiply the value of cable resistance by the factor given in Table I3 of the *On Site Guide*. In this case the factor is 1.20 and therefore the cable resistance under fault conditions will be:

$$0.585\,\Omega \times 1.20 = 0.702\,\Omega$$

The total earth fault loop impedance is therefore:

$$Z_S = 0.5\,\Omega + 0.702\,\Omega = 1.202\,\Omega$$

Table 3.1 Table I1 of the IET *On Site Guide*: value of resistance/metre for copper and aluminium conductors and of $R_1 + R_2$ per metre at 20°C in milliohms/metres

Cross-sectional area (mm²)		Resistance/metre or $(R_1 + R_2)$/metre (mΩ/m)	
Phase conductor	**Protective conductor**	**Copper**	**Aluminium**
1	–	18.10	
1	1	36.20	
1.5	–	12.10	
1.5	1	30.20	
1.5	1.5	24.20	
2.5	–	7.41	
2.5	1	25.51	
2.5	1.5	19.51	
2.5	2.5	14.82	
4	–	4.61	
4	1.5	16.71	

(continued)

Table 3.1 (continued)

Cross-sectional area (mm²)		**Resistance/metre or ($R_1 + R_2$)/metre (mΩ/m)**	
Phase conductor	**Protective conductor**	**Copper**	**Aluminium**
4	2.5	12.02	
4	4	9.22	
6	–	3.08	
6	2.5	10.49	
6	4	7.69	
6	6	6.16	
10	–	1.83	
10	4	6.44	
10	6	4.91	
10	10	3.66	
16	–	1.15	1.91
16	6	4.23	–
16	10	2.98	–
16	16	2.30	3.82
25	–	0.727	1.20
25	10	2.557	–
25	16	1.877	–
25	25	1.454	2.40
35	–	0.524	0.87
35	16	1.674	2.78
35	25	1.251	2.07
35	35	1.048	1.74
50	–	0.387	0.64
50	25	1.114	1.84
50	35	0.911	1.51
50	50	0.774	1.28

Reproduced from the IET *On Site Guide* by kind permission of the Institution of Engineering and Technology.

Values of $R_1 + R_2$ have been calculated for copper and aluminium conductors and are given in Table I1 of the *On Site Guide* as shown in Table 3.1 of this book.

The maximum permitted value given in Table 41.3 of the IET Regulations for a 20 A MCB protecting a socket outlet is 2.19 Ω, as shown in Table 3.2. The circuit earth fault loop impedance is less than this value and therefore the protective device will operate within the required disconnection time of 0.4 s.

Table 3.2 Maximum earth fault loop impedance Z_s MCB Type B maximum measured earth fault loop impedance (Ω) when overcurrent protective device is MCB Type B to BS EN 60898

	MCB rating (A)							
	6	10	16	20	25	32	40	50
For 0.4 s disconnection Z_S (Ω)	7.28	4.37	2.73	2.19	1.75	1.37	1.09	0.87

Protective conductor size

The CPC forms an integral part of the total earth fault loop impedance, so it is necessary to check that the cross-section of this conductor is adequate. If the cross-section of the CPC complies with Table 54.7 of the IET Regulations, there is no need to carry out further checks. Where line and protective conductors are made from the same material, Table 54.7 tells us that:

- for line conductors equal to or less than 16 mm^2, the protective conductor should equal the line conductor;
- for line conductors greater than 16 mm^2 but less than 35 mm^2, the protective conductor should have a cross-sectional area of 16 mm^2;
- for line conductors greater than 35 mm^2, the protective conductor should be half the size of the line conductor.

However, where the conductor cross-section does not comply with this table, then the formula given in IET Regulation 543.1.3 must be used:

$$S = \frac{\sqrt{I^2 t}}{k} \text{ (mm}^2\text{)}$$

where

S = cross-sectional area in mm^2

I = value of maximum fault current in amperes

t = operating time of the protective device

k = a factor for the particular protective conductor (see Tables 54.2 to 54.4 of the IET Regulations).

Example 1

A 230 V ring main circuit of socket outlets is wired in 2.5 mm single PVC copper cables in a plastic conduit with a separate 1.5 mm CPC. An earth fault loop impedance test identifies Z_s as 1.15 Ω. Verify that the 1.5 mm CPC meets the requirements of IET Regulation 543.1.3 when the protective device is a 30 A semi-enclosed fuse.

$$I = \text{Maximum fault current} = \frac{V}{Z_s} \text{(A)}$$

$$\therefore \frac{230}{1.15} = 200 \text{ A}$$

(continued)

Example 1 (continued)

t = maximum operating time of the protective device for a circuit not exceeding 32 A is 0.4 s from IET Regulation 411.3.2.2. From Fig. 4.14(b) you can see that the time taken to clear a fault of 200 A is about 0.4 s

k = 115 (from Table 54.3)

$$S = \frac{\sqrt{I^2 t}}{k} (\text{mm}^2)$$

$$S = \frac{\sqrt{(200\,\text{A})^2 \times 0.4\,\text{s}}}{115} = 1.10\,\text{mm}^2$$

A 1.5 mm^2 CPC is acceptable since this is the nearest standard-size conductor above the minimum cross-sectional area of 1.10 mm^2 found by calculation.

Example 2

A TN supply feeds a domestic immersion heater wired in 2.5 mm^2 PVC insulated copper cable and incorporates a 1.5 mm^2 CPC. The circuit is correctly protected with a 15 A semi-enclosed fuse to BS 3036. Establish by calculation that the CPC is of an adequate size to meet the requirements of IET Regulation 543.1.3. The characteristics of the protective device are given in IET Regulation Fig 3A2(a) of Appendix 3.

For final circuits less than 32 A the maximum operating time of the protective device is 0.4 s. From Fig 3A2(a) it can be seen that a current of about 90 A will trip the 15 A fuse in 0.4 s. The small insert table on the top right of Fig 3A2(a) of the IET Regulations gives the value of the prospective fault current required to operate the device within the various disconnection times given.

So, in this case the table states that 90 A will trip a 15 A semi-enclosed fuse in 0.4 s

$$\therefore I = 90\,\text{A}$$

$$t = 0.4\,\text{s}$$

$$k = 115 \text{ (from Table 54.3)}$$

$$S = \frac{\sqrt{I^2 t}}{k} (\text{mm}^2) \text{ (from IEE Regulation 543.1.3)}$$

$$S = \frac{\sqrt{(90\,\text{A})^2 \times 0.4\,\text{s}}}{115} = 0.49\,\text{mm}^2$$

The CPC of the cable is greater than 0.49 mm^2 and is therefore suitable. If the protective conductor is a separate conductor, that is, it does not form part of a cable as in this example and is not enclosed in a wiring system as in Example 1, the cross-section of the protective conductor must be not less than 2.5 mm^2 where mechanical protection is provided or 4.0 mm^2 where mechanical protection is ***not*** provided in order to comply with IET Regulation 544.2.3.

Visit the companion website for more on this topic.

Additional protection: RCDs

When it is required to provide the very best protection from electric shock and fire risk, earth fault protection devices are incorporated into the installation. The object of the regulations concerning these devices (411.3.2 to 411.3.3) is to remove an earth fault current very quickly, less than 0.4 s for all final circuits not exceeding 32 A, and limit the voltage which might appear on any exposed metal parts under fault conditions to not more than 50 V. They will continue to provide adequate protection throughout the life of the installation even if the earthing conditions deteriorate. This is in direct contrast to the protection provided by overcurrent devices, which require a low-resistance earth loop impedance path.

The regulations recognize RCDs as 'additional protection' in the event of failure of the provision for basic protection, fault protection or carelessness by the users of the installation (IET Regulation 415.1.1).

The basic circuit for a single-phase RCD is shown in Fig. 3.33. The load current is fed through two equal and opposing coils wound on to a common transformer core. The phase and neutral currents in a healthy circuit produce equal and opposing fluxes in the transformer core, which induces no voltage in the tripping coil. However, if more current flows in the line conductor than in the neutral conductor as a result of a fault between live and earth, an out-of-balance flux will result in an e.m.f. being induced in the trip coil which will open the double-pole switch and isolate the load. Modern RCDs have tripping sensitivities between 10 and 30 mA, and therefore a faulty circuit can be isolated before the lower lethal limit to human beings (about 50 mA) is reached.

Consumer units are now supplied which incorporate one or more RCDs, as shown in the *On Site Guide* Figs 3.6.3 (iv) to 3.6.3 (v).

Definition

RCBOs give the combined protection of an MCB and an RCD in one device.

RCBOs (residual current circuit breakers with overload protection) combine RCD protection and MCB protection into one unit.

In a split-board consumer unit, about half of the total number of final circuits are protected by the RCD. A fault on any one final circuit will trip out all of the RCD protected circuits, which may cause inconvenience.

The RCBO gives the combined protection of an MCB plus RCD for each final circuit so protected and, in the event of a fault occurring, only the faulty circuit is interrupted. This arrangement is shown in the OSG at Figs 2.1(i) to 2.1(iii) and in this book at Figs 3.34 to 3.36.

Finally, it should perhaps be said that a foolproof method of giving protection to people or animals who simultaneously touch both live and neutral has yet to be devised. The ultimate safety of an installation depends upon the skill and experience of the electrical contractor and the good sense of the user.

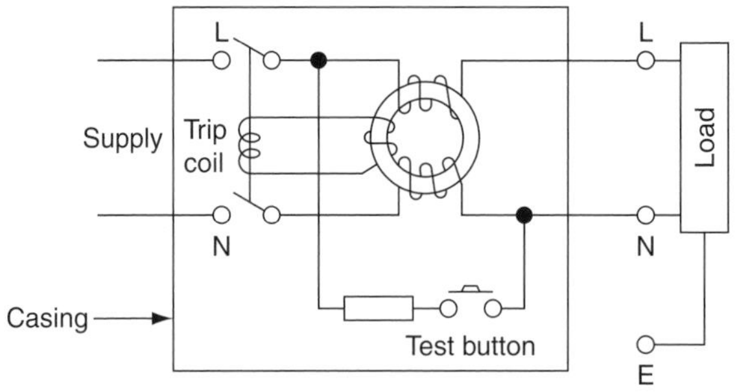

Figure 3.33 Construction of an RCD.

Electricity supply systems

The British government agreed on 1 January 1995 that the electricity supplies in the United Kingdom would be harmonized with those of the rest of Europe. Thus the voltages used previously in low-voltage supply systems of 415 V and 240 V have become 400 V for three-phase supplies and 230 V for single-phase supplies. The Electricity Supply Regulations 1988 have also been amended to permit a range of variations from the new declared nominal voltage. From January 1995 the permitted tolerance is the nominal voltage +10% or −6%. Previously it was ±6%. This gives a voltage range of 216 to 253 V for a nominal voltage of 230 V and 376 to 440 V for a nominal voltage of 400 V (IET Regulation Appendix 2 par. 14).

It is further proposed that the tolerance levels will be adjusted to ±10% of the declared nominal voltage. All European Union countries will adjust their voltages to comply with a nominal voltage of 230 V single-phase and 400 V three-phase.

The supply to a domestic, commercial or small industrial consumer's installation is usually protected at the incoming service cable position with a 100 A high breaking capacity (HBC) fuse. The maximum, that is, worst case value of external earth fault loop impedance outside of the consumer's domestic installation is:

0.8 Ω for cable sheath earth supplies (TN-S system)
0.35 Ω for protective multiple earthing (PME) supplies (TN-C-S system)
21.0 Ω excluding the consumer's earth electrode for no earth supplies (TT system).

The maximum, that is, worst case value of prospective short-circuit current is 16 kA at the supply terminals (see *Electricians Guide to the Building Regulations* Part P, Chapter 3).

Other items of equipment at this position are the energy meter and the consumer's distribution unit, providing the protection for the final circuits and the earthing arrangements for the installation.

An efficient and effective earthing system is essential to allow protective devices to operate. The limiting values of earth fault loop impedance are given in Tables 41.2 to 41.4 and Chapter 54 of the IET Regulations. Wiring Systems of Part 3 gives details of the earthing arrangements to be incorporated into the supply system to meet the requirements of the regulations. Five systems are described in the definitions but only the TN-S, TN-C-S and TT systems are suitable for public supplies.

A system consists of an electrical installation connected to a supply. Systems are classified by a capital letter designation.

The supply earthing

The supply earthing arrangements are indicated by the first letter, where T means one or more points of the supply are directly connected to earth and I means the supply is not earthed or one point is earthed through a fault-limiting impedance.

The installation earthing

The installation earthing arrangements are indicated by the second letter, where T means the exposed conductive parts are connected directly to earth and N

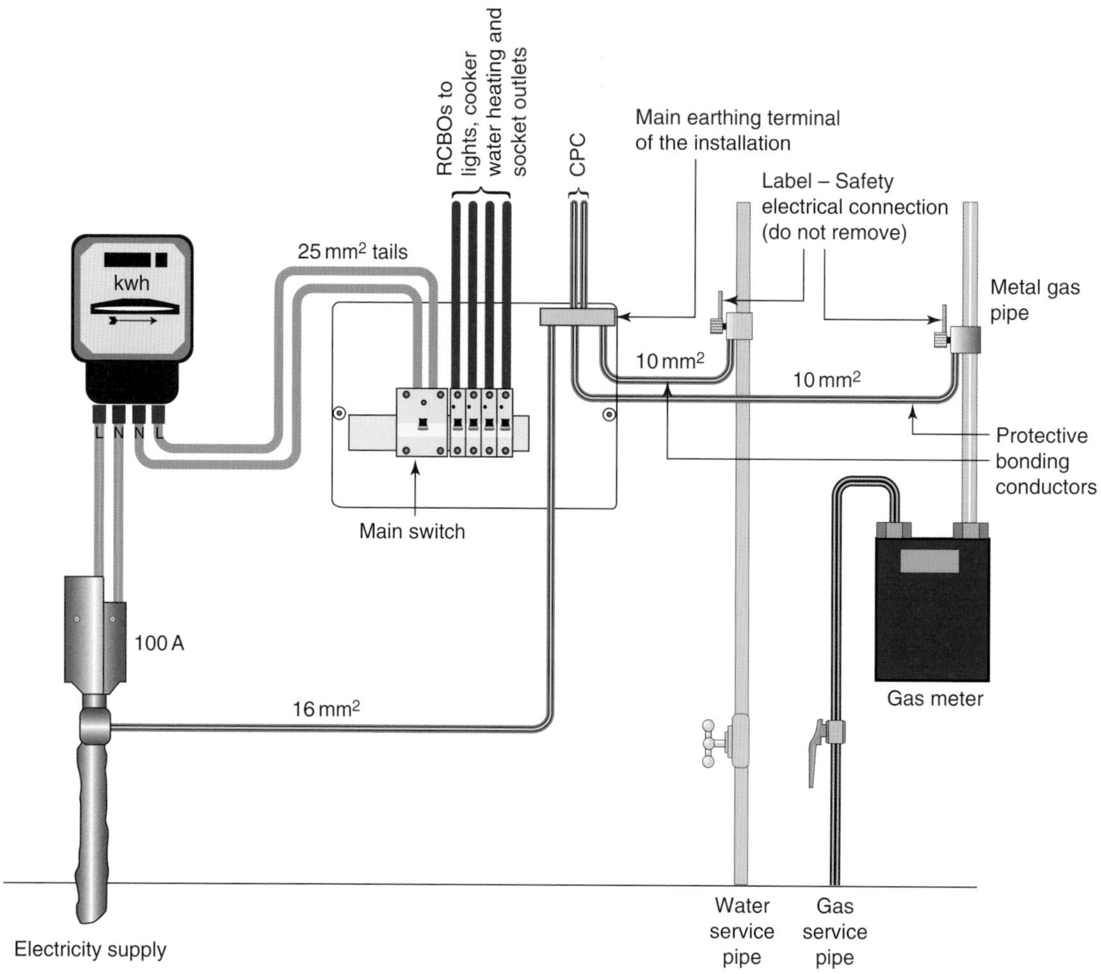

Figure 3.34 Cable sheath earth supplies (TN-S system) showing earthing and bonding arrangements.

means the exposed conductive parts are connected directly to the earthed point of the source of the electrical supply.

The earthed supply conductor

The earthed supply conductor arrangements are indicated by the third letter, where S means a separate neutral and protective conductor and C means that the neutral and protective conductors are combined in a single conductor.

Cable sheath earth supply (TN-S system)

This is one of the most common types of supply system to be found in the United Kingdom where the electricity companies' supply is provided by underground cables. The neutral and protective conductors are separate throughout the system. The protective earth conductor (PE) is the metal sheath and armour of the underground cable, and this is connected to the consumer's main earthing terminal. All extraneous conductive parts of the installation, gas pipes, water pipes and any lightning protective system are connected to the protective conductor via the main earthing terminal of the installation. The arrangement is shown in Fig. 3.34, and in Fig. 3.8 of the IET Regulations.

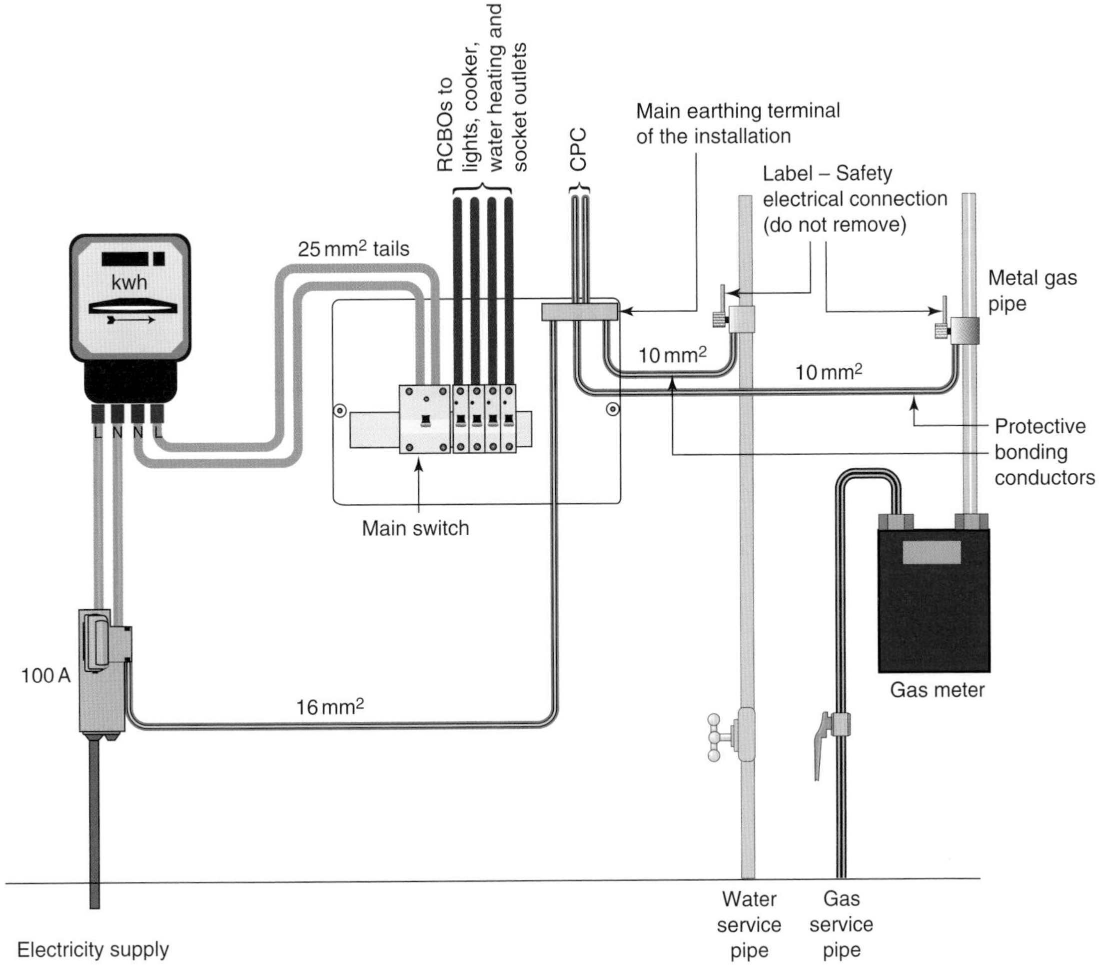

Figure 3.35 Protective multiple earthing supply (TN-C-S system) showing earthing and bonding arrangements.

Protective multiple earthing supplies (TN-C-S system)

This type of underground supply is becoming increasingly popular to supply new installations in the United Kingdom. It is more commonly referred to as protective multiple earthing (PME). The supply cable uses a combined protective earth and neutral conductor (PEN conductor). At the supply intake point a consumer's main earthing terminal is formed by connecting the earthing terminal to the neutral conductor. All extraneous conductive parts of the installation, gas pipes, water pipes and any lightning protective system are then connected to the main earthing terminals. Thus phase to earth faults are effectively converted into phase to neutral faults. The arrangement is shown in Fig. 3.35, and in Fig. 3.9 of the IET Regulations.

No earth provided supplies (TT system)

This is the type of supply more often found when the installation is fed from overhead cables. The supply authorities do not provide an earth terminal and the installation's circuit protective conductors (CPCs) must be connected to earth via an earth electrode provided by the consumer. IET Regulation 542.2.3 lists the type of earth rod, earth plate or earth tapes recognized by BS 7671. An effective earth connection is sometimes difficult to obtain and in most cases a residual current device (RCD) is provided when this type of supply is used. The arrangement is shown in Fig. 3.36.

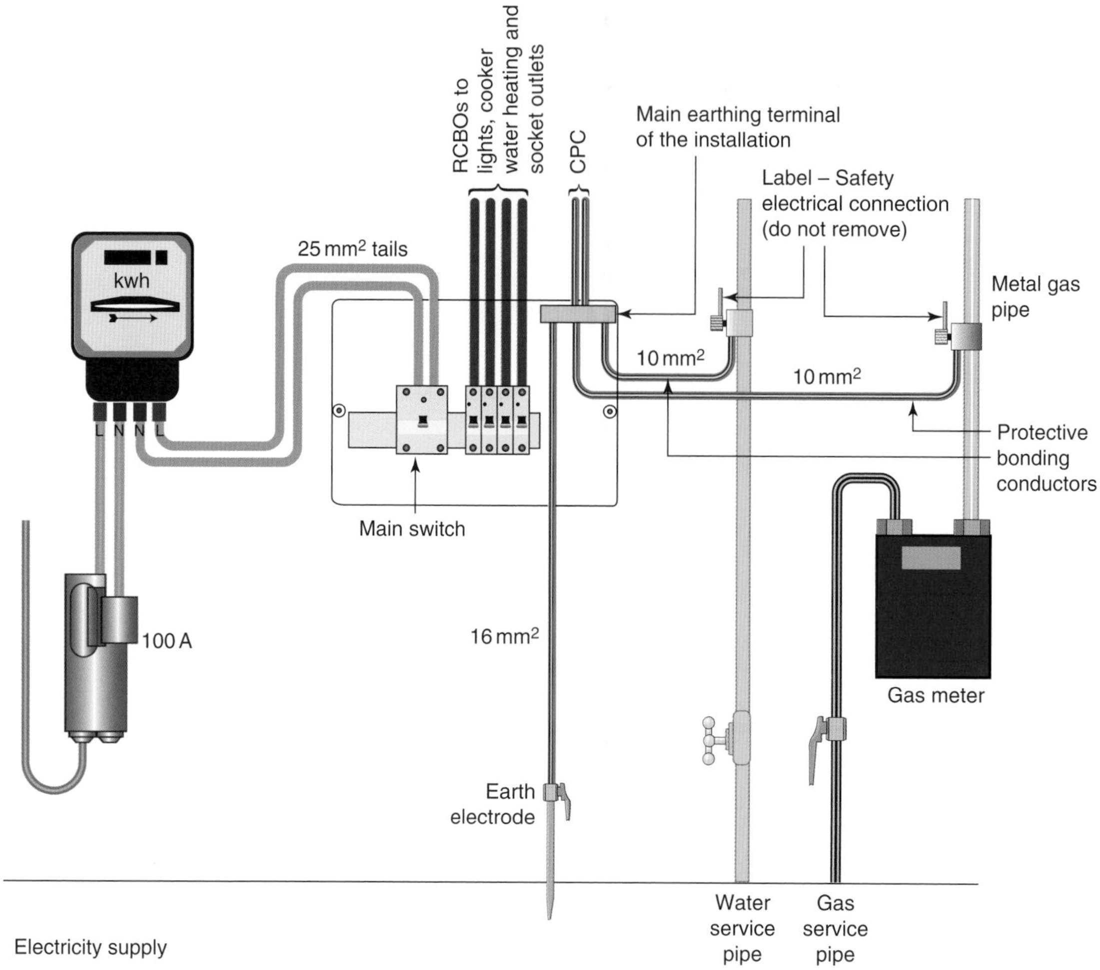

Figure 3.36 No supply earth provided (TT systems) showing earthing and bonding arrangements.

Figures 3.34, 3.35 and 3.36 show the layout of a typical domestic service position for these three supply systems. There are two other systems of supply, the TN-C and IT systems, but they do not comply with the supply regulations and therefore cannot be used for public supplies. Their use is restricted to private generating plants. For this reason I shall not include them here but they can be seen in Part 3 of the IET Regulations.

Figures 3.34 to 3.36 show circuits protected by RCBOs. The use of RCBOs will minimize inconvenience and disruption in the event of a fault occurring, because only the faulty circuit will be disconnected by the RCBO. However, alternative consumer unit arrangements using MCBs and RCDs are permissible and are shown in the *On Site Guide* at Section 3.6.3.

Conductor size calculations

The size of a cable to be used for an installation depends upon:

- the current rating of the cable under defined installation conditions; and
- the maximum permitted drop in voltage as defined by IET Regulation 525.

The factors which influence the current rating are:

1 *Design current*: cable must carry the full load current.
2 *Type of cable*: PVC, MICC, copper conductors or aluminium conductors.

3 *Installed conditions*: clipped to a surface or installed with other cables in a trunking.
4 *Surrounding temperature*: cable resistance increases as temperature increases and insulation may melt if the temperature is too high.
5 *Type of protection*: for how long will the cable have to carry a fault current?

IET Regulation 525 states that the drop in voltage from the supply terminals to the fixed current-using equipment must not exceed 3% for lighting circuits and 5% for other uses of the mains voltage. That is a maximum of 6.9 V for lighting and 11.5 V for other uses on a 230 V installation. The volt drop for a particular cable may be found from:

$$VD = \text{Factor} \times \text{Design current} \times \text{Length of run}$$

The factor is given in the tables of Appendix 4 of the IET Regulations and Appendix 6 of the *On Site Guide.* They are also given in Table 3.3 in this book.

The cable rating, denoted I_t, may be determined as follows:

$$I_t = \frac{\text{Current rating of protective device}}{\text{Any applicable correction factors}}$$

The cable rating must be chosen to comply with IET Regulation 433.1. The correction factors which may need applying are given below as:

Ca the ambient or surrounding temperature correction factor, which is given in Tables 4B1 and 4B2 of Appendix 4 of the IET Regulations. They are also shown in Table 3.3 of this book.

Cg the grouping correction factor given in Tables 4C1 to 4C6 of the IET Regulations and Table 6C of the *On Site Guide*.

Cf the 0.725 correction factor to be applied when semi-enclosed fuses protect the circuit as described in item 5.1.1 of the preface to Appendix 4 of the IET Regulations.

Ci the correction factor to be used when cables are enclosed in thermal insulation. IET Regulation 523.6.6 gives us three possible correction values:

1 Where one side of the cable is in contact with thermal insulation we must read the current rating from the column in the table which relates to reference method A (see Table 3.4).
2 Where the cable is *totally* surrounded over a length greater than 0.5 m we must apply a factor of 0.5.

COMPANION WEBSITE Visit the companion website for more on this topic.

Table 3.3 Ambient temperature correction factors

Table 4B1 of IET Regulations and F1 of the *On Site Guide* Correction factors for ambient temperature where protection is against short-circuit and overload										
Type of insulation	**Operating temperature**	**Ambient temperature (°C)**								
		25	30	35	40	45	50	55	60	65
Thermoplastic (general purpose PVC)	70°C	1.03	1.0	0.94	0.87	0.79	0.71	0.61	0.50	N/A

Adapted from the IET *On Site Guide* by kind permission of the Institution of Engineering and Technology.

Table 3.4 Current-carrying capacity of cables

Multicore cables having thermoplastic (PVC) or thermosetting insulation, non-armoured COPPER CONDUCTORS Table 4D2A of IET Regulations and Table F5(i) of the *On Site Guide* Ambient temperature: 30°C. Conductor operating temperature: 70°C Current-carrying capacity (amperes): BS 6004, BS 7629								
Conductor cross-sectional area	**Reference Method A (enclosed in conduit in an insulated wall, etc.)**		**Reference Method B (enclosed in conduit on a wall or ceiling, or in trunking)**		**Reference Method C (clipped direct)**		**Reference Method E (on a perforated cable tray) or in free air**	
	Two cables single-phase a.c. or d.c.	**Three or four cables three-phase a.c.**	**One two-core cable, single-phase a.c. or d.c.**	**One three-core cable or one four-core cable, three-phase a.c.**	**One two-core cable, single-phase a.c. or d.c.**	**One three-core cable or one four-core cable, three-phase a.c.**	**One two-core cable, single-phase a.c. or d.c.**	**One three-core cable or one four-core cable, three-phase a.c.**
1	**2**	**3**	**4**	**5**	**6**	**7**	**8**	**9**
mm²	**A**	**A**	**A**	**A**	**A**	**A**	**A**	**A**
1	11	10	13	11.5	15	13.5	17	14.5
1.5	14	13	16.5	15	19.5	17.5	22	18.5
2.5	18.5	17.5	23	20	27	24	30	25
4	25	23	30	27	37	32	40	34
6	32	29	38	34	46	41	51	43
10	43	39	52	46	63	57	70	60
16	57	52	69	62	85	76	94	80
25	75	68	90	80	112	96	119	101
35	92	83	111	99	138	119	148	126
50	110	99	133	118	168	144	180	153
70	139	125	168	149	213	184	232	196
95	167	150	201	179	258	223	282	238

Adapted from the IET *On Site Guide* by kind permission of the Institution of Engineering and Technology.

3 Where the cable is *totally* surrounded over a short length, the appropriate factor given in Table 52.2 of the IET Regulations or Table F2 of the *On Site Guide* should be applied.

Note: A cable should preferably *not* be installed in thermal insulations.

Having calculated the cable rating, I_t the smallest cable should be chosen from the appropriate table which will carry that current. This cable must also meet the voltage drop (IET Regulation 525) and this should be calculated as described earlier. When the calculated value is less than 3% for lighting and 5% for other uses of the mains voltage the cable may be considered suitable. If the calculated value is greater than this value, the next larger cable size must be tested until **a cable is found which meets both the current rating and voltage drop criteria**.

Example

A house extension has a total load of 6 kW installed some 18 m away from the mains consumer unit for lighting. A PVC insulated and sheathed twin and earth cable will provide a sub-main to this load and be clipped to the side of the ceiling joists over much of its length in a roof space which is anticipated to reach 35°C in the summer and where insulation is installed up to the top of the joists. Calculate the minimum cable size if the circuit is to be protected by a type B MCB to BS EN 60898. Assume a TN-S supply, that is, a supply having a separate neutral and protective conductor throughout.

Let us solve this question using only the tables given in the *On Site Guide*. The tables in the regulations will give the same values, but this will simplify the problem because we can refer to Tables 3.3, 3.4 and 3.5 in this book which give the relevant *On Site Guide* tables.

$$\text{Design current, } I_b = \frac{\text{Power}}{\text{Volts}} = \frac{6000\,\text{W}}{230\,\text{V}} = 26.09\,\text{A}$$

Nominal current setting of the protection for this load $I_n = 32$ A.

The cable rating I_t is given by:

$$I_t = \frac{\text{Current rating of protective device } (I_n)}{\text{The product of the correction factors}}$$

The correction factors to be included in this calculation are:

Ca ambient temperature; as shown in Table 3.3 the correction factor for 35°C is 0.94.

Cg grouping factors need not be applied.

Cf, since protection is by MCB no factor need be applied.

Ci thermal insulation demands that we assume installed Method A (see Table 3.4).

The design current is 26.09 A and we will therefore choose a 32 A MCB for the nominal current setting of the protective device, I_n.

$$\text{Cable rating, } I_t = \frac{32}{0.94} = 34.04\,\text{A}$$

(continued)

Key fact

Volt drop

Maximum permissible volt drop on 230 V supplies:

- 3% for lighting = 6.9 V
- 5% for other uses = 11.5 V

IET Regulation 525 and Table 4A6 of Appendix 4

Example (continued)

From column 2 in Table 3.4, a 10 mm cable, having a rating of 43 A, is required to carry this current.

Now test for volt drop: from Table 3.5 the volt drop per ampere per metre for a 10 mm cable is 4.4 mV. So the volt drop for this cable length and load is equal to:

$$4.4 \times 10^{-3}\ \text{V/Am} \times 26.09\,\text{A} \times 18\,\text{m} = 2.06\,\text{V}$$

Since this is less than the maximum permissible value for a lighting circuit of 6.9 V, a 10 mm cable satisfies the current and drop in voltage requirements when the circuit is protected by an MCB. This cable is run in a loft that gets hot in summer and has thermal insulation touching one side of the cable. We must, therefore, use installed reference Method A of Table 3.4. If we were able to route the cable under the floor, clipped direct or in conduit or trunking on a wall, we may be able to use a 6 mm cable for this load. You can see how the current-carrying capacity of a cable varies with the installed method by looking at Table 3.4. Compare the values in column 2 with those in column 6. When the cable is clipped direct onto a wall or surface the current rating is higher because the cable is cooler. If the alternative route was longer, you would need to test for volt drop before choosing the cable. These are some of the decisions which the electrical contractor must make when designing an installation which meets the requirements of the customer and the IET Regulations.

If you are unsure of the standard fuse and MCB rating of protective devices, you can refer to Fig. 3A4 of Appendix 3 of the IET Regulations.

COMPANION WEBSITE @ Visit the companion website for more on this topic.

Cable size for standard domestic circuits

Appendix 4 of the IET Regulations (BS 7671) and Appendix F of the IET *On Site Guide* contain tables for determining the current carrying capacities of conductors which we looked at in the last section. However, for standard domestic circuits, Table 3.6 gives a guide to cable size.

In this table, I am assuming a standard 230 V domestic installation, having a sheathed earth or PME supply terminated in a 100 A HBC fuse at the mains position. Final circuits are fed from a consumer unit, having Type B MCB protection and wired in PVC insulated and sheathed cables with copper conductors having a grey thermoplastic PVC outer sheath or a white thermosetting cable with LSF (low smoke and fume properties). I am also assuming that the surrounding temperature throughout the length of the circuit does not exceed 30°C and the cables are run singly and clipped to a surface.

Wiring systems and enclosures

The final choice of a wiring system must rest with those designing the installation and those ordering the work, but whatever system is employed, good workmanship by competent persons and the use of proper materials is essential for compliance with the regulations (IET Regulation 134.1.1). The necessary

Table 3.5 Voltage drop in cables factor

Table 4D2B of IET Regulations and F5(ii) of the On Site Guide			
Voltage drop (per ampere per metre)			**Conductor operating temperature: 70°C**
Conductor cross-sectional area (mm^2) 1	**Two-core cable, d.c. (mV/A/m) 2**	**Two-core cable, single-phase a.c. (mV/A/m) 3**	**Three- or four-core cable, three-phase (mV/A/m) 4**
1	44	44	38
1.5	29	29	25
2.5	18	18	15
4	11	11	9.5
6	7.3	7.3	6.4
10	4.4	4.4	3.8
16	2.8	2.8	2.4
		z	z
25	1.75	1.75	1.50
35	1.25	1.25	1.10
50	0.93	0.94	0.81
70	0.63	0.65	0.57
95	0.46	0.5	0.43

Adapted from the IET *On Site Guide* by kind permission of the Institution of Engineering and Technology.

Table 3.6 Cable size for standard domestic circuits

Type of final circuit	Cable size (twin and earth)	MCB rating, Type B (A)	Maximum floor area covered by circuit (m^2)	Maximum length of cable run (m)
Fixed lighting	1.0	6	–	40
Fixed lighting	1.5	6	–	60
Immersion heater	2.5	16	–	30
Storage radiator	2.5	16	–	30
Cooker (oven only)	2.5	16	–	30
13 A socket outlets (radial circuit)	2.5	20	50	30
13 A socket outlets (ring circuit)	2.5	32	100	90
13 A socket outlets (radial circuit)	4.0	32	75	35
Cooker (oven and hob)	6.0	32	–	40
Shower (up to 7.5kw)	6.0	32	–	40
Shower (up to 9.6kw)	10	40	–	40

skills can be acquired by an electrical trainee who has the correct attitude and dedication to his or her craft.

Definition

Cables can be considered to be constructed in three parts: the *conductor*, which must be of a suitable cross-section to carry the load current; the *insulation*, which has a colour or number code for identification; and the *outer sheath*, which may contain some means of providing protection from mechanical damage.

Most cables can be considered to be constructed in three parts: the **conductor**, which must be of a suitable cross-section to carry the load current; the **insulation**, which has a colour or number code for identification; and the **outer sheath**, which may contain some means of providing protection from mechanical damage.

The conductors of a cable are made of either copper or aluminium and may be stranded or solid. Solid conductors are only used in fixed wiring installations and may be shaped in larger cables. Stranded conductors are more flexible and conductor sizes from 4.0 to 25 mm^2 contain seven strands. A 10 mm^2 conductor, for example, has seven 1.35 mm diameter strands which collectively make up the 10 mm^2 cross-sectional area of the cable. Conductors above 25 mm^2 have more than seven strands, depending upon the size of the cable. Flexible cords have multiple strands of very fine wire conductors, as fine as one strand of human hair. This gives the cable its very flexible quality.

New wiring colours

Twenty-eight years ago the United Kingdom agreed to adopt the European colour code for flexible cords, that is, brown for live or phase conductor, blue for the neutral conductor and green combined with yellow for earth conductors. However, no similar harmonization was proposed for non-flexible cables used for fixed wiring. These were to remain as red for live or phase conductor, black for the neutral conductor and green combined with yellow for earth conductors.

On 31 March 2004, the IET published Amendment No. 2 to BS 7671: 2001 which specified new cable core colours for all fixed wiring in UK electrical installations. These new core colours 'harmonize' the United Kingdom with the practice in mainland Europe.

Fixed cable core colours up to 2006

- *Single-phase* supplies red line conductors, black neutral conductors, and green combined with yellow for earth conductors.

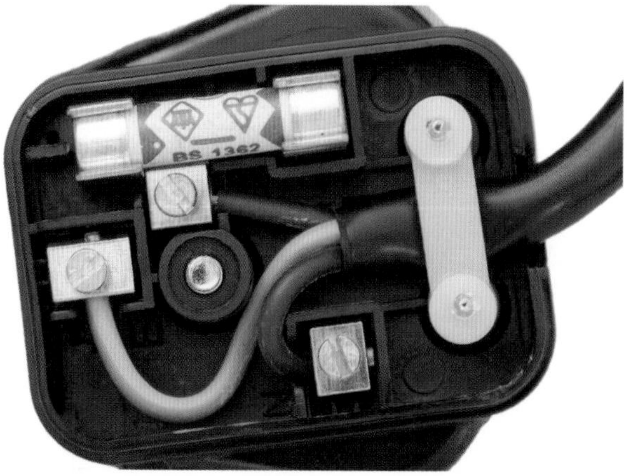

Figure 3.37 Wiring colours are standard in the UK.

- *Three-phase* supplies red, yellow and blue line conductors, black neutral conductors and green combined with yellow for earth conductors.

These core colours could not be used after 31 March 2006.

New (harmonized) fixed cable core colours

- *Single-phase* supplies brown line conductors, blue neutral conductors and green combined with yellow for earth conductors (just like flexible cords).
- *Three-phase* supplies brown, black and grey line conductors, blue neutral conductors and green combined with yellow for earth conductors.

Cable core colours used from 31 March 2004 onwards.

Extensions or alterations to existing *single-phase* installations do not require marking at the interface between the old and new fixed wiring colours. However, a warning notice must be fixed at the consumer unit or distribution fuse board which states:

> *Caution – this installation has wiring colours to two versions of BS 7671. Great care should be taken before undertaking extensions, alterations or repair that all conductors are correctly identified.*

Alterations to *three-phase* installations must be marked at the interface L1, L2, L3 for the lines and N for the neutral. Both new and old cables must be marked. These markings are preferred to coloured tape and a caution notice is again required at the distribution board. Appendix 7 of BS 7671: 2008 deals with harmonized cable core colours.

PVC insulated and sheathed cables

Domestic and commercial installations use this cable, which may be clipped direct to a surface, sunk in plaster, or installed in conduit or trunking. It is the simplest and least expensive cable. Figure 3.38 shows a sketch of a twin and earth cable.

The conductors are covered with a colour-coded PVC insulation and then contained singly or with others in a PVC outer sheath.

PVC/SWA cable

PVC insulated steel wire armour cables are used for wiring underground between buildings, for main supplies to dwellings, rising sub-mains and industrial installations. They are used where some mechanical protection of the cable conductors is required.

Safety first

PVC cables

- PVC cables should not be installed when the surrounding temperature is below 0°C.
- The PVC insulation becomes brittle at low temperatures and may be damaged during installation.

IET Regulation 522.1.2

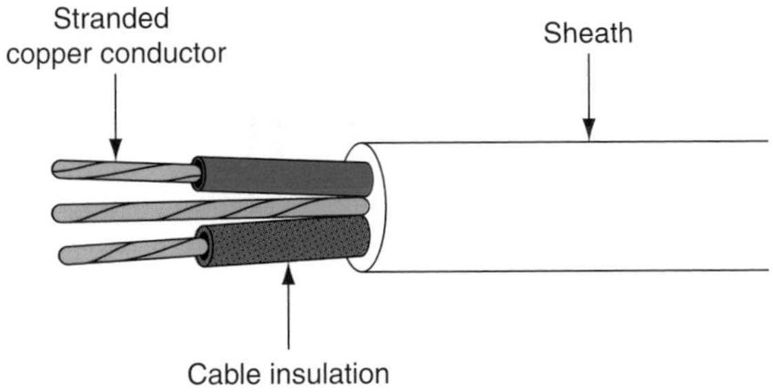

Figure 3.38 A twin and earth PVC insulated and sheathed cable.

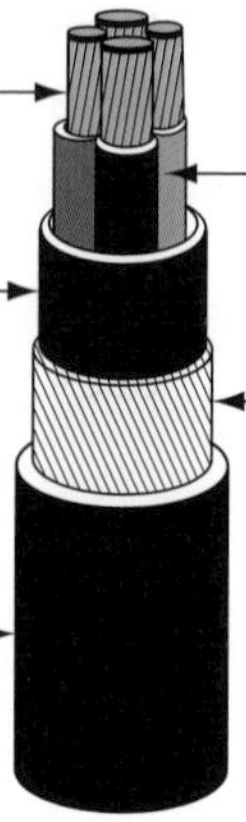

Figure 3.39 A four-core PVC/SWA cable.

Visit the companion website for more on this topic.

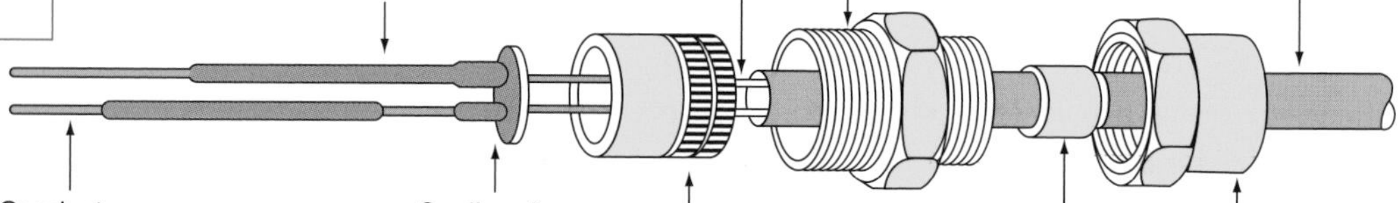

Figure 3.40 MI cable with terminating seal and gland.

The conductors are covered with colour-coded PVC insulation and then contained either singly or with others in a PVC sheath (see Fig. 3.39). Around this sheath is placed an armour protection of steel wires twisted along the length of the cable, and a final PVC sheath covering the steel wires protects them from corrosion. The armour sheath also provides the circuit protective conductor (CPC) and the cable is simply terminated using a compression gland.

MI cable

A mineral insulated (MI) cable has a seamless copper sheath which makes it waterproof and fire- and corrosion-resistant. These characteristics often make it the only cable choice for hazardous or high-temperature installations such as oil refineries and chemical works, boiler houses and furnaces, petrol pump and fire alarm installations.

The cable has a small overall diameter when compared to alternative cables and may be supplied as bare copper or with a PVC oversheath. It is colour-coded orange for general electrical wiring, white for emergency lighting or red for fire alarm wiring. The copper outer sheath provides the CPC, and the cable is terminated with a pot and sealed with compound and a compression gland (see Fig. 3.40).

The copper conductors are embedded in a white powder, magnesium oxide, which is non-ageing and non-combustible, but which is hygroscopic, which means that it readily absorbs moisture from the surrounding air, unless adequately terminated. The termination of an MI cable is a complicated process requiring the electrician to demonstrate a high level of practical skill and expertise for the termination to be successful.

FP 200 cable

FP 200 cable is similar in appearance to an MI cable in that it is a circular tube, or the shape of a pencil, and is available with a red or white sheath. However, it is much simpler to use and terminate than an MI cable.

The cable is available with either solid or stranded conductors that are insulated with 'insudite', a fire-resistant insulation material. The conductors are then screened by wrapping an aluminium tape around the insulated conductors, that is, between the insulated conductors and the outer sheath. This aluminium tape screen is applied metal side down and in contact with the bare CPC.

The sheath is circular and made of a robust thermoplastic low-smoke, zero-halogen material.

FP 200 is available in 2, 3, 4, 7, 12 and 19 cores with a conductor size range from 1.0 to 4.0 mm.

The cable is as easy to use as a PVC insulated and sheathed cable. No special terminations are required; the cable may be terminated through a grommet into a knock-out box or terminated through a simple compression gland.

The cable is a fire-resistant cable, primarily intended for use in fire alarms and emergency lighting installations, or it may be embedded in plaster.

High-voltage overhead cables

Suspended from cable towers or pylons, overhead cables must be light, flexible and strong.

The cable is constructed of stranded aluminium conductors formed around a core of steel-stranded conductors (see Fig. 3.41). The aluminium conductors carry the current and the steel core provides the tensile strength required to suspend the cable between pylons. The cable is not insulated since it is placed out of reach and insulation would only add to the weight of the cable.

Optical fibre cables

The introduction of fibre-optic cable systems and digital transmissions will undoubtedly affect future cabling arrangements and the work of the electrician.

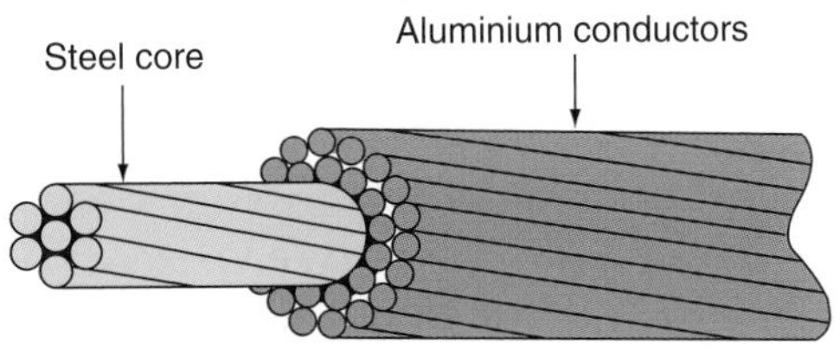

Figure 3.41 132 kV overhead cable construction.

Visit the companion website for more on this topic.

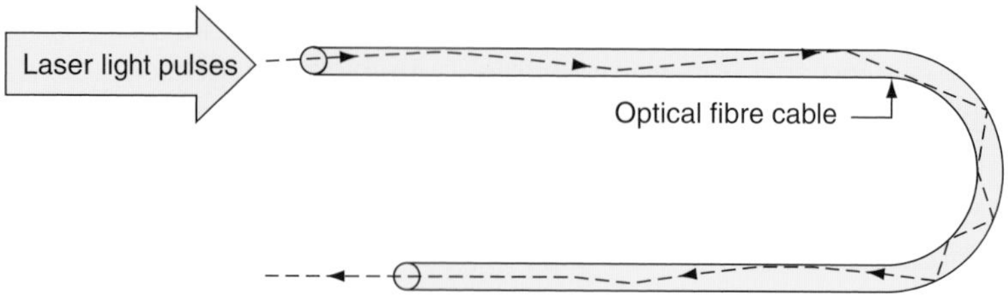

Figure 3.42 Digital pulses of laser light down an optical fibre cable.

Networks based on the digital technology currently being used so successfully by the telecommunications industry are very likely to become the long-term standard for computer systems. Fibre-optic systems dramatically reduce the number of cables required for control and communications systems, and this will in turn reduce the physical room required for these systems. Fibre-optic cables are also immune to electrical noise when run parallel to mains cables and, therefore, the present rules of segregation and screening may change in the future. There is no spark risk if the cable is accidentally cut and, therefore, such circuits are intrinsically safe. Intrinsic safety is described later in this chapter under the heading 'Hazardous area installations'.

Definition

Optical fibre cables are communication cables made from optical-quality plastic, the same material from which spectacle lenses are manufactured. The energy is transferred down the cable as digital pulses of laser light, as against current flowing down a copper conductor in electrical installation terms.

Optical fibre cables are communication cables made from optical-quality plastic, the same material from which spectacle lenses are manufactured. The energy is transferred down the cable as digital pulses of laser light, as against current flowing down a copper conductor in electrical installation terms. The light pulses stay within the fibre-optic cable because of a scientific principle known as 'total internal refraction' which means that the laser light bounces down the cable and when it strikes the outer wall it is always deflected inwards and, therefore, does not escape out of the cable, as shown in Fig. 3.42.

The cables are very small because the optical quality of the conductor is very high and signals can be transmitted over great distances. They are cheap to produce and lightweight because these new cables are made from high-quality plastic and not high-quality copper. Single-sheathed cables are often called 'simplex' cables and twin-sheathed cables 'duplex', that is, two simplex cables together in one sheath. Multi-core cables are available containing up to 24 single fibres.

Fibre-optic cables look like steel wire armour cables (but of course they are lighter) and should be installed in the same way, and given the same level of protection, as SWA cables. Avoid tight-radius bends if possible and kinks at all costs. Cables are terminated in special joint boxes which ensure cable ends are cleanly cut and butted together to ensure the continuity of the light pulses. Fibre-optic cables are Band I circuits when used for data transmission and must therefore be segregated from other mains cables to satisfy the IET Regulations.

The testing of fibre-optic cables requires special instruments to measure the light attenuation (i.e. light loss) down the cable. Finally, when working with fibre-optic cables, electricians should avoid direct eye contact with the low-energy laser light transmitted down the conductors.

Component parts of an electrical circuit

For a piece of electrical equipment to work efficiently and effectively it must be correctly connected to an electrical circuit. So what is an electrical circuit?

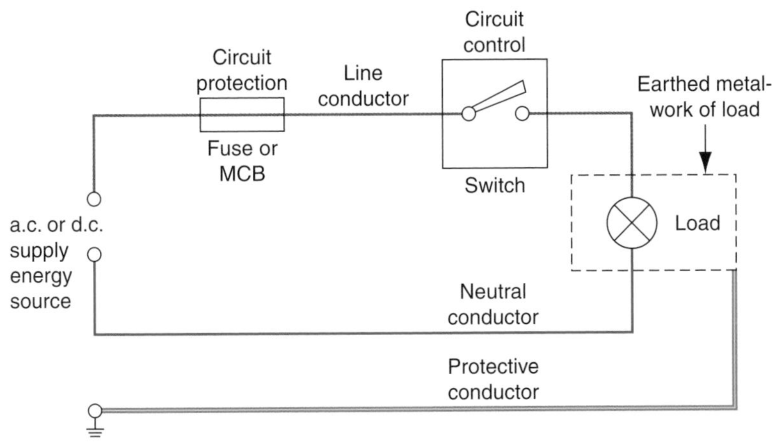

Figure 3.43 Component parts of an electrical circuit.

An electrical circuit has the following five components, as shown in Fig. 3.43:

1 A source of electrical energy. This might be a battery giving a d.c. (direct current) supply or the mains supply which is a.c. (alternating current).
2 A source of circuit protection. This might be a fuse or circuit-breaker which will protect the circuit from 'overcurrent'.
3 The circuit conductors or cables. These carry voltage and current to power the load.
4 A means to control the circuit. This might be a simple on/off switch but it might also be a dimmer or a thermostat.
5 A load. This is something which needs electricity to make it work. It might be an electric lamp, an electrical appliance, an electric motor or an iPod.

Choosing an appropriate wiring system

An electrical installation is made up of many different electrical circuits, lighting circuits, power circuits, single-phase domestic circuits and three-phase industrial or commercial circuits.

Whatever the type of circuit, the circuit conductors are contained within cables or enclosures.

Part 5 of the IET Regulations tells us that electrical equipment and materials must be chosen so that they are suitable for the installed conditions, taking into account temperature, the presence of water, corrosion, mechanical damage, vibration or exposure to solar radiation. Therefore, PVC insulated and sheathed cables are suitable for domestic installations but, for a cable requiring mechanical protection and suitable for burying underground, a PVC/SWA cable would be preferable. These two types of cable are shown in Figs 3.38 and 3.39.

MI cables are waterproof, heatproof and corrosion-resistant with some mechanical protection. These qualities often make it the only cable choice for hazardous or high-temperature installations such as oil refineries, chemical works, boiler houses and petrol pump installations. An MI cable with terminating gland and seal is shown in Fig. 3.40.

Wiring systems and enclosures

The final choice of a wiring system must rest with those designing the installation and those ordering the work, but whatever system is employed, good workmanship by competent persons and the use of proper materials is essential for compliance with the IET Regulations (IET Regulation 134.1.1). The necessary skills can be acquired by an electrical trainee who has the correct attitude and dedication to his craft.

PVC insulated and sheathed cable installations

PVC insulated and sheathed wiring systems are used extensively for lighting and socket installations in domestic dwellings. Mechanical damage to the cable caused by impact, abrasion, penetration, compression or tension must be minimized during installation (IET Regulation 522.6.1). The cables are generally fixed using plastic clips incorporating a masonry nail, which means the cables can be fixed to wood, plaster or brick with almost equal ease. Cables should be run horizontally or vertically, not diagonally, down a wall. All kinks should be removed so that the cable is run straight and neatly between clips fixed at equal distances providing adequate support for the cable so that it does not become damaged by its own weight (IET Regulation 522.8.4 and Table D1 of the *On Site Guide*). Table D1 of the *On Site Guide* is shown in Table 3.7. Where cables are bent, the radius of the bend should not cause the conductors to be damaged (IET Regulation 522.8.3 and Table D5 of the *On Site Guide*).

Terminations or joints in the cable may be made in ceiling roses, junction boxes, or behind sockets or switches, provided that they are enclosed in a non-ignitable material, are properly insulated and are mechanically and electrically secure (IET Regulation 526). All joints must be accessible for inspection, testing and maintenance when the installation is completed (IET Regulation 526.3).

The 3rd Amendment to the 17th Edition of the IET Regulations introduced a new Regulation 521.200.1, requiring robust support of wiring systems in escape routes so that they will not become compromised or collapse in the event of a fire.

Where PVC insulated and sheathed cables are concealed in walls, floors or partitions, they must be provided with a box incorporating an earth terminal at each outlet position. PVC cables do not react chemically with plaster, as do some cables, and consequently PVC cables may be buried under plaster. Further protection by channel or conduit is only necessary if mechanical protection from nails or screws is required or to protect them from the plasterer's trowel. However, IET Regulation 522.6.101 now tells us that where PVC cables are to be embedded in a wall or partition at a depth of less than 50 mm they should be run along one of the permitted routes shown in Fig. 3.45. Figure 3.44 shows a typical PVC installation. To identify the most probable cable routes, IET Regulation 522.6.201 tells us that outside a zone formed by a 150 mm border all around a wall edge, cables can only be run horizontally or vertically to a point or accessory if they are contained in a substantial earthed enclosure, such as a conduit, which can withstand nail penetration, as shown in Fig. 3.45.

Where the accessory or cable is fixed to a wall which is less than 100 mm thick, protection must also be extended to the reverse side of the wall if a position can be determined.

Where none of this protection can be complied with, the cable must be given additional protection with a 30 mA RCD (IET Regulation 522.6.201).

Table 3.7 Spacing of cable supports

Spacings of supports for cables in accessible positions								
Overall diameter of cable* (mm)	**Maximum spacings of clips**							
	Non-armoured thermosetting, thermoplastic or lead sheathed cables				**Armoured cables**		**Mineral insulated copper sheathed or aluminium sheathed cables**	
	Generally		**In caravans**					
	Horizontal (mm)**	**Vertical** (mm)**	**Horizontal** (mm)**	**Vertical** (mm)**	**Horizontal** (mm)**	**Vertical** (mm)**	**Horizontal** (mm)**	**Vertical** (mm)**
1	**2**	**3**	**4**	**5**	**6**	**7**	**8**	**9**
Not exceeding 9	250	400			–	–	600	800
Exceeding 9 and not exceeding 15	300	400	250 (for all sizes)	400 (for all sizes)	350	450	900	1200
Exceeding 15 and not exceeding 20	350	450			400	550	1500	2000
Exceeding 20 and not exceeding 40	400	550			450	600	–	–

Note: For the spacing of supports for cables having an overall diameter exceeding 40 mm, and for single-core cables having conductors of cross-sectional area 300 mm^2 and larger, the manufacturer's recommendations should be observed.
*For flat cables taken as the dimension of the major axis.
**The spacings stated for horizontal runs may be applied also to runs at an angle of more than 30° from the vertical. For runs at an angle of 30° or less from the vertical, the vertical spacings are applicable.
Adapted from the IET *On Site Guide* by kind permission of the Institution of Engineering and Technology.

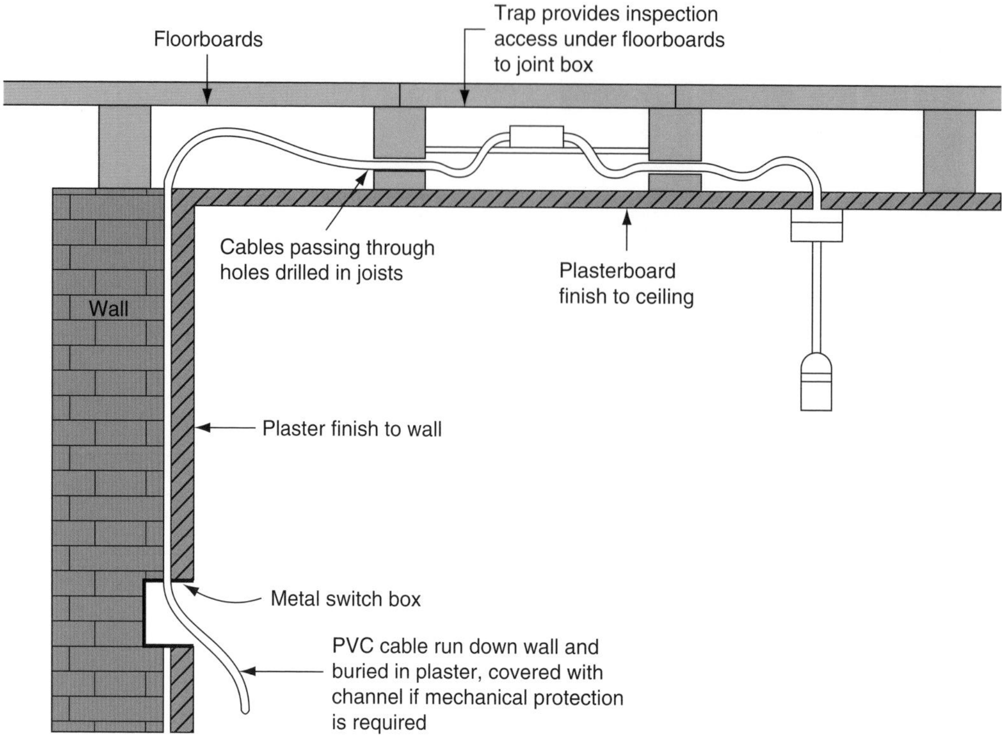

Figure 3.44 A concealed PVC sheathed wiring system.

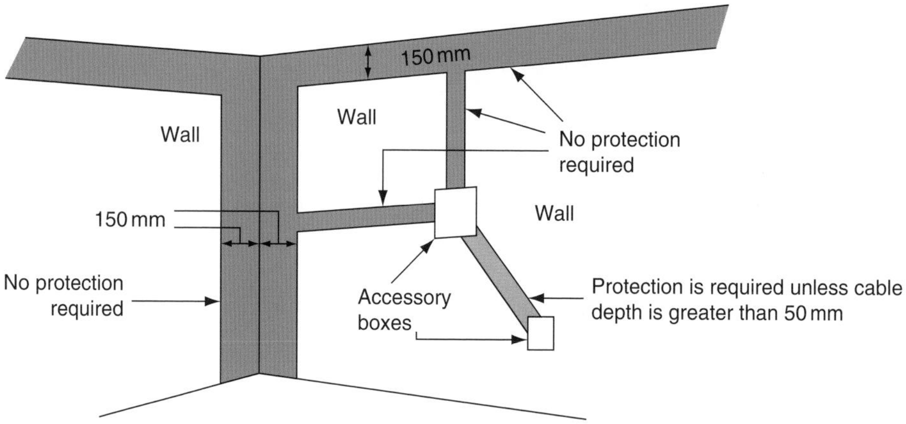

Figure 3.45 Permitted cable routes.

Where cables pass through walls, floors and ceilings the hole should be made good with incombustible material such as mortar or plaster to prevent the spread of fire (IET Regulations 527.1.2 and 527.2.1). Cables passing through metal boxes should be bushed with a rubber grommet to prevent abrasion of the cable. Holes drilled in floor joists through which cables are run should be 50 mm below the top or 50 mm above the bottom of the joist to prevent damage to the cable by nail penetration (IET Regulation 522.6.200), as shown in Fig. 3.46. PVC cables should not be installed when the surrounding temperature is below 0°C or when the cable temperature has been below 0°C for the previous 24 hours because the insulation becomes brittle at low temperatures and may be damaged during installation.

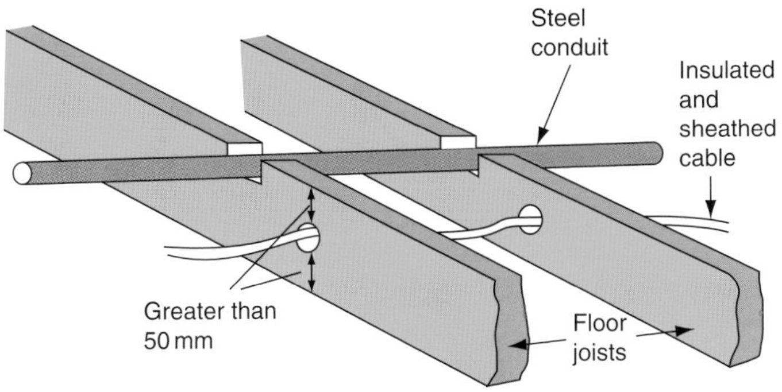

Notes:
1. Maximum diameter of hole should be 0.25 × joist depth.
2. Holes on centre line in a zone between 0.25 and 0.4 × span.
3. Maximum depth of notch should be 0.125 × joist depth.
4. Notches on top in a zone between 0.1 and 0.25 × span.
5. Holes in the same joist should be at least 3 diameters apart.

Figure 3.46 Correct installation of cables in floor joists.

Try this

Definitions

- In the margin write down a short definition of a 'competent person'.

Conduit installations

A **conduit** is a tube, channel or pipe in which insulated conductors are contained. The conduit, in effect, replaces the PVC outer sheath of a cable, providing mechanical protection for the insulated conductors. A conduit installation can be rewired easily or altered at any time, and this flexibility, coupled with mechanical protection, makes conduit installations popular for commercial and industrial applications. There are three types of conduit used in electrical installation work: steel, PVC and flexible.

Definition

A *conduit* is a tube, channel or pipe in which insulated conductors are contained.

Steel conduit

Steel conduits are made to a specification defined by BS 4568 and are either heavy gauge welded or solid drawn. Heavy gauge is made from a sheet of steel welded along the seam to form a tube and is used for most electrical installation work. Solid drawn conduit is a seamless tube which is much more expensive and only used for special gas-tight, explosion-proof or flameproof installations.

Conduit is supplied in 3.75 m lengths and typical sizes are 16, 20, 25 and 32 mm. Conduit tubing and fittings are supplied in a black enamel finish for internal use or hot galvanized finish for use on external or damp installations. A wide range of fittings is available and the conduit is fixed using saddles or pipe hooks, as shown in Fig. 3.47.

Top tip

Cutting materials

After cutting any metals, conduit or trunking, etc. make sure that you remove any burrs with a file in order to avoid cut fingers.

Metal conduits are threaded with stocks and dies and bent using special bending machines. The metal conduit is also utilized as the CPC and, therefore, all connections must be screwed up tightly and all burrs removed so that cables will not be damaged as they are drawn into the conduit. Metal conduits containing a.c. circuits must contain phase and neutral conductors in the same conduit to prevent eddy currents from flowing, which would result in the metal conduit becoming hot (IET Regulations 521.5.1, 522.8.1 and 522.8.11).

PVC conduit

PVC conduit used on typical electrical installations is heavy gauge standard impact tube manufactured to BS 4607. The conduit size and range of fittings are the same as those available for metal conduit. PVC conduit is most often

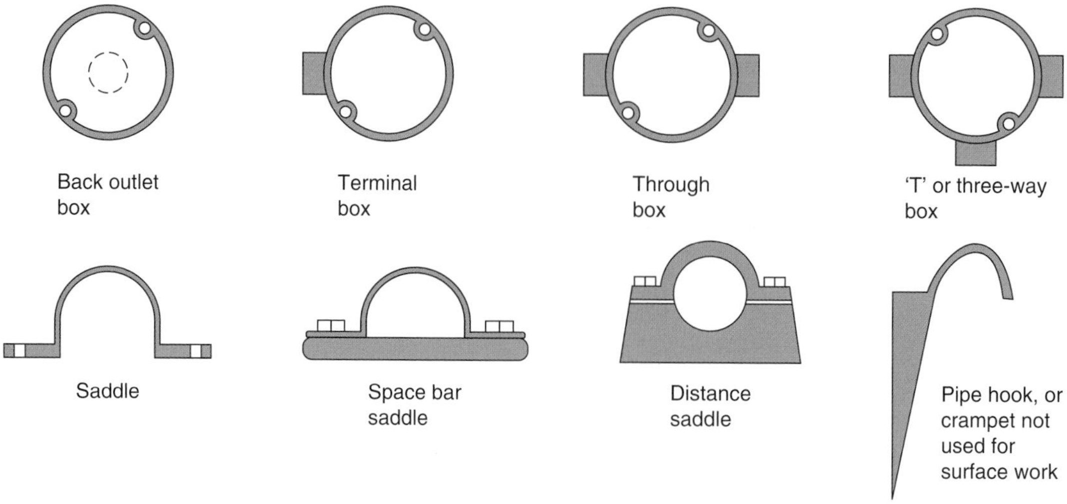

Figure 3.47 Conduit fittings and saddles.

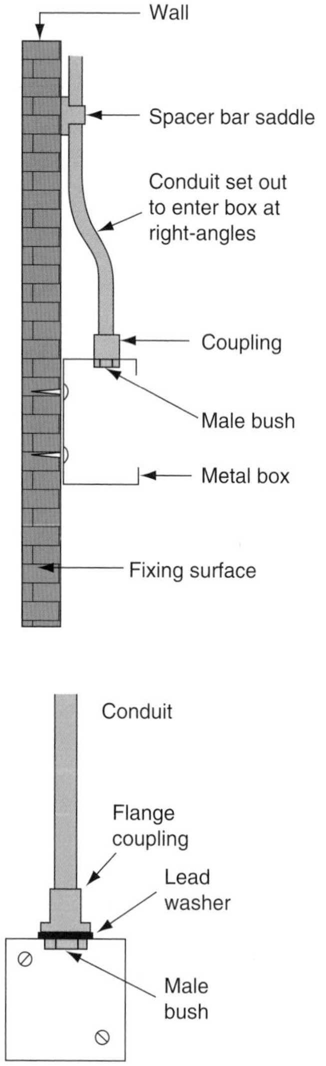

Figure 3.48 Terminating conduits.

joined by placing the end of the conduit into the appropriate fitting and fixing with a PVC solvent adhesive. PVC conduit can be bent by hand using a bending spring of the same diameter as the inside of the conduit. The spring is pushed into the conduit to the point of the intended bend and the conduit then bent over the knee. The spring ensures that the conduit keeps its circular shape. In cold weather, a little warmth applied to the point of the intended bend often helps to achieve a more successful bend.

The advantages of a PVC conduit system are that it may be installed much more quickly than steel conduit and is non-corrosive, but it does not have the mechanical strength of steel conduit. Since PVC conduit is an insulator it cannot be used as the CPC and a separate earth conductor must be run to every outlet. It is not suitable for installations subjected to temperatures below 25°C or above 60°C. Where luminaires are suspended from PVC conduit boxes, precautions must be taken to ensure that the lamp does not raise the box temperature or that the mass of the luminaire supported by each box does not exceed the maximum recommended by the manufacturer (IET Regulations 522.1 and 522.2). PVC conduit also expands much more than metal conduit and so long runs require an expansion coupling to allow for conduit movement and to help prevent distortion during temperature changes.

All conduit installations must be erected first before any wiring is installed (IET Regulation 522.8.2). The radius of all bends in conduit must not cause the cables to suffer damage, and therefore the minimum radius of bends given in Table D5 of the *On Site Guide* applies (IET Regulation 522.8.3). All conduits should terminate in a box or fitting and meet the boxes or fittings at right-angles, as shown in Fig. 3.48. Any unused conduit-box entries should be blanked off and all boxes covered with a box lid, fitting or accessory to provide complete enclosure of the conduit system. Conduit runs should be separate from other services, unless intentionally bonded, to prevent arcing from occurring from a faulty circuit within the conduit, which might cause the pipe of another service to become punctured.

When drawing cables into conduit they must first be *run off* the cable drum. That is, the drum must be rotated as shown in Fig. 3.49 and not allowed to *spiral off*, which will cause the cable to twist.

Cables should be fed into the conduit in a manner which prevents any cable from crossing over and becoming twisted inside the conduit. The cable insulation must not be damaged on the metal edges of the draw-in box. Cables can be pulled in on a draw wire if the run is a long one. The draw wire itself may be drawn in on a fish tape, which is a thin spring steel or plastic tape.

A limit must be placed on the number of bends between boxes in a conduit run and the number of cables which may be drawn into a conduit to prevent the cables from being strained during wiring. Appendix E of the *On Site Guide* gives a guide to the cable capacities of conduits and trunking.

Flexible conduit

Flexible conduit manufactured to BS 731-1: 1993 is made of interlinked metal spirals often covered with a PVC sleeving. The tubing must not be relied upon to provide a continuous earth path and, consequently, a separate CPC must be run either inside or outside the flexible tube (IET Regulation 543.2.7).

Flexible conduit is used for the final connection to motors so that the vibrations of the motor are not transmitted throughout the electrical installation and to allow for modifications to be made to the final motor position and drive belt adjustments.

Definition

Flexible conduit manufactured to BS 731-1: 1993 is made of interlinked metal spirals often covered with a PVC sleeving.

Conduit capacities

Single PVC insulated conductors are usually drawn into the installed conduit to complete the installation. Having decided upon the type, size and number of cables required for a final circuit, it is then necessary to select the appropriate size of conduit to accommodate those cables.

The tables in Appendix E of the *On Site Guide* describe a 'factor system' for determining the size of conduit required to enclose a number of conductors. The tables are shown in Tables 3.8 and 3.9. The method is as follows:

- Identify the cable factor for the particular size of conductor; see Table 3.8.
- Multiply the cable factor by the number of conductors, to give the sum of the cable factors.
- Identify the appropriate part of the conduit factor table given by the length of run and number of bends (see Table 3.9).
- The correct size of conduit to accommodate the cables is that conduit which has a factor equal to or greater than the sum of the cable factors.

Definition

Single PVC insulated conductors are usually drawn into the installed conduit to complete the installation.

Table 3.8 Conduit cable factors

Cable factors for use in conduit in long straight runs over 3 m, or runs of any length incorporating bends		
Type of conductor	**Conductor cross-sectional area (mm²)**	**Cable factor**
Solid or stranded	1	16
	1.5	22
	2.5	30
	4	43
	6	58
	10	105
	16	145
	25	217

The inner radius of a conduit bend should be not less than 2.5 times the outside diameter of the conduit. Adapted from the IET *On Site Guide* by kind permission of the Institution of Engineering and Technology.

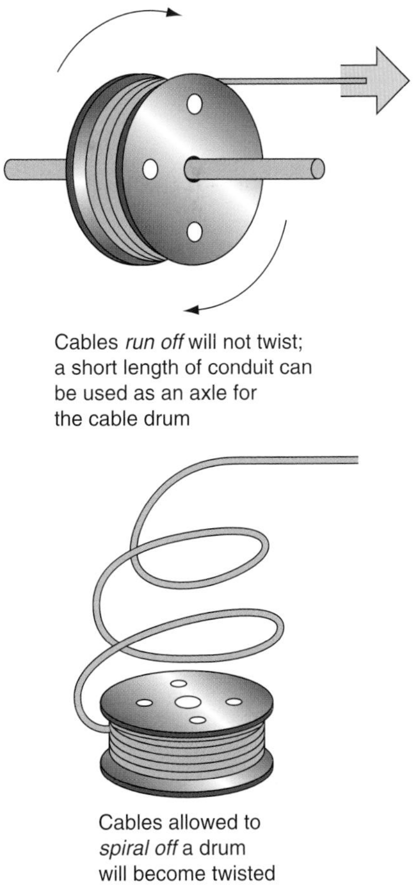

Figure 3.49 Running off cable from a drum.

Table 3.9 Conduit cable factors

Cable factors for runs incorporating bends and long straight runs																				
Length of run (m)	Conduit diameter (mm)																			
	16	20	25	32	16	20	25	32	16	20	25	32	16	20	25	32	16	20	25	32
	Straight				One bend				Two bends				Three bends				Four bends			
1	Covered by				188	303	543	947	177	286	514	900	158	256	463	818	130	213	388	692
1.5					182	294	528	923	167	270	487	857	143	233	422	750	111	182	333	600
2	Tables				177	286	514	900	158	256	463	818	130	213	388	692	97	159	292	529
2.5					171	278	500	878	150	244	442	783	120	196	358	643	86	141	260	474
3	A and B				167	270	487	857	143	233	422	750	111	182	333	600				
3.5	179	290	521	911	162	263	475	837	136	222	404	720	103	169	311	563				
4	177	286	514	900	158	256	463	818	130	213	388	692	97	159	292	529				
4.5	174	282	507	889	154	250	452	800	125	204	373	667	91	149	275	500				
5	171	278	500	878	150	244	442	783	120	196	358	643	86	141	260	474				
6	167	270	487	857	143	233	422	750	111	182	333	600								
7	162	263	475	837	136	222	404	720	103	169	311	563								
8	158	256	463	818	130	213	388	692	97	159	292	529								
9	154	250	452	800	125	204	373	667	91	149	275	500								
10	150	244	442	783	120	196	358	643	86	141	260	474								

Additional factors: For 38 mm diameter use 1.4 × (32 mm factor).
For 50 mm diameter use 2.6 × (32 mm factor).
For 63 mm diameter use 4.2 × (32 mm factor).

Adapted from the IET *On Site Guide* by kind permission of the Institution of Engineering and Technology.

Example 1

Six 2.5 mm^2 PVC insulated cables are to be run in a conduit containing two bends between boxes 10 m apart. Determine the minimum size of conduit to contain these cables.

From Table 3.8:

$$\begin{aligned}\text{The factor for one 2.5mm}^2\text{ cable} &= 30\\ \text{The sum of the cable factors} &= 6 \times 30\\ &= 180\end{aligned}$$

From Table 3.9, a 25 mm conduit, 10 m long and containing two bends, has a factor of 260. A 20 mm conduit containing two bends only has a factor of 141 which is less than 180, the sum of the cable factors, and, therefore, 25 mm conduit is the minimum size to contain these cables.

Example 2

Ten 1.0 mm^2 PVC insulated cables are to be drawn into a plastic conduit which is 6 m long between boxes and contains one bend. A 4.0 mm PVC insulated CPC is also included. Determine the minimum size of conduit to contain these conductors.

From Table 3.8:

$$\begin{aligned}\text{The factor for one 1.0mm cable} &= 16\\ \text{The factor for one 4.0mm cable} &= 43\\ \text{The sum of the cable factors} &= (10 \times 16) + (1 \times 43)\\ &= 203\end{aligned}$$

From Table 3.9, a 20 mm conduit, 6 m long and containing one bend, has a factor of 233. A 16 mm conduit containing one bend only has a factor of 143 which is less than 203, the sum of the cable factors, and, therefore, 20 mm conduit is the minimum size to contain these cables.

Trunking installations

A **trunking** is an enclosure provided for the protection of cables which is normally square or rectangular in cross-section, having one removable side. Trunking may be thought of as a more accessible conduit system, and for industrial and commercial installations it is replacing the larger conduit sizes. A trunking system can have great flexibility when used in conjunction with conduit; the trunking forms the background or framework for the installation, with conduits running from the trunking to the point controlling the current-using apparatus. When an alteration or extension is required it is easy to drill a hole in the side of the trunking and run a conduit to the new point. The new wiring can then be drawn through the new conduit and the existing trunking to the supply point.

Definition

A *trunking* is an enclosure provided for the protection of cables which is normally square or rectangular in cross-section, having one removable side. Trunking may be thought of as a more accessible conduit system.

Trunking is supplied in 3 m lengths and various cross-sections measured in millimetres from 50 × 50 up to 300 × 150. Most trunking is available in either steel or plastic.

Metallic trunking

Metallic trunking is formed from mild steel sheet, coated with grey or silver enamel paint for internal use or a hot-dipped galvanized coating where damp conditions might be encountered and made to a specification defined by BS EN 500 85. A wide range of accessories is available, such as 45° bends, 90° bends, tee and four-way junctions, for speedy on-site assembly. Alternatively, bends may be fabricated in lengths of trunking, as shown in Fig. 3.50. This may be necessary or more convenient if a bend or set is non-standard, but it does take more time to fabricate bends than merely to bolt on standard accessories.

Definition

Metallic trunking is formed from mild steel sheet, coated with grey or silver enamel paint for internal use or a hot-dipped galvanized coating where damp conditions might be encountered.

The trunking installation must be treated as an exposed conductive part and be properly earthed in accordance with Regulation 411.3.1.1. Care must be taken to provide reliable earth continuity and an adequate earth fault current path by making all joints electrically and mechanically secure.

When fabricating bends the trunking should be supported with wooden blocks for sawing and filing, in order to prevent the sheet-steel from vibrating or becoming deformed. Fish-plates must be made and riveted or bolted to the trunking to form a solid and secure bend. When manufactured bends are used, the continuity of the earth path must be ensured across the joint by making all fixing screw connections very tight, or fitting a separate copper strap between the trunking and the standard bend. If an earth continuity test on the trunking is found to be unsatisfactory, an insulated CPC must be installed inside the trunking. The size of the protective conductor will be determined by the largest cable contained in the trunking, as described in Table 54.7 of the IET Regulations. If the circuit conductors are less than 16 mm^2, then a 16 mm^2 CPC will be required.

Top tip

Sawing metal

When sawing metals, conduit trunking or tray, a safe and effective speed is 50 strokes per minute.

Non-metallic trunking

Trunking and trunking accessories are also available in high-impact PVC. The accessories are usually secured to the lengths of trunking with a PVC solvent adhesive. PVC trunking, like PVC conduit, is easy to install and is non-corrosive. A separate CPC will need to be installed and non-metallic trunking may require more frequent fixings because it is less rigid than metallic trunking. All trunking

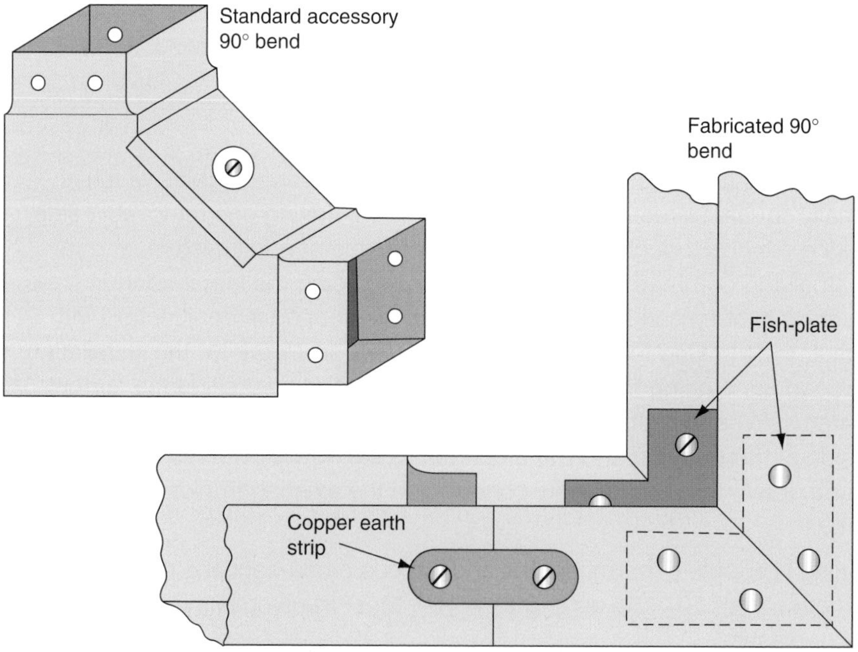

Figure 3.50 Alternative trunking bends.

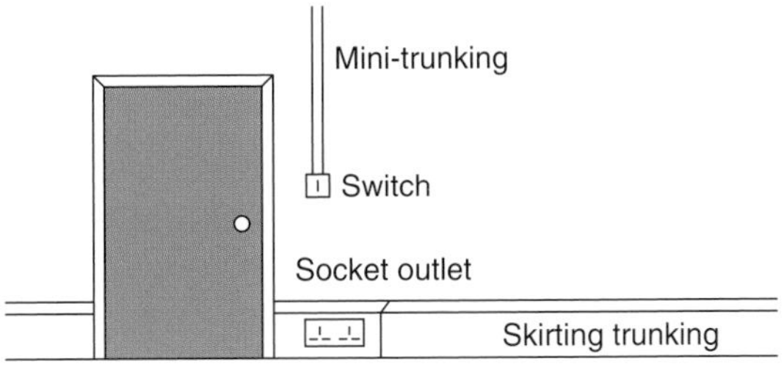

Figure 3.51 Typical installation of skirting trunking and mini-trunking.

fixings should use round-headed screws to prevent damage to cables since the thin sheet construction makes it impossible to countersink screw heads.

Mini-trunking

Mini-trunking is very small PVC trunking, ideal for surface wiring in domestic and commercial installations such as offices. The trunking has a cross-section of 16 × 16 mm, 25 × 16 mm, 38 × 16 mm or 38 × 25 mm and is ideal for switch drops or for housing auxiliary circuits such as telephone or audio equipment wiring. The modern square look in switches and sockets is complemented by the mini-trunking which is very easy to install (see Fig. 3.51).

> **Definition**
>
> *Mini-trunking* is very small PVC trunking, ideal for surface wiring in domestic and commercial installations such as offices.

Skirting and dado trunking

Skirting trunking is a trunking manufactured from PVC or steel in the shape of a skirting board and is frequently used in commercial buildings such as hospitals, laboratories and offices. The trunking is fitted around the walls of a room at either the skirting board level or at the working surface level when it is called dado trunking and contains the wiring for socket outlets and telephone points which are mounted on the lid, as shown in Figs. 3.51 and 3.53.

> **Definition**
>
> *Skirting trunking* is a trunking manufactured from PVC or steel in the shape of a skirting board and is frequently used in commercial buildings such as hospitals, laboratories and offices.

Where any trunking passes through walls, partitions, ceilings or floors, short lengths of lid should be fitted so that the remainder of the lid may be removed later without difficulty. Any damage to the structure of the buildings must be made good with mortar, plaster or concrete in order to prevent the spread of fire. Fire barriers must be fitted inside the trunking every 5 m, or at every floor level or room-dividing wall if this is a shorter distance, as shown in Fig. 3.52(a).

Where trunking is installed vertically, the installed conductors must be supported so that the maximum unsupported length of non-sheathed cable does not exceed 5 m. Figure 3.52(b) shows cables woven through insulated pin supports, which is one method of supporting vertical cables.

PVC insulated cables are usually drawn into an erected conduit installation or laid into an erected trunking installation. Table E4 of the *On Site Guide* only gives factors for conduits up to 32 mm in diameter, which would indicate that conduits larger than this are not in frequent or common use. Where a cable enclosure greater than 32 mm is required because of the number or size of the conductors, it is generally more economical and convenient to use trunking.

Trunking capacities

The **ratio** of the space occupied by all the cables in a conduit or trunking to the whole space enclosed by the conduit or trunking is known as the **space factor**. Where sizes and types of cable and trunking are not covered by the tables in the *On Site Guide*, a space factor of 45% must not be exceeded. This means that the cables must not fill more than 45% of the space enclosed by the trunking. The tables take this factor into account.

> **Definition**
>
> The *ratio* of the space occupied by all the cables in a conduit or trunking to the whole space enclosed by the conduit or trunking is known as the *space factor*.

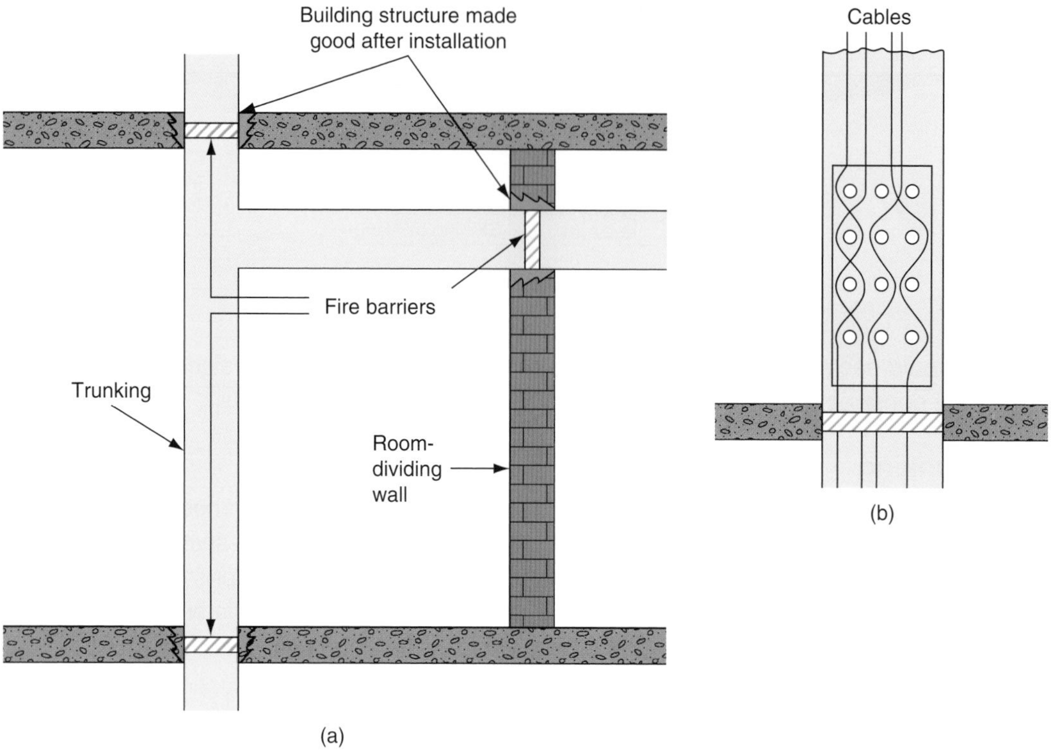

Figure 3.52 Installation of trunking: (a) fire barriers in trunking, and (b) cable supports in vertical trunking.

Table 3.10 Trunking cable factors

Cable factors for trunking			
Type of conductor	**Conductor cross-sectional area (mm^2)**	**PVC BS 6004 Cable factor**	**Thermosetting BS 7211 Cable factor**
Solid	1.5	8.0	8.6
	2.5	11.9	11.9
Stranded	1.5	8.6	9.6
	2.5	12.6	13.9
	4	16.6	18.1
	6	21.2	22.9
	10	35.3	36.3
	16	47.8	50.3
	25	73.9	75.4

Notes: These factors are for metal trunking and may be optimistic for plastic trunking where the cross-sectional area available may be significantly reduced from the nominal by the thickness of the wall material. The provision of spare space is advisable; however, any circuits added at a later date must take into account grouping. Appendix 4, BS 7671.
Adapted from the IET *On Site Guide* by kind permission of the Institution of Engineering and Technology.

To calculate the size of trunking required to enclose a number of cables:

- Identify the cable factor for the particular size of conductor (see Table 3.10).
- Multiply the cable factor by the number of conductors to give the sum of the cable factors.

Table 3.11 Trunking cable factors

Factors for trunking			
Dimensions of trunking (mm × mm)	**Factor**	**Dimensions of trunking (mm × mm)**	**Factor**
50 × 38	767	200 × 100	8572
50 × 50	1037	200 × 150	13001
75 × 25	738	200 × 200	17429
75 × 38	1146	225 × 38	3474
75 × 50	1555	225 × 50	4671
75 × 75	2371	225 × 75	7167
100 × 25	993	225 × 100	9662
100 × 38	1542	225 × 150	14652
100 × 50	2091	225 × 200	19643
100 × 75	3189	225 × 225	22138
100 × 100	4252	300 × 38	4648
150 × 38	2999	300 × 50	6251
150 × 50	3091	300 × 75	9590
150 × 75	4743	300 × 100	12929
150 × 100	6394	300 × 150	19607
150 × 150	9697	300 × 200	26285
200 × 38	3082	300 × 225	29624
200 × 50	4145	300 × 300	39428
200 × 75	6359		

Space factor – 45% with trunking thickness taken into account.
Adapted from the IET *On Site Guide* by kind permission of the Institution of Engineering and Technology.

- Consider the factors for trunking shown in Table 3.11. The correct size of trunking to accommodate the cables is that trunking which has a factor equal to, or greater than, the sum of the cable factors.

Example 3

Calculate the minimum size of trunking required to accommodate the following single-core PVC cables:

20 × 1.5 mm solid conductors
20 × 2.5 mm solid conductors
21 × 4.0 mm stranded conductors
16 × 6.0 mm stranded conductors

From Table 3.10, the cable factors are:

for 1.5 mm solid cable – 8.0
for 2.5 mm solid cable – 11.9

(continued)

Example 3 (continued)

for 4.0 mm stranded cable – 16.6

for 6.0 mm stranded cable – 21.2

The sum of the cable terms is:

$(20 \times 8.0) + (20 \times 11.9) + (21 \times 16.6) + (16 \times 21.2) = 1085.8$. From Table 3.11, 75 × 38 mm trunking has a factor of 1146 and, therefore, the minimum size of trunking to accommodate these cables is 75 × 38 mm, although a larger size, say, 75 × 50 mm, would be equally acceptable if this was more readily available as a standard stock item.

Segregation of circuits

Where an installation comprises a mixture of low-voltage and very low-voltage circuits such as mains lighting and power, fire alarm and telecommunication circuits, they must be separated or *segregated* to prevent electrical contact (IET Regulation 528.1).

For the purpose of these regulations various circuits are identified by one of two bands as follows:

Band I: telephone, radio, bell, call and intruder alarm circuits, emergency circuits for fire alarm and emergency lighting.

Band II: mains voltage circuits.

When Band I circuits are insulated to the same voltage as Band II circuits, they may be drawn into the same compartment.

When trunking contains rigidly fixed metal barriers along its length, the same trunking may be used to enclose cables of the separate bands without further precautions, provided that each band is separated by a barrier, as shown in Fig. 3.53.

Multi-compartment PVC trunking cannot provide band segregation since there is no metal screen between the bands. This can only be provided in PVC trunking if screened cables are drawn into the trunking.

Cable tray installations

Cable tray is a sheet-steel channel with multiple holes. The most common finish is hot-dipped galvanized but PVC-coated tray is also available. It is used

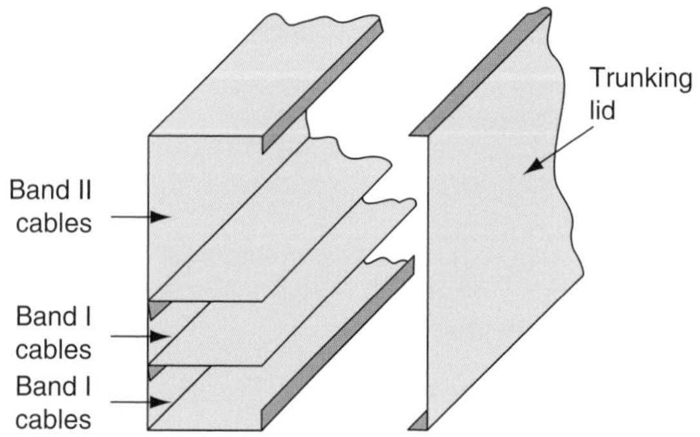

Figure 3.53 Segregation of cables in trunking.

extensively on large industrial and commercial installations for supporting MI and SWA cables which are laid on the cable tray and secured with cable ties through the tray holes.

Cable tray should be adequately supported during installation by brackets which are appropriate for the particular installation. The tray should be bolted to the brackets with round-headed bolts and nuts, with the round head inside the tray so that cables drawn along the tray are not damaged.

The tray is supplied in standard widths from 50 to 900 mm, and a wide range of bends, tees and reducers is available. Figure 3.54 shows a factory-made 90° bend at B. The tray can also be bent using a cable tray bending machine to create bends such as that shown at A in Fig. 3.54. The installed tray should be securely bolted with round-headed bolts where lengths or accessories are attached, so that there is a continuous earth path which may be bonded to an electrical earth. The whole tray should provide a firm support for the cables, and therefore the tray fixings must be capable of supporting the weight of both the tray and cables.

Definition

Cable tray is a sheet-steel channel with multiple holes. The most common finish is hot-dipped galvanized but PVC-coated tray is also available. It is used extensively on large industrial and commercial installations for supporting MI and SWA cables which are laid on the cable tray and secured with cable ties through the tray holes.

Cable basket installations

Cable basket is becoming very popular for commercial and industrial installations. It is made from steel wire into a basket channel with sides.

Cable basket allows maximum airflow around the cables which are laid into the basket without fixing.

Cable basket requires similar installation techniques to cable tray and should be adequately supported.

PVC/SWA cable installations

Steel wire armoured PVC insulated cables are now extensively used on industrial installations and often laid on cable tray. This type of installation has the advantage of flexibility, allowing modifications to be made speedily as the need arises. The cable has a steel wire armouring giving mechanical protection

Definition

Steel wire armoured PVC insulated cables are now extensively used on industrial installations and often laid on cable tray.

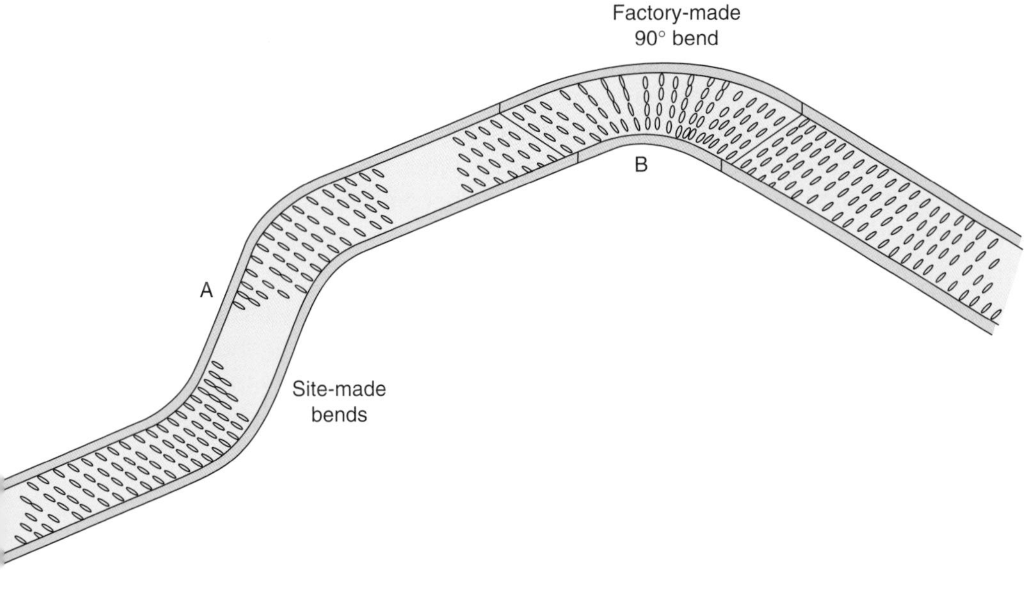

Figure 3.54 Cable tray with bends.

and permitting it to be laid directly in the ground or in ducts, or it may be fixed directly or laid on a cable tray. Figure 3.39 shows a PVC/SWA cable.

It should be remembered that when several cables are grouped together the current rating will be reduced according to the correction factors given in Appendix 4 (Table 4C1) of the IET Regulations.

The cable is easy to handle during installation, is pliable and may be bent to a radius of eight times the cable diameter. The PVC insulation would be damaged if installed in ambient temperatures over 70°C or below 0°C, but once installed the cable can operate at low temperatures.

The cable is terminated with a simple gland which compresses a compression ring onto the steel wire armouring to provide the earth continuity between the switchgear and the cable.

MI cable installations

Mineral insulated cables are available for general wiring as:

- light-duty MI cables for voltages up to 600 V and sizes from 1.0 to 10 mm^2;
- heavy-duty MI cables for voltages up to 1000 V and sizes from 1.0 to 150 mm^2.

Figure 3.40 shows an MI cable and termination.

The cables are available with bare sheaths or with a PVC oversheath. The cable sheath provides sufficient mechanical protection for all but the most severe situations, where it may be necessary to fit a steel sheath or conduit over the cable to give extra protection, particularly near floor level in some industrial situations.

The cable may be laid directly in the ground, in ducts, on cable tray or clipped directly to a structure. It is not affected by water, oil or the cutting fluids used in engineering and can withstand very high temperatures or even fire. The cable diameter is small in relation to its current-carrying capacity and it should last indefinitely if correctly installed because it is made from inorganic materials. These characteristics make the cable ideal for Band I emergency circuits, boiler houses, furnaces, petrol stations and chemical plant installations.

The cable is supplied in coils and should be run off during installation and not spiralled off, as described in Fig. 3.49 for conduit. The cable can be work hardened if over-handled or over-manipulated. This makes the copper outer sheath stiff and may result in fracture. The outer sheath of the cable must not be penetrated, otherwise moisture will enter the magnesium oxide insulation and lower its resistance. To reduce the risk of damage to the outer sheath during installation, cables should be straightened and formed by hammering with a hide hammer or a block of wood and a steel hammer. When bending MI cables the radius of the bend should not cause the cable to become damaged and clips should provide adequate support (IET Regulation 522.8.5); see Table 3.7.

The cable must be prepared for termination by removing the outer copper sheath to reveal the copper conductors. This can be achieved by using a rotary stripper tool or, if only a few cables are to be terminated, the outer sheath can be removed with side cutters, peeling off the cable in a similar way to peeling the skin from a piece of fruit with a knife. When enough conductor has been revealed, the outer sheath must be cut off square to facilitate the fitting of the sealing pot, and this can be done with a ringing tool. All excess magnesium oxide powder must be wiped from the conductors with a clean cloth. This is to prevent moisture from penetrating the seal by capillary action.

Cable ends must be terminated with a special seal to prevent the entry of moisture. Figure 3.40 shows a brass screw-on seal and gland assembly, which allows termination of the MI cables to standard switchgear and conduit fittings. The sealing pot is filled with a sealing compound, which is pressed in from one side only to prevent air pockets from forming, and the pot closed by crimping home the sealing disc with an MI crimping tool such as that shown in Fig. 4.1. Such an assembly is suitable for working temperatures up to 105°C. Other compounds or powdered glass can increase the working temperature up to 250°C.

The conductors are not identified during the manufacturing process and so it is necessary to identify them after the ends have been sealed. A simple continuity or polarity test, as described later in this chapter, can identify the conductors which are then sleeved or identified with coloured markers.

Connection of MI cables can be made directly to motors, but to absorb the vibrations a 360° loop should be made in the cable just before the termination. If excessive vibration is expected, the MI cable should be terminated in a conduit through box and the final connection made by flexible conduit.

Copper MI cables may develop a green incrustation or patina on the surface, even when exposed to normal atmospheres. This is not harmful and should not be removed. However, if the cable is exposed to an environment which might encourage corrosion, an MI cable with an overall PVC sheath should be used.

Water-heating circuits

A small, single-point over-sink-type water heater may be considered as a permanently connected appliance and so may be connected to a ring circuit through a fused connection unit. A water heater of the immersion type is usually rated at a maximum of 3 kW, and could be considered as a permanently connected appliance, fed from a fused connection unit. However, many immersion heating systems are connected into storage vessels of about 150 litres in domestic installations, and the *On Site Guide* states that immersion heaters fitted to vessels in excess of 15 litres should be supplied by their own circuit (OSG Appendix H5).

Therefore, immersion heaters must be wired on a separate radial circuit when they are connected to water vessels which hold more than 15 litres. Figure 3.55 shows the wiring arrangements for an immersion heater. Every switch must be a double-pole (DP) switch and out of reach of anyone using a fixed bath or shower when the immersion heater is fitted to a vessel in a bathroom.

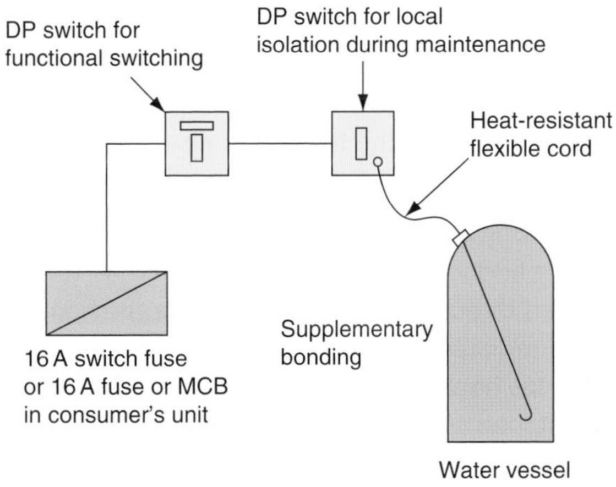

Figure 3.55 Immersion heater wiring.

Supplementary protective bonding to pipework *will only be required* as an addition to fault protection (IET Regulation 415.2) if the immersion heater vessel is in a bathroom that *does not have*:

- all circuits protected by a 30 mA RCD; **and**
- main protective bonding (IET Regulation 701.415.2) as shown in Fig. 3.6.

Electric space-heating circuits

Electrical heating systems can be broadly divided into two categories: unrestricted local heating and off-peak heating.

Unrestricted local heating may be provided by portable electric radiators which plug into the socket outlets of the installation. Fixed heaters that are wall mounted or inset must be connected through a fused connection and incorporate a local switch, either on the heater itself or as a part of the fuse connecting unit. Heating appliances where the heating element can be touched must have a DP switch which disconnects all conductors. This requirement includes radiators which have an element inside a silica-glass sheath.

Off-peak heating systems may provide central heating from storage radiators, ducted warm air or underfloor heating elements. All three systems use the thermal storage principle, whereby a large mass of heat-retaining material is heated during the off-peak period and allowed to emit the stored heat throughout the day. The final circuits of all off-peak heating installations must be fed from a separate supply controlled by an electricity board time clock.

Figure 3.56 A room radiator.

When calculating the size of cable required to supply a single-storage radiator, it is good practice to assume a current demand equal to 3.4 kW at each point. This will allow the radiator to be changed at a future time with the minimum disturbance to the installation. Each radiator must have a 20 A DP means of isolation adjacent to the heater and the final connection should be via a flex outlet. See Fig. 3.57 for wiring arrangements.

Ducted warm air systems have a centrally sited thermal storage heater with a high storage capacity. The unit is charged during the off-peak period, and a fan drives the stored heat in the form of warm air through large air ducts to outlet grilles in the various rooms. The wiring arrangements for this type of heating are shown in Fig. 3.58.

The single-storage heater is heated by an electric element embedded in bricks and rated between 6 and 15 kW depending upon its thermal capacity. A radiator of this capacity must be supplied on its own circuit, in cable capable of carrying the maximum current demand and protected by a fuse or miniature circuit-breaker (MCB) of 30, 45 or 60 A as appropriate. At the heater position, a DP switch must be installed to terminate the fixed heater wiring. The flexible cables used for the final connection to the heaters must be of the heat-resistant type.

Floor-warming installations use the thermal storage properties of concrete. Special cables are embedded in the concrete floor screed during construction. When current is passed through the cables they become heated, the concrete absorbs this heat and radiates it into the room. The wiring arrangements are shown in Fig. 3.59. Once heated, the concrete will give off heat for a long time after the supply is switched off and is, therefore, suitable for connection to an off-peak supply.

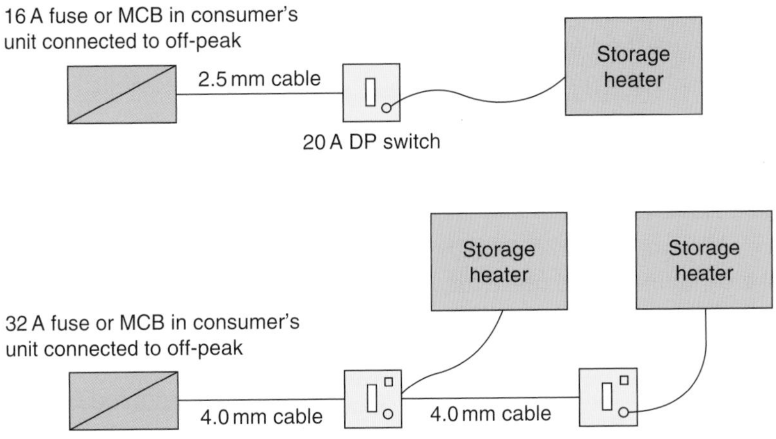

Figure 3.57 Possible wiring arrangements for storage heaters.

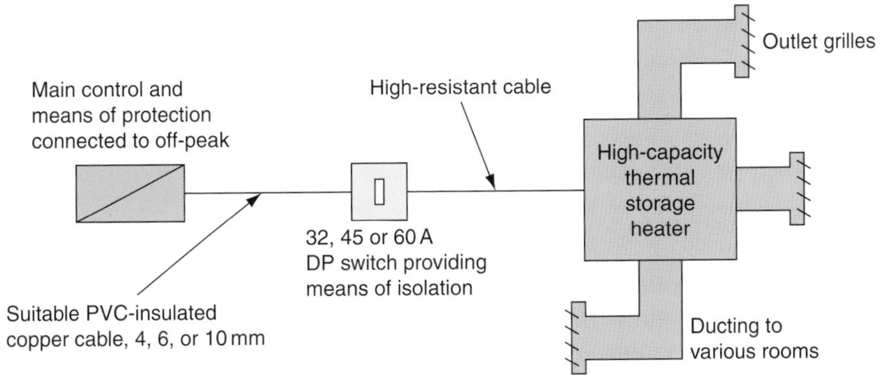

Figure 3.58 Ducted warm air heating system.

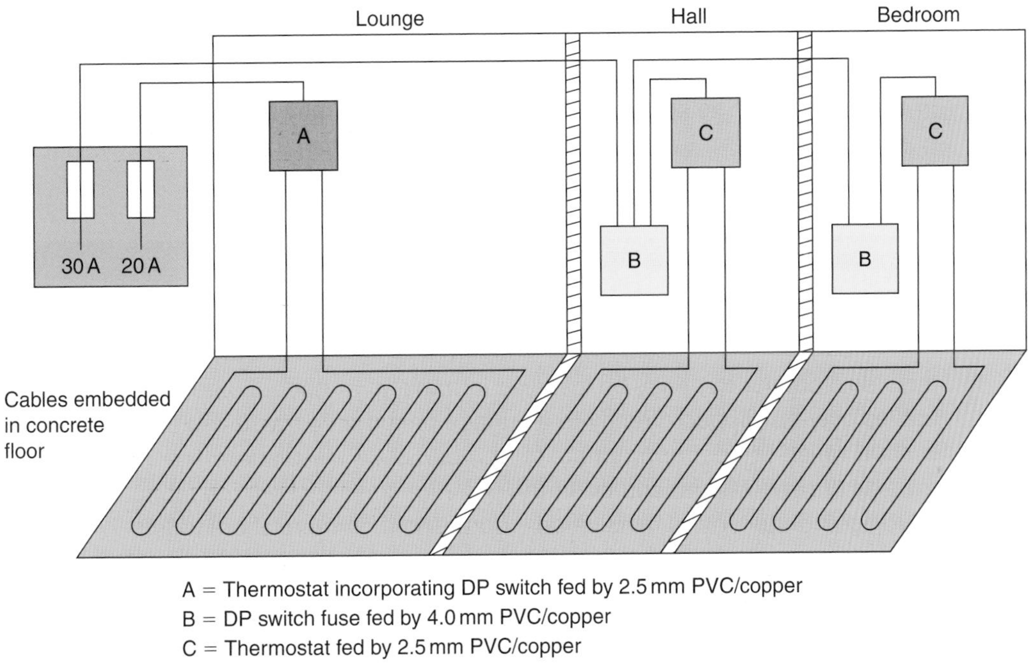

Figure 3.59 Floor-warming installations.

Underfloor heating cables installed in bathrooms or shower rooms must incorporate an earthed metallic sheath or be covered by an earthed metallic grid connected to the protective conductor of the supply circuit (IET Regulation 701.753).

Cooker circuit

A cooker with a rating above 3 kW must be supplied on its own circuit but since it is unlikely that in normal use every heating element will be switched on at the same time, a diversity factor may be applied in calculating the cable size, as detailed in the *On Site Guide.*

Consider, as an example, a cooker with the following elements fed from a cooker control unit incorporating a 13 A socket:

$$\begin{aligned}4 \times 2\text{ kW fast-boiling rings} &= 8000\text{ W}\\ 1\times 2\text{ kW grill} &= 2000\text{ W}\\ 1\times 2\text{ kW oven} &= 2000\text{ W}\\ \text{Total loading} &= 12\,000\text{ W}\end{aligned}$$

When connected to 230 V

$$\text{Current rating} = \frac{12000}{230} = 52.17\text{ A}$$

Applying the diversity factor of Table 1A

$$\begin{aligned}\text{Total current rating} &= 52.17\text{ A}\\ \text{First 10 amperes} &= 10\text{ A}\\ 30\%\text{ of }42.17\text{A} &= 12.65\text{ A}\\ \text{Socket outlet} &= 5\text{ A}\\ \text{Assessed current demand} &= 10 + 12.65 + 5 = 27.65\text{ A}\end{aligned}$$

Therefore, a cable capable of carrying 27.65 A may be used safely rather than a 52.17 A cable.

A cooking appliance must be controlled by a switch separate from the cooker but in a readily accessible position. Where two cooking appliances are installed in one room, such as split-level cookers, one switch may be used to control both appliances provided that neither appliance is more than 2 m from the switch (*On Site Guide*, Appendix 8).

Support and fixing methods for electrical equipment

Individual conductors may be installed in trunking or conduit and individual cables may be clipped directly to a surface or laid on a tray using the wiring system which is most appropriate for the particular installation. The installation method chosen will depend upon the contract specification, the fabric of the building and the type of installation – domestic, commercial or industrial.

It is important that the wiring systems and fixing methods are appropriate for the particular type of installation and compatible with the structural materials used in the building construction. The electrical installation must be compatible with the installed conditions, must not damage the fabric of the building, or weaken load-bearing girders or joists.

The installation designer must ask him- herself the following questions:

- Does this wiring system meet the contract specification?
- Is the wiring system compatible with this particular installation?
- Do I need to consider any special regulations such as those required by agricultural and horticultural installations, swimming pools or flameproof installations?
- Will this type of electrical installation be aesthetically acceptable and compatible with the other structural materials?

The installation electrician must ask him- herself the following questions:

- Am I using materials and equipment which meet the relevant British Standards and the contract specification?
- Am I using an appropriate fixing method for this wiring system or piece of equipment?
- Will the structural material carry the extra load that my conduits and cables will place upon it?
- Will my fixings and fittings weaken the existing fabric of the building?
- Will the electrical installation interfere with other supplies and services?
- Will all terminations and joints be accessible upon completion of the erection period? (IET Regulations 513.1 and 526.3.)
- Will the materials being used for the electrical installation be compatible with the intended use of the building?
- Am I working safely and efficiently and in accordance with the IET Regulations (BS 7671)?

A domestic installation usually calls for a PVC insulated and sheathed wiring system. These cables are generally fixed using plastic clips incorporating a masonry nail which means that the cables can be fixed to wood, plaster or brick with almost equal ease.

Cables must be run straight and neatly between clips fixed at equal distances which provide adequate support for the cable so that it does not become damaged by its own weight (IET Regulation 522.8.4), shown in Table 3.7.

A commercial or industrial installation might call for a conduit or trunking wiring system. A conduit is a tube, channel or pipe in which insulated conductors are contained. The conduit, in effect, replaces the PVC outer sheath of a cable, providing mechanical protection for the insulated conductors. A conduit installation can be rewired easily or altered at any time and this flexibility, coupled with mechanical protection, makes conduit installations popular for commercial and industrial applications. Steel conduits and trunking are, however, much heavier than single cables and, therefore, need substantial and firm fixings and supports. A wide range of support brackets is available for fixing conduit, trunking and tray installations to the fabric of a commercial or industrial installation. Some of these are shown in Fig. 3.60.

When a heavier or more robust fixing is required to support cabling or equipment, a nut and bolt or screw fixing is called for. Wood screws may be screwed directly into wood but when fixing to stone, brick or concrete it is first necessary to drill a hole in the masonry material which is then plugged with a material (usually plastic) to which a screw can be secured.

For the most robust fixing to masonry materials, an expansion bolt such as that made by Rawlbolt should be used.

For lightweight fixings to hollow partitions or plasterboard, a spring toggle can be used. Plasterboard cannot support a screw fixing directly into itself but the

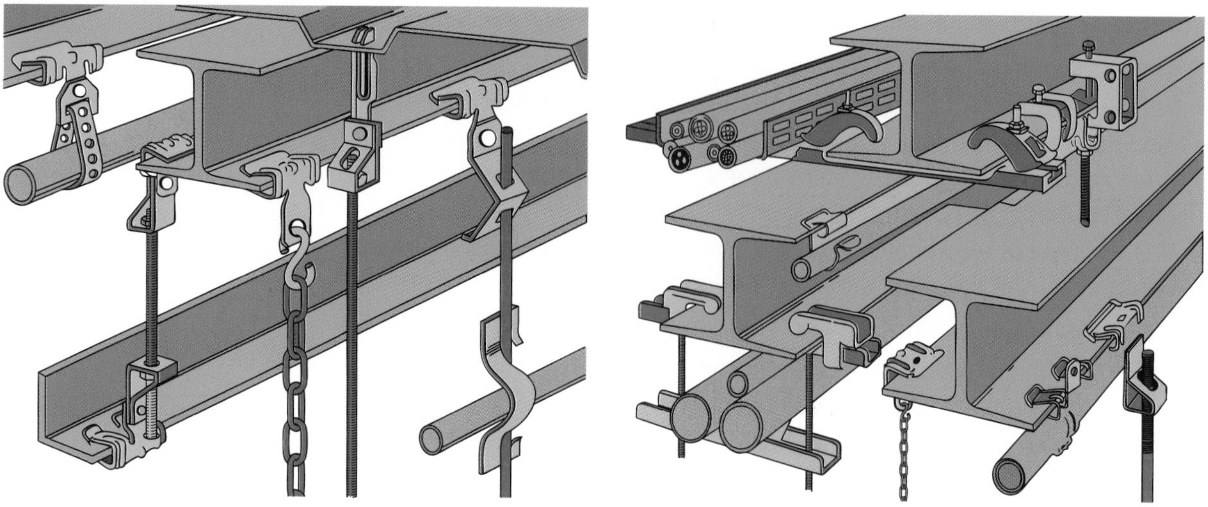

Figure 3.60 Some manufactured girder supports for electrical equipment.

spring toggle spreads the load over a larger area, making the fixing suitable for light loads.

Let us look in a little more detail at individual joining, support and fixing methods.

Joining materials

Plastic can be joined with an appropriate solvent. Metal may be welded, brazed or soldered, but the most popular method of on-site joining of metal on electrical installations is by nuts and bolts or rivets.

A nut and bolt joint may be considered a temporary fastening since the parts can easily be separated if required by unscrewing the nut and removing the bolt. A rivet is a permanent fastening since the parts riveted together cannot be easily separated.

Two pieces of metal joined by a bolt and nut and by a machine screw and nut are shown in Fig. 3.61. The nut is tightened to secure the joint. When joining trunking or cable trays, a round-head machine screw should be used with the head inside to reduce the risk of damage to cables being drawn into the trunking or tray.

Thin sheet material such as trunking is often joined using a pop riveter. Special rivets are used with a hand tool, as shown in Fig. 3.62. Where possible, the parts to be riveted should be clamped and drilled together with a clearance hole for the rivet. The stem of the rivet is pushed into the nose bush of the riveter until the alloy sleeve of the rivet is flush with the nose bush (a). The rivet is then placed in the hole and the handles squeezed together (b). The alloy sleeve is compressed and the rivet stem will break off when the rivet is set and the joint complete (c). To release the broken-off stem piece, the nose bush is turned upward and the handles opened sharply. The stem will fall out and is discarded (d).

Bracket supports

Conduit and trunking may be fixed directly to a surface such as a brick wall or concrete ceiling, but where cable runs are across girders or other steel framework, spring steel clips may be used but support brackets or clips often require manufacturing.

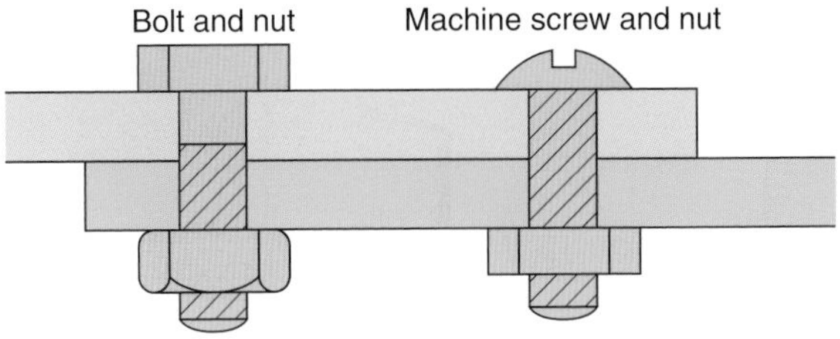

Figure 3.61 Joining of metal.

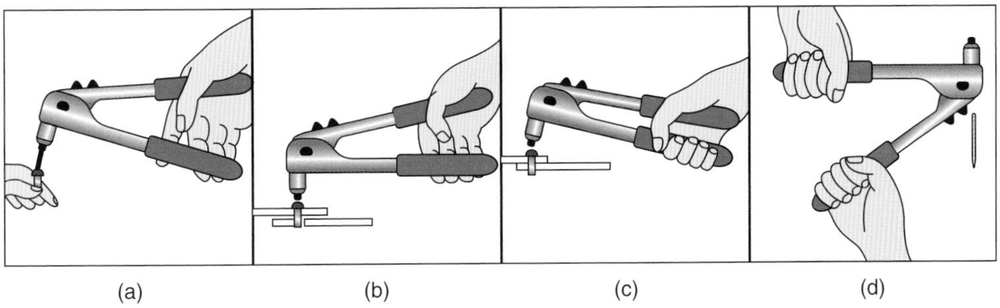

Figure 3.62 Metal joining with pop rivets.

COMPANION WEBSITE Visit the companion website for more on this topic.

The brackets are usually made from flat iron, which is painted after manufacturing to prevent corrosion. They may be made on-site by the electrician or, if many brackets are required, the electrical contractor may make a working sketch with dimensions and have the items manufactured by a blacksmith or metal fabricator.

The type of bracket required will be determined by the installation, but Fig. 3.63 gives some examples of brackets which may be modified to suit particular circumstances.

Fixing methods

PVC insulated and sheathed wiring systems are usually fixed with PVC clips in order to comply with IET Regulations 522.8.3 and 4 and shown in Table 3.7. The clips are supplied in various sizes to hold the cable firmly, and the fixing nail is a hardened masonry nail. Figure 3.64 shows a cable clip of this type. The use of a masonry nail means that fixings to wood, plaster, brick or stone can be made with equal ease.

When heavier cables, trunking, conduit or luminaires have to be fixed, a screw fixing is often needed. Wood screws may be screwed directly into wood but when fixing to brick, stone, plaster or concrete it is necessary to drill a hole in the masonry material, which is then plugged with a material to which the screw can be secured.

Plastic plugs

A **plastic plug** is made of a hollow plastic tube split up to half its length to allow for expansion. Each size of plastic plug is colour coded to match a wood screw size.

A hole is drilled into the masonry, using a masonry drill of the same diameter, to the length of the plastic plug (see Fig. 3.65). The plastic plug is inserted into the hole and tapped home until it is level with the surface of the masonry. Finally, the

Definition

A *plastic plug* is made of a hollow plastic tube split up to half its length to allow for expansion. Each size of plastic plug is colour coded to match a wood screw size.

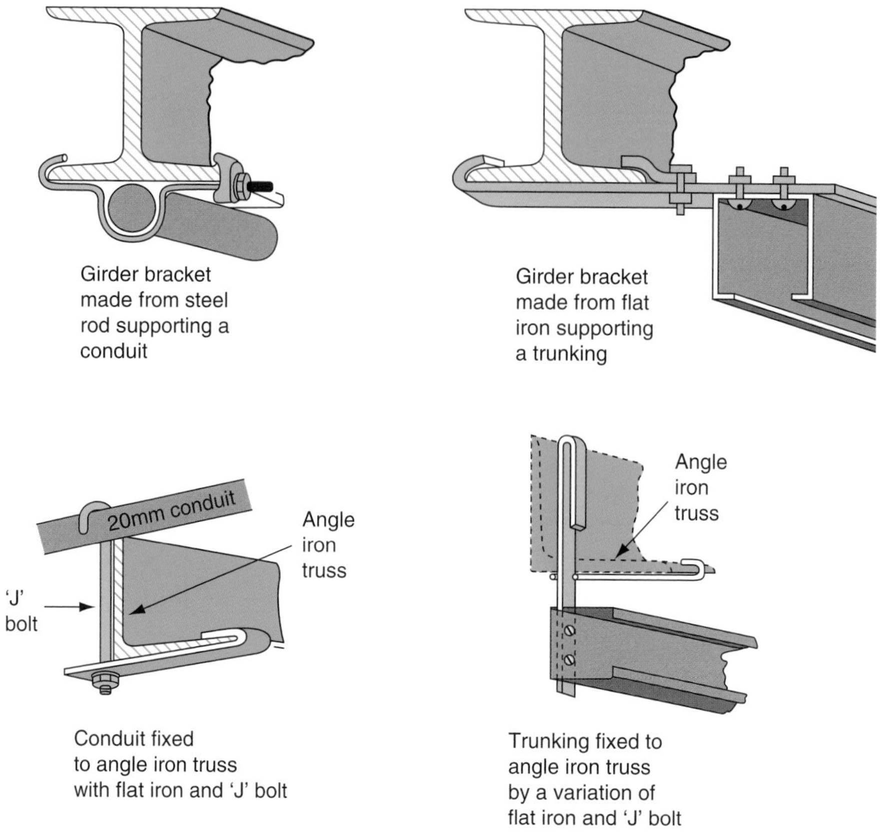

Figure 3.63 Bracket supports for conduits and trunking.

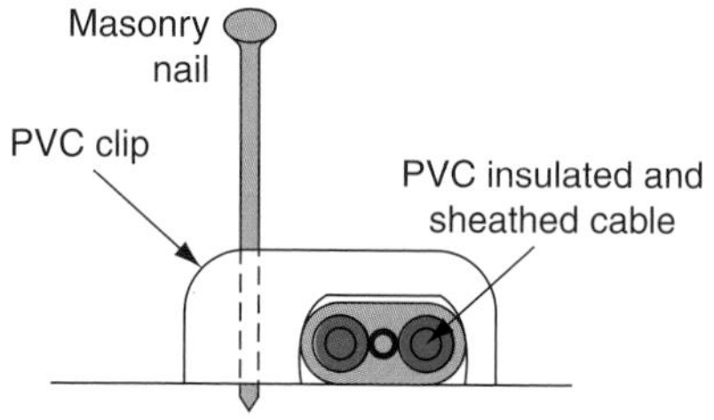

Figure 3.64 PVC insulated and sheathed cable clip.

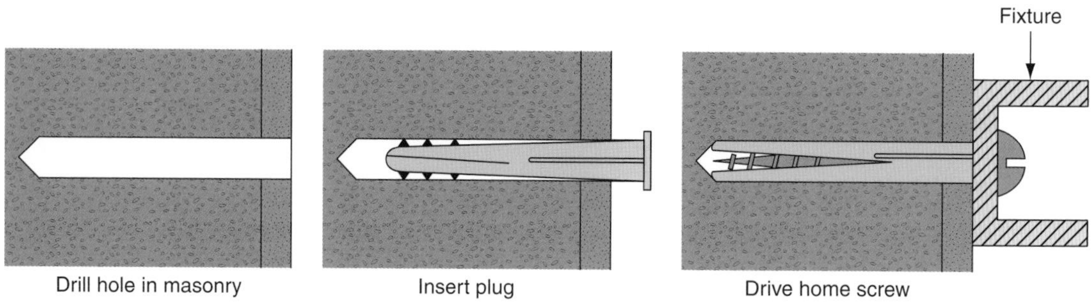

Figure 3.65 Screw fixing to plastic plug.

Visit the companion website for more on this topic.

fixing screw is driven into the plastic plug until it becomes tight and the fixture is secure.

Definition

The most well-known *expansion bolt* is made by Rawlbolt and consists of a split iron shell held together by a steel ferrule at one end and a spring wire clip at the other end. Tightening the bolt draws up an expanding bolt inside the split iron shell, forcing the iron to expand and grip the masonry. Rawlbolts are for heavy-duty masonry fixings.

Expansion bolts

The most well-known **expansion bolt** is made by Rawlbolt and consists of a split iron shell held together by a steel ferrule at one end and a spring wire clip at the other end. Tightening the bolt draws up an expanding bolt inside the split iron shell, forcing the iron to expand and grip the masonry. Rawlbolts are for heavy-duty masonry fixings (see Fig. 3.66).

A hole is drilled in the masonry to take the iron shell and ferrule. The iron shell is inserted with the spring wire clip end first so that the ferrule is at the outer surface. The bolt is passed through the fixture, located in the expanding nut and tightened until the fixing becomes secure.

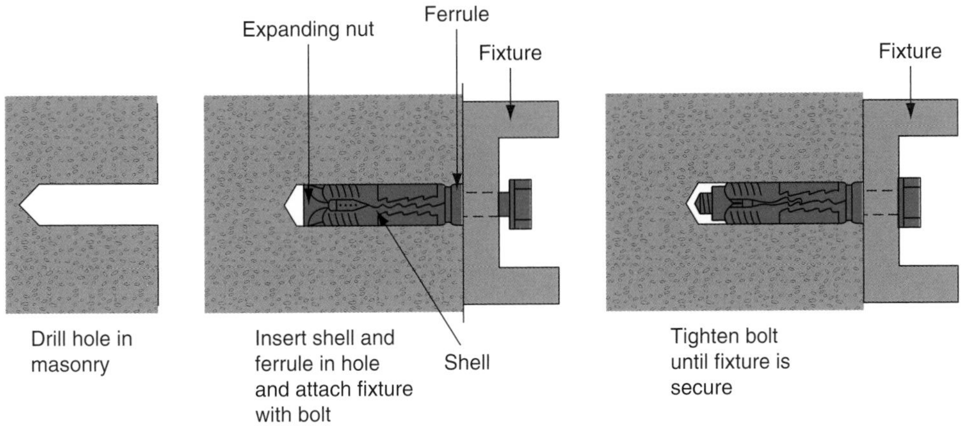

Figure 3.66 Expansion bolt fixing.

Visit the companion website for more on this topic.

Spring toggle bolts

A **spring toggle bolt** provides one method of fixing to hollow partition walls which are usually faced with plasterboard and a plaster skimming. Plasterboard and plaster wall or ceiling surfaces are not strong enough to support a load fixed directly into the plasterboard, but the spring toggle spreads the load over a larger area, making the fixing suitable for light loads (see Fig. 3.67).

A hole is drilled through the plasterboard and into the cavity. The toggle bolt is passed through the fixture and the toggle wings screwed into the bolt. The toggle wings are compressed and passed through the hole in the plasterboard and into the cavity where they spring apart and rest on the cavity side of the plasterboard. The bolt is tightened until the fixing becomes firm. The bolt of the spring toggle cannot be removed after fixing without the loss of the toggle wings. If it becomes necessary to remove and refix the fixture a new toggle bolt will have to be used.

Top tip

Inserting rawl plugs

When placing rawl plugs into masonry or tile surfaces, ensure the plug is pushed several millimetres past the surface.

Definition

A *spring toggle bolt* provides one method of fixing to hollow partition walls which are usually faced with plasterboard and a plaster skimming.

Special situations or hazardous electrical locations

All electrical installations and installed equipment must be safe to use and free from the dangers of electric shock, but some installations or locations require special consideration because of the inherent dangers of the installed conditions. The danger may arise because of the corrosive or explosive nature of the atmosphere, because the installation must be used in damp or low-temperature conditions or because there is a need to provide additional mechanical protection for the electrical system. Part 7 of the IET Regulations deals with these special installations or locations. In this section we will consider some of the installations which require special consideration.

Safety first

Fire

If your electrical activities cause damage to the fabric of the building then:

- the openings remaining must be sealed according to the degree of fire resistance demonstrated by the original building material.

IET Regulation 527.2

Antistatic precautions

Static electricity is a voltage charge which builds up to many thousands of volts between two surfaces when they rub together. A dangerous situation occurs when the static charge has built up to a potential capable of striking an arc through the air gap separating the two surfaces.

Static charges build up in a thunderstorm. A lightning strike is the discharge of the thunder-cloud, which might have built up to a voltage of 100 MV, to the

Definition

Static electricity is a voltage charge which builds up to many thousands of volts between two surfaces when they rub together.

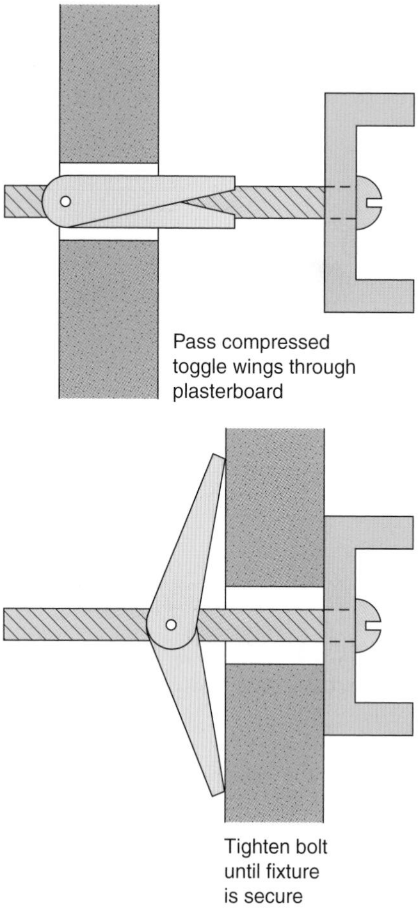

Figure 3.67 Spring toggle bolt fixing.

general mass of earth which is at 0 V. Lightning discharge currents are of the order of 20 kA, hence the need for lightning conductors on vulnerable buildings in order to discharge the energy safely.

Static charge builds up between any two insulating surfaces or between an insulating surface and a conducting surface, but it is not apparent between two conducting surfaces.

Definition

Static charge builds up between any two insulating surfaces or between an insulating surface and a conducting surface, but it is not apparent between two conducting surfaces.

A motor car moving through the air builds up a static charge which sometimes gives the occupants a minor shock as they step out and touch the door handle.

Static electricity also builds up in modern offices and similar carpeted areas. The combination of synthetic carpets, man-made footwear materials and dry, air-conditioned buildings contributes to the creation of static electrical charges building up on people moving about these buildings. Individuals only become aware of the charge if they touch earthed metalwork, such as a stair banister rail, before the static electricity has been dissipated. The effect is a sensation of momentary shock.

The precautions against this problem include using floor coverings that have been 'treated' to increase their conductivity or that contain a proportion of natural fibres that have the same effect. The wearing of leather-soled footwear also reduces the likelihood of a static charge persisting, as does increasing the humidity of the air in the building.

A nylon overall and nylon bed sheets build up static charge which is the cause of the 'crackle' when you shake them. Many flammable liquids have the same properties as insulators, and therefore liquids, gases, powders and paints moving through pipes build up a static charge.

Petrol pumps, operating theatre oxygen masks and car spray booths are particularly at risk because a spark in these situations may ignite the flammable liquid, powder or gas.

So how do we protect ourselves against the risks associated with static electrical charges? I said earlier that a buildup of static charge is not apparent between two conducting surfaces, and this gives a clue to the solution. Bonding surfaces together with protective bonding conductors prevents a buildup of static electricity between the surfaces. If we use large-diameter pipes, we reduce the flow rates of liquids and powders and, therefore, we reduce the buildup of static charge. Hospitals use cotton sheets and uniforms, and use protective bonding extensively in operating theatres. Rubber, which contains a proportion of graphite, is used to manufacture antistatic trolley wheels and surgeons' boots. Rubber constructed in this manner enables any buildup of static charge to 'leak' away. Increasing humidity also reduces static charge because the water droplets carry away the static charge, thus removing the hazard.

Avoiding shutdown of IT equipment

Every modern office now contains computers, and many systems are linked together or networked. Most computer systems are sensitive to variations or distortions in the mains supply and many computers incorporate filters which produce high-protective conductor currents of around 2 or 3 mA. This is clearly not a fault current, but is typical of the current which flows in the circuit protective conductor of IT equipment under normal operating conditions. IET Regulations 543.7.1 and 4 deal with the earthing requirements for the installation of equipment having high-protective conductor currents. IET Guidance Note

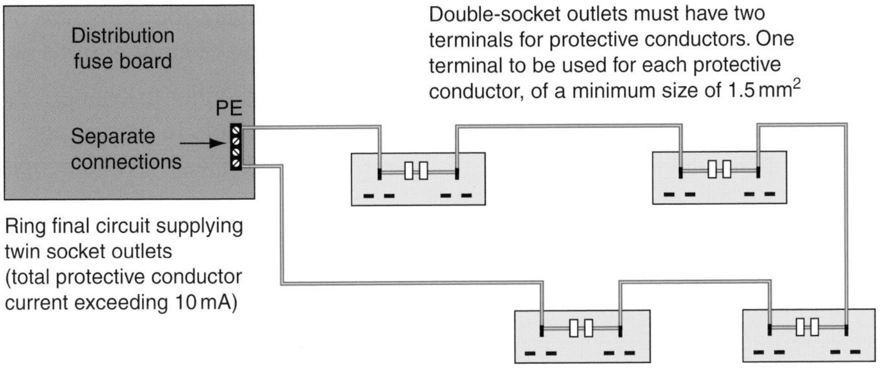

Figure 3.68 Recommended method of connecting IT equipment to socket outlets.

7 recommends that IT equipment should be connected to double sockets as shown in Fig. 3.68.

Surge protection

A transient over-voltage, or surge, is a voltage spike of very short duration. It may be caused by a lightning strike or a switching action on the system. It sends a large voltage spike for a few microseconds down the mains supply which is sufficient to damage sensitive electronic equipment. Supplies to computer circuits must be 'clean' and 'secure'. Mainframe computers and computer networks are sensitive to mains distortion or interference, which is referred to as 'noise'. Noise is mostly caused by switching an inductive circuit which causes a transient spike, or by brush gear making contact with the commutator segments of an electric motor. These distortions in the mains supply can cause computers to 'crash' or provoke errors, and are shown in Fig. 3.69.

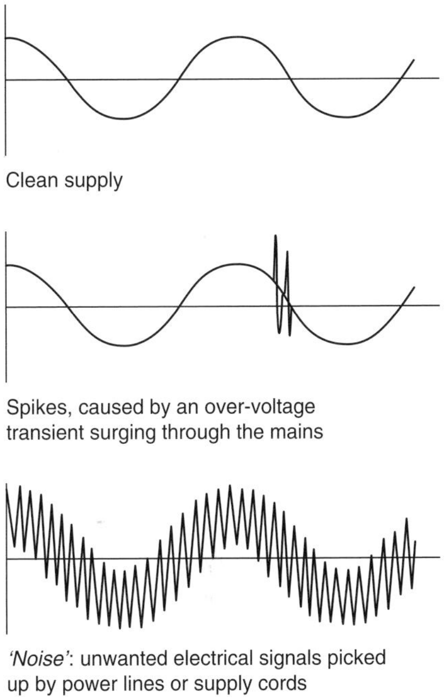

Figure 3.69 Distortions in the a.c. mains supply.

To avoid this, a 'clean' supply is required for the computer network. This can be provided by taking the ring or radial circuits for the computer supplies from a point as close as possible to the intake position of the electrical supply to the building. A clean earth can also be taken from this point, which is usually one core of the cable and not the armour of an SWA cable, and distributed around the final wiring circuit. Alternatively, the computer supply can be cleaned by means of a filter such as that shown in Fig. 3.70.

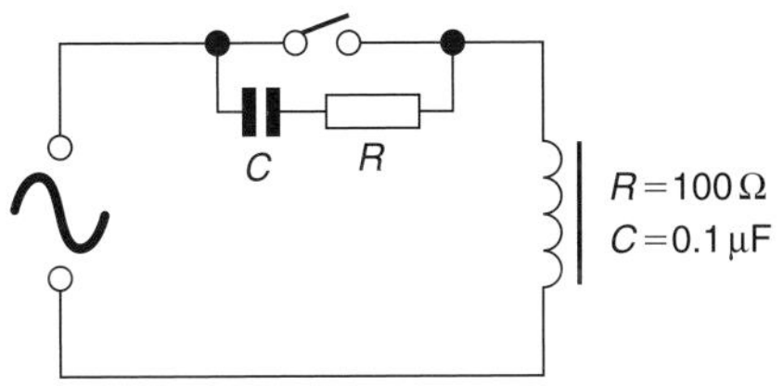

Figure 3.70 A simple noise suppressor.

A surge protection device is a device intended to limit transient over-voltages, and to divert damaging surge currents away from sensitive equipment.

Section 534 of the IET Regulations contains the requirements for the installation of surge protective devices (SPDs) to limit transient over-voltages where required by section 443 of BS 7671:2008 or where specified by the designer.

Secure supplies

The mains electrical supply in the United Kingdom is extremely reliable and secure. However, the loss of supply to a mainframe computer or computer network for even a second can cause the system to 'crash', and hours, or even days, of work can be lost.

One solution to this problem is to protect 'precious' software systems with an uninterruptible power supply (UPS). A **UPS** is essentially a battery supply electronically modified to provide a clean and secure a.c. supply. The UPS is plugged into the mains supply and the computer systems are plugged into the UPS.

Definition

A UPS is essentially a battery supply electronically modified to provide a clean and secure a.c. supply.

A UPS to protect a small network of, say, six PCs is physically about the size of one PC hard drive and is usually placed under or at the side of an operator's desk.

It is best to dedicate a ring or radial circuit to the UPS and either to connect the computer equipment permanently or use non-standard outlets to discourage the unauthorized use and overloading of these special supplies by, for example, kettles.

Finally, remember that most premises these days contain some computer equipment and systems. Electricians intending to isolate supplies for testing or modification should *first check and then check again* before they finally isolate the supply in order to avoid loss or damage to computer systems.

Damage to electronic devices by 'over-voltage'

The use of electronic circuits in all types of electrical equipment has increased considerably over recent years. Electronic circuits and components can now be found in leisure goods, domestic appliances, motor starting and control circuits, discharge lighting, emergency lighting, alarm circuits and special-effects lighting systems. All electronic circuits are low-voltage circuits carrying very small currents.

Electrical installation circuits usually carry in excess of 1A and often carry hundreds of amperes. Electronic circuits operate in the milliampere or even microampere range. The test instruments used on electronic circuits must have a *high impedance* so that they do not damage the circuit when connected to take readings.

The use of an insulation resistance test as described by the IET Regulations must be avoided with any electronic equipment. The working voltage of this instrument can cause total devastation to modern electronic equipment. When carrying out an insulation resistance test as part of the prescribed series of tests for an electrical installation, all electronic equipment must first be disconnected or damage will result.

Any resistance measurements made on electronic circuits must be achieved with a battery-operated ohmmeter, with a high impedance to avoid damaging the electronic components.

Risks associated with high-frequency or large capacitive circuits

Definition

Induction heating processes use high-frequency power to provide very focused heating in industrial processes.

Induction heating processes use high-frequency power to provide very focused heating in industrial processes.

The induction heater consists of a coil of large cross-section. The work-piece or object to be heated is usually made of ferrous metal and is placed inside the coil. When the supply is switched on, eddy currents are induced into the work-piece and it heats up very quickly so that little heat is lost to conduction and convection.

The frequency and size of the current in the coil determines where the heat is concentrated in the work-piece:

- the higher the current, the greater is the surface penetration;
- the longer the current is applied, the deeper the penetration;
- the higher the frequency, the less is the depth of heat penetration.

For shallow penetration, high-frequency, high-current, short time application is typically used for tool tempering. Other applications are brazing and soldering industrial and domestic gas boiler parts.

When these machines are not working they look very harmless but when they are working they operate very quietly and there is no indication of the intense heat that they are capable of producing. Domestic and commercial microwave ovens operate at high frequency. The combination of risks of high frequency and intense heating means that before any maintenance, repair work or testing is carried out, the machine must first be securely isolated and no one should work on these machines unless they have received additional training to enable them to do so safely.

Industrial wiring systems are very inductive because they contain many inductive machines and circuits, such as electric motors, transformers, welding plants and discharge lighting. The inductive nature of the industrial load causes the current to lag behind the voltage and creates a bad power factor. Power factor is the percentage of current in an alternating current circuit that can be used as energy for the intended purpose. A power factor of, say, 0.7 indicates that 70% of the current supplied is usefully employed by the industrial equipment.

An inductive circuit, such as that produced by an electric motor, induces an electromagnetic force which opposes the applied voltage and causes the current waveform to lag the voltage waveform. Magnetic energy is stored up in the load during one half cycle and returned to the circuit in the next half cycle. If a capacitive circuit is employed, the current leads the voltage since the capacitor stores energy as the current rises and discharges it as the current falls. So here we have the idea of a solution to the problem of a bad power factor created by inductive industrial loads. Power factor and power factor improvement is discussed below.

The **power factor** at which consumers take their electricity from the local electricity supply authority is outside the control of the supply authority. The power factor of the consumer is governed entirely by the electrical plant and equipment that is installed and operated within the consumer's buildings. Domestic consumers do not have a bad power factor because they use very little inductive equipment. Most of the domestic load is neutral and at unity power factor.

Definition

The *power factor* of the consumer is governed entirely by the electrical plant and equipment that is installed and operated within the consumer's buildings.

Electricity supply authorities discourage the use of equipment and installations with a low power factor because they absorb part of the capacity of the generating plant and the distribution network to no useful effect. They, therefore, penalize industrial consumers with a bad power factor through a maximum demand tariff, metered at the consumer's intake position. If the power factor falls below a datum level of between 0.85 and 0.9 then extra charges are incurred. In this way industrial consumers are encouraged to improve their power factor.

Power factor improvement of most industrial loads is achieved by connecting capacitors to either:

- individual items of equipment;
- banks of capacitors which may be connected to the main busbars of the installation at the intake position.

Definition

Power factor improvement of most industrial loads is achieved by connecting capacitors to either:

- individual items of equipment;
- banks of capacitors.

The method used will depend upon the utilization of the installed equipment by the industrial or commercial consumer. If the load is constant then banks of capacitors at the mains intake position would be indicated. If the load is variable then power factor correction equipment could be installed adjacent to the machine or piece of equipment concerned.

Power factor correction by capacitors is the most popular method because of the following:

- they require no maintenance;
- capacitors are flexible and additional units may be installed as an installation or system is extended;
- capacitors may be installed adjacent to individual pieces of equipment or at the mains intake position. Equipment may be placed on the floor or fixed high up and out of the way.

Capacitors store charge and must be disconnected before the installation or equipment is tested in accordance with Section 6 of the IET Regulations BS 7671.

Small power factor correction capacitors, as used in discharge lighting, often incorporate a high-value resistor connected across the mains terminals. This discharges the capacitor safely when not in use. Banks of larger capacity capacitors may require discharging to make them safe when not in use. To discharge a capacitor safely and responsibly it must be discharged slowly over a period in excess of five 'time-constants' through a suitable discharge resistor.

Capacitors and time constants are discussed in Chapter 3 of *Advanced Electrical Installation Work* 8th Edition by Trevor Linsley under the subheading 'Energy stored in a capacitor'.

Presence of storage batteries

Definition

Since an emergency occurring in a building may cause the mains supply to fail, the *emergency lighting* should be supplied from a source which is independent from the main supply.

Since an emergency occurring in a building may cause the mains supply to fail, the **emergency lighting** should be supplied from a source which is independent from the main supply. A battery's ability to provide its output instantly makes it a very satisfactory source of standby power. In most commercial, industrial and public service buildings housing essential services, the alternative power supply would be from batteries, but generators may also be used. Generators can have a large capacity and duration, but a major disadvantage is the delay of time while the generator runs up to speed and takes over the load. In some premises a delay of more than five seconds is considered unacceptable, and in these cases a battery supply is required to supply the load until the generator can take over.

The emergency lighting supply must have an adequate capacity and rating for the specified duration of time (IET Regulation 313.2). BS 5266 and BS EN 1838 state that after a battery is discharged by being called into operation for its specified duration of time, it should be capable of once again operating for the specified duration of time following a recharge period of no longer than 24 hours. The duration of time for which the emergency lighting should operate will be specified by a statutory authority but is normally one to three hours. The British Standard states that escape lighting should operate for a minimum of one hour. Standby lighting operation time will depend upon financial considerations and the importance of continuing the process or activity within the premises after the mains supply has failed.

The contractor installing the emergency lighting should provide a test facility which is simple to operate and secure against unauthorized interference. The emergency lighting installation must be segregated completely from any other wiring, so that a fault on the main electrical installation cannot damage the emergency lighting installation (IET Regulation 528.1).

The batteries used for the emergency supply should be suitable for this purpose. Motor vehicle batteries are not suitable for emergency lighting applications, except in the starter system of motor-driven generators. The fuel supply to a motor-driven generator should be checked. The battery room of a central battery system must be well ventilated and, in the case of a motor-driven generator, adequately heated to ensure rapid starting in cold weather.

The British Standard recommends that the full load should be carried by the emergency supply for at least one hour in every six months. After testing, the emergency system must be carefully restored to its normal operative state. A record should be kept of each item of equipment and the date of each test by a qualified or responsible person. It may be necessary to produce the record as evidence of satisfactory compliance with statutory legislation to a duly authorized person.

Self-contained units are suitable for small installations of up to about 12 units. The batteries contained within these units should be replaced about every five years, or as recommended by the manufacturer.

Storage batteries are secondary cells. A secondary cell has the advantage of being rechargeable. If the cell is connected to a suitable electrical supply, electrical energy is stored on the plates of the cell as chemical energy. When the cell is connected to a load, the chemical energy is converted to electrical energy.

> **Definition**
>
> *Storage batteries* are secondary cells. A secondary cell has the advantage of being rechargeable. If the cell is connected to a suitable electrical supply, electrical energy is stored on the plates of the cell as chemical energy. When the cell is connected to a load, the chemical energy is converted to electrical energy.

A lead-acid cell is a secondary cell. Each cell delivers about 2 V, and when six cells are connected in series a 12 V battery is formed.

A lead-acid battery is constructed of lead plates which are deeply ribbed to give maximum surface area for a given weight of plate. The plates are assembled in groups, with insulating separators between them. The separators are made of a porous insulating material, such as wood or ebonite, and the whole assembly is immersed in a dilute sulphuric acid solution in a plastic container.

The capacity of a cell to store charge is a measure of the total quantity of electricity which it can cause to be displaced around a circuit after being fully charged. It is stated in ampere-hours, abbreviation Ah, and calculated at the 10-hour rate which is the steady load current which would completely discharge the battery in 10 hours. Therefore, a 50 Ah battery will provide a steady current of 5A for 10 hours.

Maintenance of lead-acid batteries

- The plates of the battery must always be covered by dilute sulphuric acid. If the level falls, it must be topped up with distilled water.
- Battery connections must always be tight and should be covered with a thin coat of petroleum jelly.
- The specific gravity or relative density of the battery gives the best indication of its state of charge. A discharged cell will have a specific gravity of 1.150, which will rise to 1.280 when fully charged. The specific gravity of a cell can be tested with a hydrometer.
- To maintain a battery in good condition it should be regularly trickle-charged. A rapid charge or discharge encourages the plates to buckle, and may cause permanent damage. Most batteries used for standby supplies today are equipped with constant voltage chargers. The principle of these is that after the battery has been discharged by it being called into operation, the terminal voltage will be depressed and this enables a relatively large current (1–5A) to flow from the charger to recharge the battery. As the battery becomes more fully charged its voltage will rise until it reaches the constant voltage level where the current output from the charger will drop until it is just sufficient to balance the battery's internal losses. The main advantage of this system is that the battery controls the amount of charge it receives and is

therefore automatically maintained in a fully charged condition without human intervention and without the use of any elaborate control circuitry.

- The room used to charge the emergency supply storage batteries must be well ventilated because the charged cell gives off hydrogen and oxygen, which are explosive in the correct proportions.

Working alone

Some working situations are so potentially hazardous that not only must PPE be worn but you must also never work alone and safe working procedures must be in place before your work begins to reduce the risk.

It is unsafe to work in isolation in the following situations:

- when working above ground;
- when working below ground;
- when working in confined spaces;
- when working close to unguarded machinery;
- when a fire risk exists;
- when working close to toxic or corrosive substances such as battery acid.

Refer back to Chapter 1 for more detail.

Hazardous area installations

The British Standards concerned with hazardous areas were first published in the 1920s and were concerned with the connection of electrical apparatus in the mining industry. Since those early days many national and international standards, as well as codes of practice, have been published to inform the manufacture, installation and maintenance of electrical equipment in all hazardous areas.

The relevant British Standards for Electrical Apparatus for Potentially Explosive Atmospheres are BS 5345, BS EN 60079 and BS EN 50014: 1998.

They define a hazardous area as 'any place in which an explosive atmosphere may occur in such quantity as to require special precautions to protect the safety of workers'. Clearly these regulations affect the petroleum industry, but they also apply to petrol filling stations.

Most flammable liquids only form an explosive mixture between certain concentration limits. Above and below this level of concentration the mix will not explode. The lowest temperature at which sufficient vapour is given off from a flammable substance to form an explosive gas–air mixture is called the **flashpoint**. A liquid which is safe at normal temperatures will require special consideration if heated to flashpoint. An area in which an explosive gas–air mixture is present is called a **hazardous area**, as defined by the British Standards, and any electrical apparatus or equipment within a hazardous area must be classified as flameproof.

Flameproof electrical equipment is constructed so that it can withstand an internal explosion of the gas for which it is certified, and prevent any spark or flame resulting from that explosion from leaking out and igniting the surrounding atmosphere. This is achieved by manufacturing flameproof equipment to a robust standard of construction. All access and connection points have wide machined flanges which damp the flame in its passage across the flange. Flanged surfaces are firmly bolted together with many recessed bolts, as shown in Fig. 3.71.

Definition

The lowest temperature at which sufficient vapour is given off from a flammable substance to form an explosive gas–air mixture is called the *flashpoint*.

Definition

An area in which an explosive gas–air mixture is present is called a *hazardous area*, as defined by the British Standards, and any electrical apparatus or equipment within a hazardous area must be classified as flameproof.

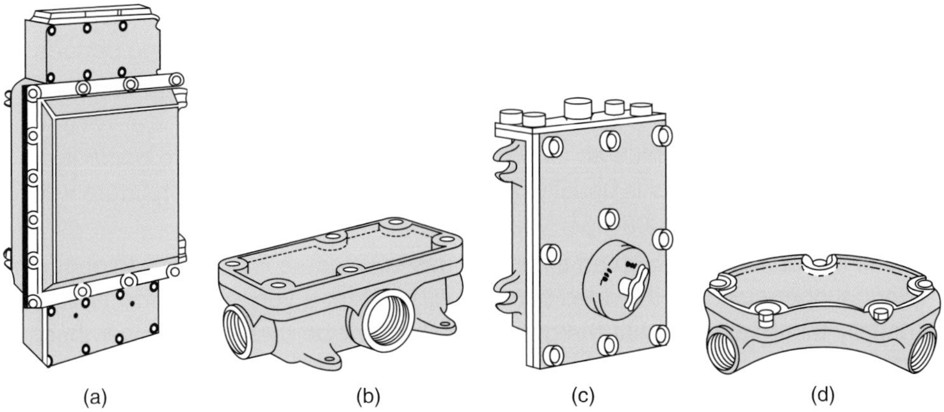

Figure 3.71 Flameproof fittings: (a) flameproof distribution board; (b) flameproof rectangular junction box; (c) double-pole switch; (d) flameproof inspection bend.

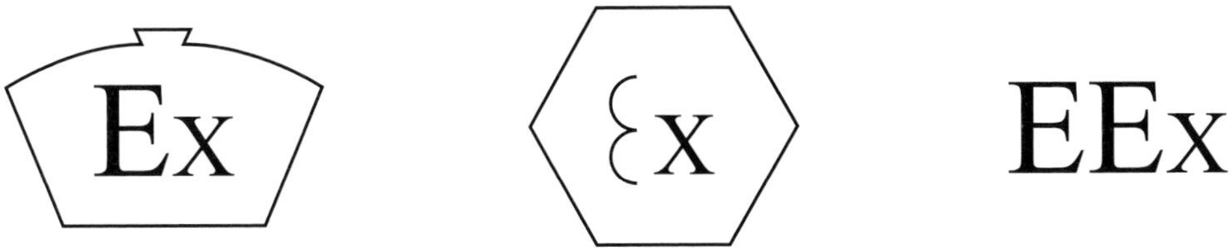

Figure 3.72 Flameproof equipment markings.

Wiring systems within a hazardous area must be to flameproof fittings using an appropriate method, such as:

- PVC cables encased in solid drawn heavy-gauge screwed steel conduit terminated at approved enclosures having wide flanges and bolted covers.
- Mineral insulated cables terminated into accessories with approved flameproof glands. These have a longer gland thread than normal MICC glands of the type shown in Fig. 3.39. Where the cable is laid underground, it must be protected by a PVC sheath and laid at a depth of not less than 500 mm.
- PVC armoured cables terminated into accessories with approved flameproof glands or any other wiring system which is approved by the British Standard. All certified flameproof enclosures will be marked Ex, indicating that they are suitable for potentially explosive situations, or EEx, where equipment is certified to the harmonized European Standard. All the equipment used in a flameproof installation must carry the appropriate markings, as shown in Fig. 3.72, if the integrity of the wiring system is to be maintained.

Flammable and explosive installations are to be found in the petroleum and chemical industries, which are classified as group II industries. Mining is classified as group I and receives special consideration from the Mining Regulations because of the extreme hazards of working underground. Petrol filling pumps must be wired and controlled by flameproof equipment to meet the requirements of the Petroleum Regulation Act 1928 and 1936 and any local licensing laws concerning the keeping and dispensing of petroleum spirit.

Hazardous area classification

The British Standard divides the risk associated with inflammable gases and vapours into three classes or zones.

1 Zone 0 is the most hazardous, and is defined as a zone or area in which an explosive gas–air mixture is *continuously present* or present for long periods. ('Long periods' is usually taken to mean that the gas–air mixture will be present for longer than 1000 h per year.)
2 Zone 1 is an area in which an explosive gas–air mixture is *likely to occur* in normal operation. (This is usually taken to mean that the gas–air mixture will be present for up to 1000 h per year.)
3 Zone 2 is an area in which an explosive gas–air mixture is *not likely* to occur in normal operation and if it does occur it will exist for a very short time. (This is usually taken to mean that the gas–air mixture will be present for fewer than 10 h per year.)

If an area is not classified as zone 0, 1 or 2, then it is deemed to be non-hazardous, so that normal industrial electrical equipment may be used.

The electrical equipment used in zone 2 will contain a minimum amount of protection. For example, normal sockets and switches cannot be installed in a zone 2 area, but oil-filled radiators may be installed if they are directly connected and controlled from outside the area. Electrical equipment in this area should be marked Ex'o' for oil-immersed or Ex'p' for powder-filled.

In zone 1 all electrical equipment must be flameproof, as shown in Fig. 3.71, and marked Ex'd' to indicate a flameproof enclosure.

Ordinary electrical equipment cannot be installed in zone 0, even when it is flameproof protected. However, many chemical and oil-processing plants are entirely dependent upon instrumentation and data transmission for their safe operation. Therefore, very low-power instrumentation and data-transmission circuits can be used in special circumstances, but the equipment must be *intrinsically safe*, and used in conjunction with a 'safety barrier' installed outside the hazardous area. Intrinsically safe equipment must be marked Ex'ia' or Ex's', specially certified for use in zone 0.

Intrinsic safety

Definition

An *intrinsically safe circuit* is one in which no spark or thermal effect is capable of causing ignition of a given explosive atmosphere.

By definition, an **intrinsically safe circuit** is one in which no spark or thermal effect is capable of causing ignition of a given explosive atmosphere. The intrinsic safety of the equipment in a hazardous area is assured by incorporating a Zener diode safety barrier into the control circuit such as that shown in Fig. 3.73. In normal operation, the voltage across a Zener diode is too low for it to conduct, but if a fault occurs, the voltage across Z_1 and Z_2 will rise, switching them on and blowing the protective fuse. Z_2 is included in the circuit as a 'back-up' in case the first Zener diode fails.

An intrinsically safe system, suitable for use in zone 0, is one in which *all* the equipment, apparatus and interconnecting wires and circuits are intrinsically safe.

Index of Protection (IP) BS EN 60529

IET Regulation 612.4.5 tells us that where barriers and enclosures have been installed to prevent direct contact with live parts, they must afford a degree of protection not less than IP2X and IP4X, but what does this mean?

The Index of Protection is a code which gives us a means of specifying the suitability of equipment for the environmental conditions in which it will be used.

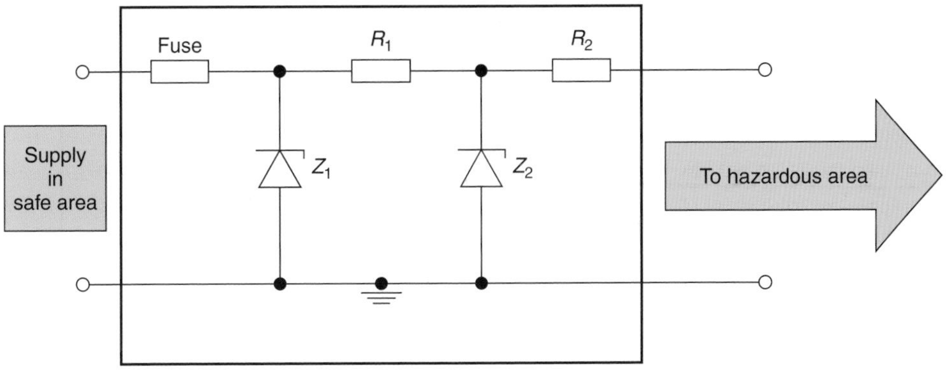

Figure 3.73 Zener safety barrier.

The tests to be carried out for the various degrees of protection are given in the British and European Standard BS EN 60529.

The code is written as IP (Index of Protection) followed by two numbers XX. The first number gives the degree of protection against the penetration of solid objects into the enclosure. The second number gives the degree of protection against water penetration. For example, a piece of equipment classified as IP45 will have barriers installed which prevent a 1 mm diameter rigid steel bar from making contact with live parts and be protected against the ingress of water from jets of water applied from any direction. Where a degree of protection is not specified, the number is replaced by an 'X' which simply means that the degree of protection is not specified although some protection may be afforded. The 'X' is used instead of '0', since '0' would indicate that no protection was given. The index of protection codes is shown in Fig. 3.74.

Appendix 5 of the IET Regulations identifies the required IP classification for electrical equipment being used in hazardous conditions and requiring water protection as follows:

IPX1 or IPX2 where water vapour occasionally condenses on electrical equipment.

IPX3 where sprayed water forms a continous film on the floor.

IPX4 where equipment may be subjected to splashed water (e.g. construction sites).

IPX5 where hosed water is regularly used (e.g. car washing).

IPX6 for seashore locations (e.g. marinas and piers).

IPX7 for locations which may become flooded, immersing equipment in water.

IPX8 electrical equipment is permanently immersed in water (e.g. swimming pools).

Appendix 5 of the IET Regulations also identifies the required IP classification to prevent dust or objects from penetrating electrical equipment as follows:

IP2X to protect against solid objects thicker than a finger, approximately 12 mm.

IP3X to prevent penetration by small objects, of which the smallest is 2.5 mm.

IP4X to prevent penetration by very small objects, of which the smallest is 1.0 mm.

IP5X where light dust penetration would not harm the electrical equipment.

IP6X where dust must not penetrate the equipment.

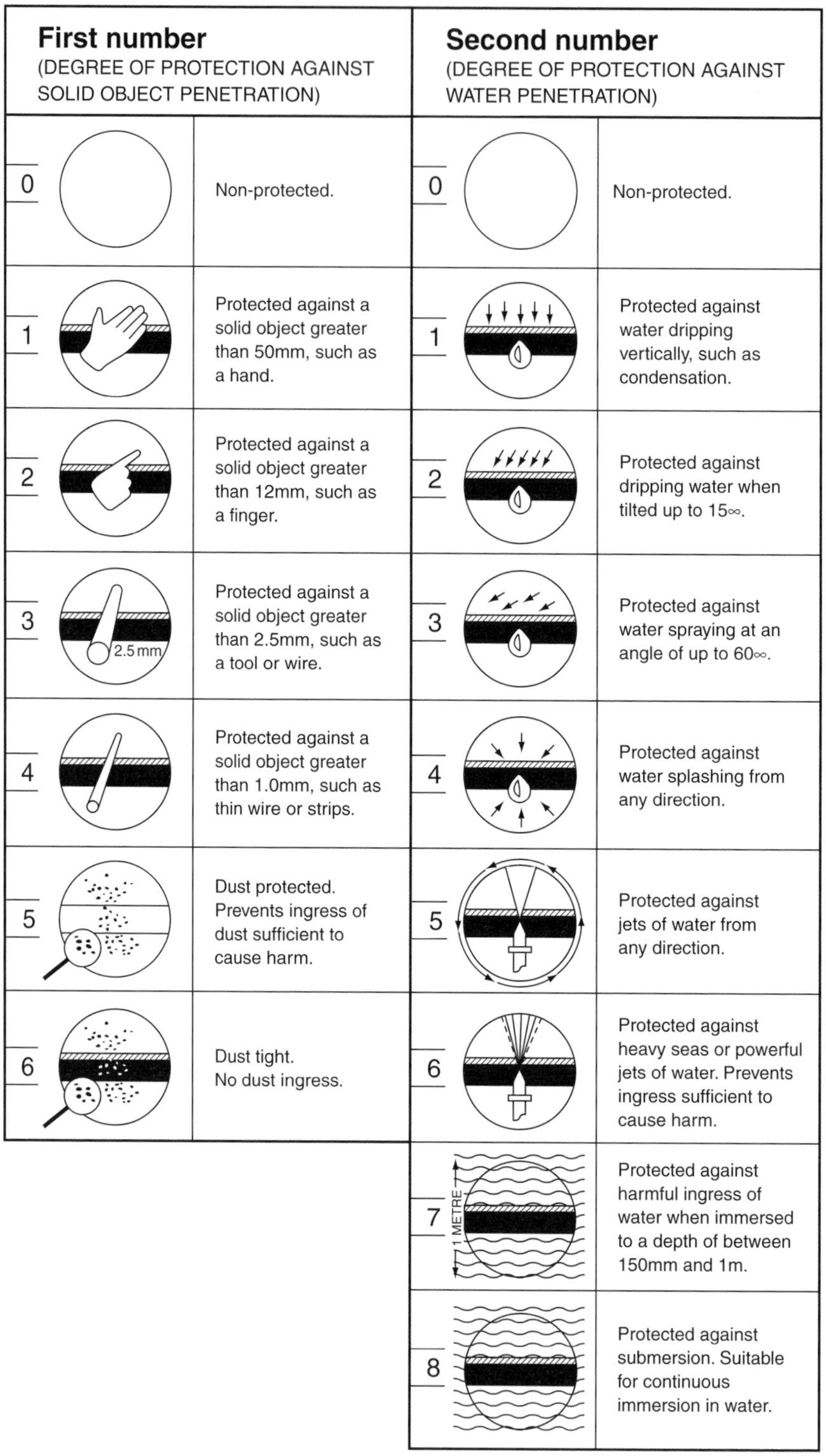

Figure 3.74 Index of protection codes.

Temporary construction site installations

Temporary electrical supplies provided on construction sites can save many man hours of labour by providing the energy required for fixed and portable tools and lighting which speeds up the completion of a project. However, construction sites are dangerous places and the temporary electrical supply which is installed

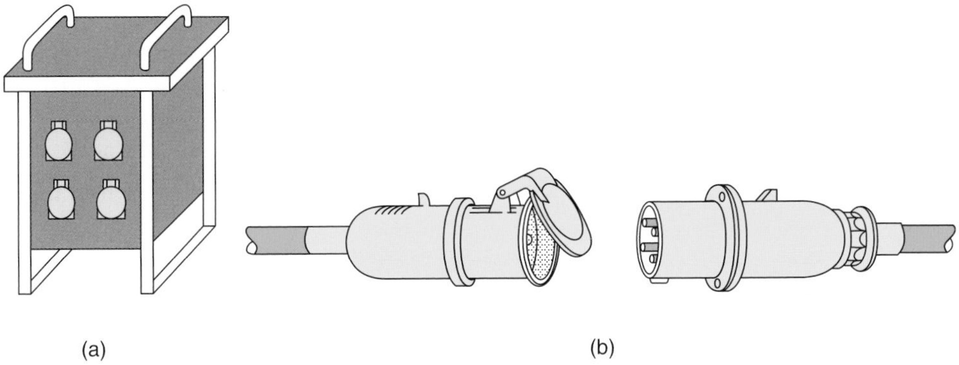

Figure 3.75 110 V distribution unit and cable connector, suitable for construction site electrical supplies: (a) reduced-voltage distribution unit incorporating industrial sockets to BS EN 60309; and (b) industrial plug and connector.

to assist the construction process must comply with all of the relevant wiring regulations for permanent installations (IET Regulation 110.1). All equipment must be of a robust construction in order to fulfil the on-site electrical requirements while being exposed to rough handling, vehicular nudging, the wind, rain and sun. All equipment socket outlets, plugs and couplers must be of the industrial type to BS EN 60439 and BS EN 60309 and specified by IET Regulation 704.511.1 as shown in Fig. 3.75.

Where an electrician is not permanently on-site, MCBs are preferred so that over-current protection devices can be safely reset by an unskilled person. The British Standards Code of Practice 1017, *The Distribution of Electricity on Construction and Building Sites*, advises that protection against earth faults may be obtained by first providing a low impedance path, so that overcurrent devices can operate quickly as described earlier in this chapter, and second by fitting an RCD in addition to the over-current protection device (IET Regulation 704.410.3.10). The seventeenth edition of the IET Regulations considers construction sites as being very special locations, devoting the whole of Section 704 to their requirements. A construction site installation should be tested and inspected in accordance with Part 6 of the IET Regulations every three months throughout the construction period.

The source of supply for the temporary installation may be from a petrol- or diesel-generating set or from the local supply company. When the local electricity company provides the supply, the incoming cable must be terminated in a non-combustible or not readily combustible waterproof and locked enclosure to prevent unauthorized access and provide metering arrangements.

IET Regulations 704.313, 704.410.3.10 and 411.8 tell us that reduced low voltage is strongly *preferred* for portable hand lamps and tools used on construction and demolition sites.

The distribution of electrical supplies on a construction site would typically be as follows:

- 400 V three-phase for supplies to major items of plant having a rating above 3.75 kW such as cranes and lifts. These supplies must be wired in armoured cables.
- 230 V single-phase for supplies to items of equipment which are robustly installed such as floodlighting towers, small hoists and site offices. These supplies must be wired in armoured cable unless run inside the site offices.

- 110V single-phase for supplies to all mobile hand tools and all mobile lighting equipment. The supply is usually provided by a reduced voltage distribution unit which incorporates splash-proof sockets fed from a centre-tapped 110V transformer. This arrangement limits the voltage to earth to 55V, which is recognized as safe in most locations. A 110V distribution unit is shown in Fig. 3.75. Edison screw lamps are used for 110V lighting supplies so that they are not interchangeable with 230V site office lamps.

Safety first

Construction sites
- low voltage or
- battery tools must be used.

There are occasions when even a 110V supply from a centre-tapped transformer is too high; for example, supplies to inspection lamps for use inside damp or confined places. In these circumstances a safety extra-low voltage (SELV) supply would be required.

Industrial plugs have a keyway which prevents a tool from one voltage from being connected to the socket outlet of a different voltage. They are also colour coded for easy identification as follows:

400V – red

230V – blue

110V – yellow

50V – white

25V – violet.

Agricultural and horticultural installations

Especially adverse installation conditions are to be encountered on agricultural and horticultural installations because of the presence of livestock, vermin, dampness, corrosive substances and mechanical damage. 17th edition of the IET Wiring Regulations considers these installations very special locations and has devoted the whole of Section 705 to their requirements. In situations accessible to livestock the electrical equipment should be of a type which is appropriate for the external influences likely to occur, and should have at least protection IP44; that is, protection against solid objects and water splashing from any direction (IET Regulation 705.512.2; see also Fig. 3.74).

In buildings intended for livestock, all fixed wiring systems must be inaccessible to the livestock and cables liable to be attacked by vermin must be suitably protected (IET Regulation 705.513.2).

PVC cables enclosed in heavy-duty PVC conduit are suitable for installations in most agricultural buildings. All exposed and extraneous metalwork must be provided with supplementary protective bonding in areas where livestock is kept (IET Regulation 705.415.2.1). In many situations, waterproof socket outlets to BS 196 must be installed. All socket outlet circuits must be protected by an RCD complying with the appropriate British Standard and the operating current must not exceed 30mA.

Cables buried on agricultural or horticultural land should be buried at a depth of not less than 600mm, or 1000mm where the ground may be cultivated, and the cable must have an armour sheath and be further protected by cable tiles. Overhead cables must be insulated and installed so that they are clear of farm machinery or placed at a minimum height of 6m to comply with IET Regulation 705.522.

Horses and cattle have a very low body resistance, which makes them susceptible to an electric shock at voltages lower than 25V r.m.s. The sensitivity of farm animals to electric shock means that they can be contained by an

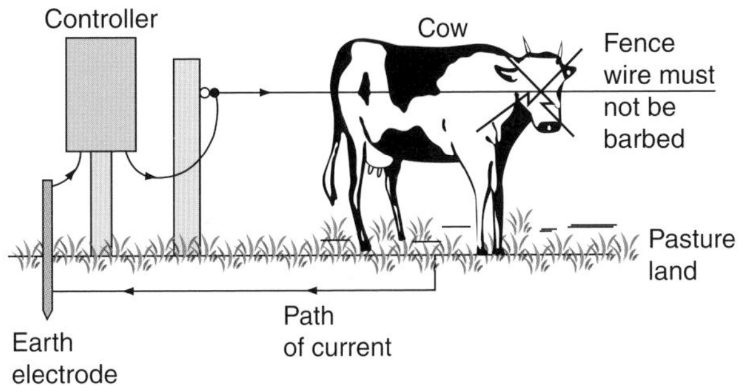

Figure 3.76 Farm animal control by electric fence.

electric fence. An animal touching the fence receives a short pulse of electricity which passes through the animal to the general mass of earth and back to an earth electrode sunk near the controller, as shown in Fig. 3.76. The pulses are generated by a capacitor–resistor circuit inside the controller which may be mains or battery operated. There must be no risk to any human coming into contact with the controller, which should be manufactured to BS 2632. The output voltage of the controller must not exceed 10 kV and the energy must not be greater than 5 J. The duration of the pulse must not be greater than 1.5 ms and the pulse must never have a frequency greater than one pulse per second. This shock level is very similar to that which can be experienced by touching a spark plug lead on a motor car. The energy levels are very low at 5 J. There are 3.6 million joules of energy in 1 kWh.

Earth electrodes connected to the earth terminal of an electric fence controller must be separate from the earthing system of any other circuit and should be situated outside the resistance area of any electrode used for protective earthing. The electric fence controller and the fence wire must be installed so that they do not come into contact with any power, telephone or radio systems, including poles. Agricultural and horticultural installations should be tested and inspected in accordance with Part 6 of the Wiring Regulations every three years.

Visit the companion website for more on this topic.

Caravans and caravan sites

The electrical installations on caravan sites, and within caravans, must comply in all respects with the wiring regulations for buildings. All the dangers which exist in buildings are present in and around caravans, including the added dangers associated with repeated connection and disconnection of the supply and the flexing of the caravan installation in a moving vehicle. 17th edition of the IET Regulations has devoted Section 721 to the electrical installation in caravans and motor caravans and Section 708 to caravan parks.

Safety first

Electric shock

- Animals and humans must be protected against electric shock.

Touring caravans must be supplied from a 16A industrial-type socket outlet adjacent to the caravan park pitch, similar to that shown in Fig. 3.75. Each socket outlet must be provided with individual over-current protection and an individual residual current circuit-breaker with a rated tripping current of 30 mA (IET Regulations 708.553.1.12 and 708.553.1.13). The distance between the caravan connector and the site socket outlet must not be more than 20 m (IET Regulation 708.530.3). These requirements are shown in Fig. 3.77.

The supply cables must be installed outside the pitch area and be buried at a depth of at least 0.6 m (IET Regulation 708.521.1.1).

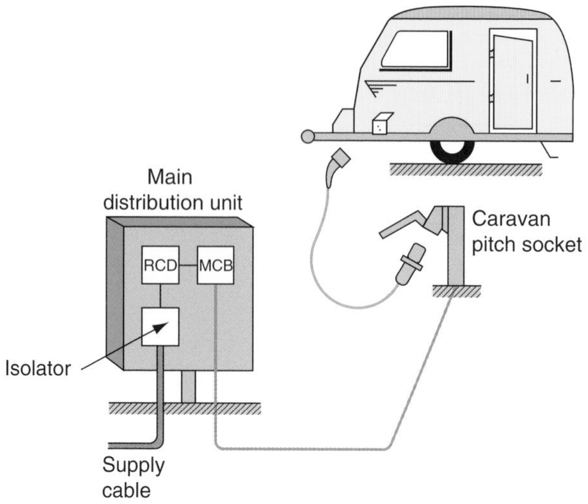

Figure 3.77 Electrical supplies to caravans.

Key fact

Caravans

- every caravan pitch must have at least one 16A industrial socket;
- each socket must have RCD and over-current protection.

IET Regulation 708.553

The caravan or motor caravan must be provided with a mains isolating switch and an RCD to break all live conductors (IET Regulation 721.411). An adjacent notice detailing how to connect and disconnect the supply safely must also be provided, as shown in IET Regulation 721.514. Electrical equipment must not be installed in fuel storage compartments (IET Regulation 721.528.3.4). Caravans flex when being towed, and therefore the installation must be wired in flexible or stranded conductors of at least 1.5 mm cross-section. The conductors must be supported on horizontal runs at least every 25 cm and the metalwork of the caravan and chassis must be bonded with 4.0 mm^2 cable.

The wiring of the extra-low-voltage battery supply must be run in such a way that it does not come into contact with the 230 V wiring system (IET Regulation 721.528.1).

The caravan should be connected to the pitch socket outlet by means of a flexible cable, no longer than 25 m and having a minimum cross-sectional area of 2.5 mm^2 or as detailed in Section 721 of the IET Regulations.

Because of the mobile nature of caravans it is recommended that the electrical installation be tested and inspected at intervals considered appropriate, preferably not less than once every three years and annually if the caravan is used frequently (IET Regulation 721.514.1).

Bathroom installations

Rooms containing a fixed bath-tub or shower basin are considered an area of increased shock risk and, therefore, additional regulations are specified in Section 701 of the IET Regulations. This is to reduce the risk of electric shock to people in circumstances where body resistance is lowered because of contact with water. The regulations can be summarized as follows:

- socket outlets must not be installed and no provision is made for connection of portable appliances unless the socket outlet can be fixed 3 metres horizontally beyond the zone 1 boundary within the bath or shower room (IET Regulation 701.512.3);
- only shaver sockets which comply with BS EN 60742, that is, those which contain an isolating transformer, may be installed in zone 2 or outside the zones in the bath or shower room (IET Regulation 701.512.3);

- all circuits serving a bath or shower room, that is, both power and lighting, must be additionally protected by an RCD having a rated maximum operating current of 30 mA (IET Regulation 701.411.3.3);
- there are restrictions as to where appliances, switchgear and wiring accessories may be installed (see 'Zones for bath and shower rooms' below).
- the 3rd Amendment to the 17th Edition of the IET Regulations (701.411.3.3) has introduced an additional requirement that any circuit 'passing through' the location must also be RCD protected.
- local supplementary protective bonding (IET Regulation 701.415.2) must be provided to all gas, water and central heating pipes in addition to metallic baths, *unless the following two requirements are both met*:
  (i) all bathroom circuits, both lighting and power, are protected by a 30 mA RCD in addition to an MCB or fuse; and
  (ii) the bath or shower is located in a building with main protective bonding in place, as described in Fig. 3.6 (IET Regulation 411.3.1.2).

Note: Local supplementary protective bonding may be an additional requirement of the Local Authority regulations in, for example, licensed premises, student accommodation and rented property.

Zones for bath and shower rooms

Locations that contain a bath or shower are divided into zones or separate areas, as shown in Fig. 3.78.

Zone 0 – the bath-tub or shower basin itself, which can contain water and is, therefore, the most dangerous zone.

Zone 1 – the next most dangerous zone in which people stand in water.

Zone 2 – the next most dangerous zone in which people might be in contact with water.

Outside zones – people are least likely to be in contact with water but are still in a potentially dangerous environment and the general IET Regulations apply.

- spaces under the bath which are accessible '*only with the use of a tool*' are outside zones
- spaces under the bath which are accessible '*without the use of a tool*' are zone 1.

Electrical equipment and accessories are restricted within the zones.

Zone 0 – being the most potentially dangerous zone, for all practical purposes no electrical equipment can be installed in this zone. However, the IET Regulations permit that where SELV fixed equipment with a rated voltage not exceeding 12 V a.c. cannot be located elsewhere, it may be installed in this zone (IET Regulation 701.55). The electrical equipment must have at least IPX7 protection against total immersion in water (IET Regulation 701.512.2).

Zone 1 – water heaters, showers and shower pumps and SELV fixed equipment may be installed in zone 1. The electrical equipment must have at least IPX4 protection against water splashing from any direction. If the electrical equipment may be exposed to water jets from, for example, commercial cleaning equipment, then the electrical equipment must have IPX5 protection. (The Index of Protection codes were discussed earlier, and are shown in Fig. 3.74.)

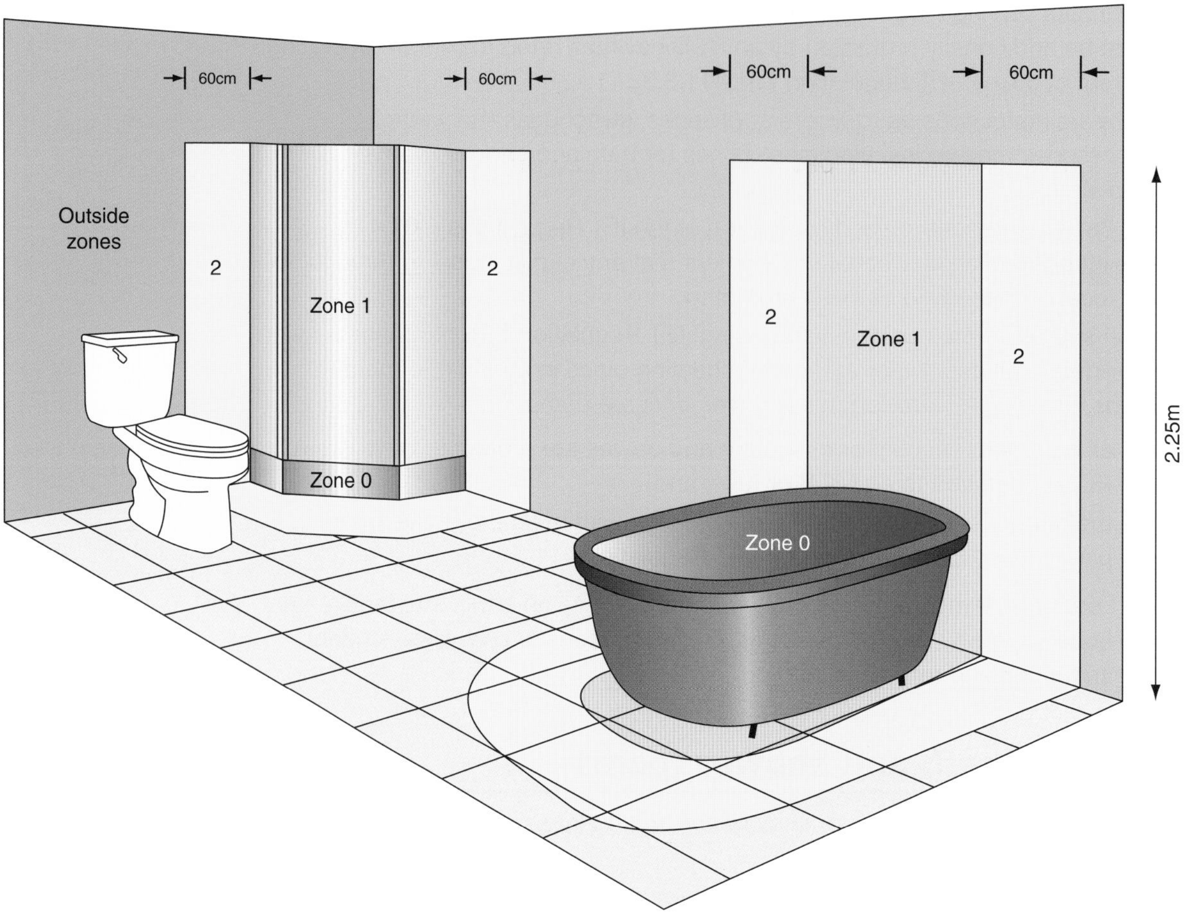

Figure 3.78 Bathroom zone dimensions. (Any window recess is zone 2.)

Zone 2 – luminaires and fans, and equipment from zone 1 plus shaver units to BS EN 60742 may be installed in zone 2. The electrical equipment must be suitable for installation in that zone according to the manufacturer's instructions and have at least IPX4 protection against splashing or IPX5 protection if commercial cleaning is anticipated.

Outside zones – appliances are allowed plus accessories except socket outlets unless the location containing the bath or shower is very big and the socket outlet can be installed at least 3 m horizontally beyond the zone 1 boundary (IET Regulation 701.512.3) and has additional RCD protection (IET Regulation 701.411.3.3).

Key fact

Bathrooms

- All bathroom circuits, both power and lighting, must have additional RCD protection.

If underfloor heating is installed in these areas it must have an overall earthed metallic grid or the heating cable must have an earthed metallic sheath, which is connected to the protective conductor of the supply circuit (IET Regulation 701.753).

Supplementary protective bonding

Modern plumbing methods make considerable use of non-metals (PTFE tape on joints, for example). Therefore, the metalwork of water and gas installations cannot be relied upon to be continuous throughout.

The IET Regulations describe the need to consider additional protection by supplementary protective bonding in situations where there is a high risk of electric shock (e.g. in kitchens and bathrooms) (IET Regulation 415.2).

In kitchens, supplementary protective bonding of hot and cold taps, sink tops and exposed water and gas pipes *is only required* if an earth continuity test proves that they are not already effectively and reliably connected to the main protective bonding, having negligible impedance, by the soldered pipe fittings of the installation. If the test proves unsatisfactory, the metalwork must be bonded using a single-core copper conductor with PVC green/yellow insulation, which will normally be 4 mm^2 for domestic installations but must comply with IET Regulation 543.1.1.

In rooms containing a fixed bath or shower, supplementary protective bonding conductors *must* be installed to reduce to a minimum the risk of an electric shock unless the following *two* conditions are met:

1. All bathroom circuits are protected by a fuse or MCB plus a 30 mA RCD.
2. The bathroom is located in a building with a main protective bonding system in place (IET Regulation 701.415.2). Such a system is shown in Fig. 3.6.

Supplementary protective bonding conductors in domestic premises will normally be of 4 mm^2 copper with PVC insulation to comply with IET Regulation 543.1.1 and must be connected between all exposed metalwork (e.g. between metal baths, bath and sink taps, shower fittings, metal waste pipes and radiators), as shown in Fig. 3.79.

The bonding connection must be made to a cleaned pipe, using a suitable bonding clip. Fixed at, or near, the connection must be a permanent label saying 'Safety electrical connection – do not remove' (IET Regulation 514.3.1), as shown in Fig. 3.80.

Communications

When we talk about good communications we are talking about transferring information from one person to another both quickly and accurately. We do this by talking to other people, looking at drawings and plans, and discussing these with colleagues from the same company and with other professionals who have an interest in the same project. The technical information used within our industry comes from many sources. The IET Regulations (BS 7671) is the 'electrician's bible' and forms the basis of all our electrical design calculations and installation methods. British Standards, European Harmonised Standards and Codes of Practice provide detailed information for every sector of the electrical industry, influencing all design-and-build considerations.

Sources of technical information

The equipment and accessories available to use in a specific situation can often be found in the very comprehensive manufacturers' catalogues and the catalogues of the major wholesalers that service the electrical industries.

All of this technical information may be distributed and retrieved by using:

- conventional drawings and diagrams which we will look at in more detail below;

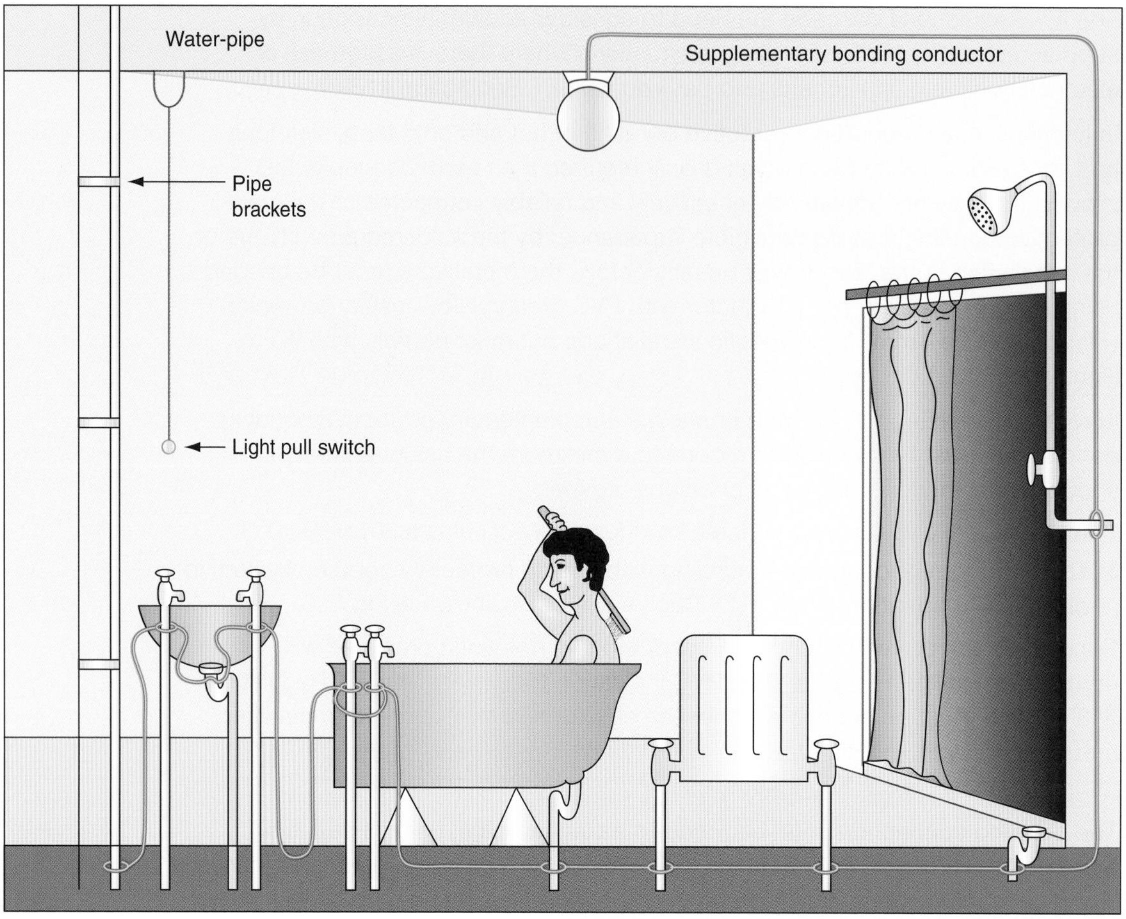

Figure 3.79 Supplementary protective bonding in bathrooms to metal pipework.

Figure 3.81 Legrand 'rapid' earth bonding clamp.

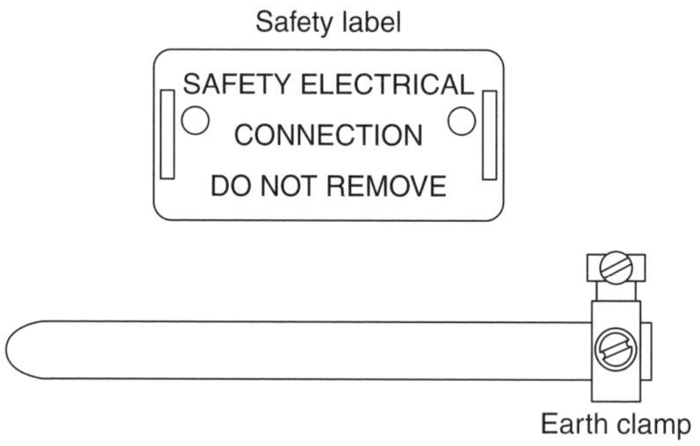

Figure 3.80 Typical earth bonding clamp.

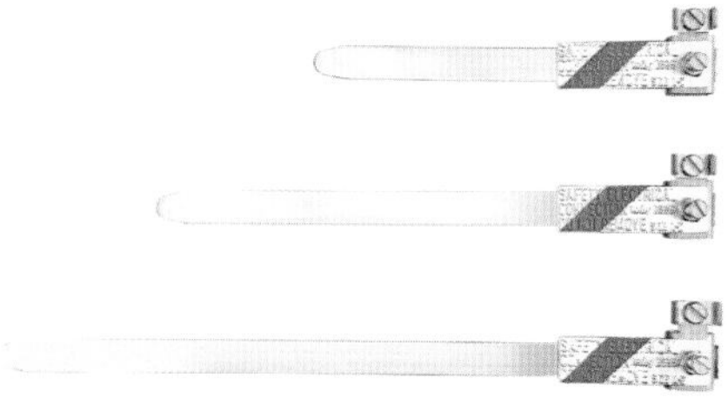

Figure 3.82 Legrand adjustable earth bonding strap.

- sketch drawing to illustrate an idea or the shape of, say, a bracket to hold a piece of electrical equipment;
- the Internet can be used to download British Standards and Codes of Practice;
- the Internet can also be used to download health and safety information from the health and safety executive at: www.gov.uk/hseor%20www.opsi.gov.uk;
- CDs, DVDs, USB memory sticks and email can be used to communicate and store information electronically;
- the facsimile (Fax) machine can be used to communicate with other busy professionals, information, say, about a project you are working on together.

If you are working at your company office with access to online computers, then technical information is only a fingertip or mouse click away. However, a construction site is a hostile environment for a laptop and so a hard copy of any data is preferable on-site.

We will look at the types of drawings and diagrams which we use within our industry to communicate technical information between colleagues and other professionals. The type of diagram to be used in any particular situation is the one which most clearly communicates the desired information.

Types of technical information

Technical information is communicated to electrical personnel in lots of different ways. It comes in the form of:

- *Specifications* – these are details of the client's requirements, usually drawn up by an architect. For example, the specification may give information about the type of wiring system to be employed or detail the types of luminaires or other equipment to be used.
- *Manufacturer's data* – if certain equipment is specified, let's say a particular type of luminaire or other piece of equipment, then the manufacturer's data sheet will give specific instructions for its assembly and fixing requirements. It is always good practice to read the data sheet before fitting the equipment. A copy of the data sheet should also be placed in the job file for the client to receive when the job is completed.
- *Reports and schedules* – a report is the written detail of something that has happened or the answer to a particular question asked by another professional person or the client. A **schedule** gives information about a programme or timetable of work: it might be a list or a chart giving details of when certain events will take place. For example, when the electricians will start to do the 'first fix' and how many days it will take.
- *User instructions* – give information about the operation of a piece of equipment. Manufacturers of equipment provide 'user instructions' and a copy should be placed in the job file for the client to receive when the project is handed over.
- *Job sheets and time sheets* – give 'on-site' information. Job sheets give information about what is to be done and are usually issued by a manager to an electrician. Time sheets are a record of where an individual worker has been spending his or her time, which job and for how long. This information is used to make up individual wages and to allocate company costs to a particular job. We will look at these again later under the subheading 'On-site documentation'.

Definition

A *schedule* is a list or programme of planned work.

Those who need or use technical information

Technical information is required by many of the professionals involved in any electrical activity, so who are the key people?

- The operative – in our case this will be the skilled electricians actually on-site, doing the job for the electrical company.
- The supervisor – he or she may have overall responsibility for a number of electricians on-site and will need the 'big picture'.
- The contractor – the main contractor takes on the responsibility of the whole project for the client. The main contractor may take on a subcontractor to

carry out some part of the whole project. On a large construction site the electrical contractor is usually the subcontractor.
- The site agent – he or she will be responsible for the smooth running of the whole project and for bringing the contract to a conclusion on schedule and within budget. The site agent may be nominated by the architect.
- The customer or client – they also are the people ordering the work to be done. They will pay the final bill that pays everyone's wages.

Drawings and diagrams

So that the electrician will know where to install the sockets, lights and equipment they will probably be provided with a site plan or layout drawing by the architect or main contractor.

Site plans or layout drawings

These are scale drawings based upon the architect's site plan of the building and show the position of the electrical equipment which is to be installed. The electrical equipment is identified by a graphical symbol. The standard symbols used by the electrical contracting industry are those recommended by the British Standard EN 60617, *Graphical Symbols for Electrical Power, Telecommunications and Electronic Diagrams*. Some of the more common electrical installation symbols are given in Fig. 3.83.

The site plan or layout drawing will be drawn to a scale, smaller than the actual size of the building, so to find the actual measurement you must measure the distance on the drawing and multiply by the scale.

For example, if the site plan is drawn to a scale of 1:100, then 10 mm on the site plan represents 1 m measured in the building.

The layout drawing or site plan of a small domestic extension is shown in Fig. 3.84. It can be seen that the mains intake position, probably a consumer unit, is situated in the store-room which also contains one light controlled by a switch at the door. The bathroom contains one lighting point controlled by a one-way pull switch at the door. The kitchen has two doors and a switch is installed at each door to control the fluorescent luminaire. There are also three double sockets situated around the kitchen. The sitting-room has a two-way switch at each door controlling the centre lighting point. Two wall lights with built-in switches are to be wired, one at each side of the window. Two double sockets and one switched socket are also to be installed in the sitting-room. The bedroom has two lighting points controlled independently by two one-way switches at the door. The wiring diagrams and installation procedures for all these circuits will be looked at in more detail shortly.

Try this

Drawings

The next time you are on-site ask your supervisor to show you the site plans. Ask them:

- How does the scale work?
- Can you put names to the equipment represented by British Standard symbols?

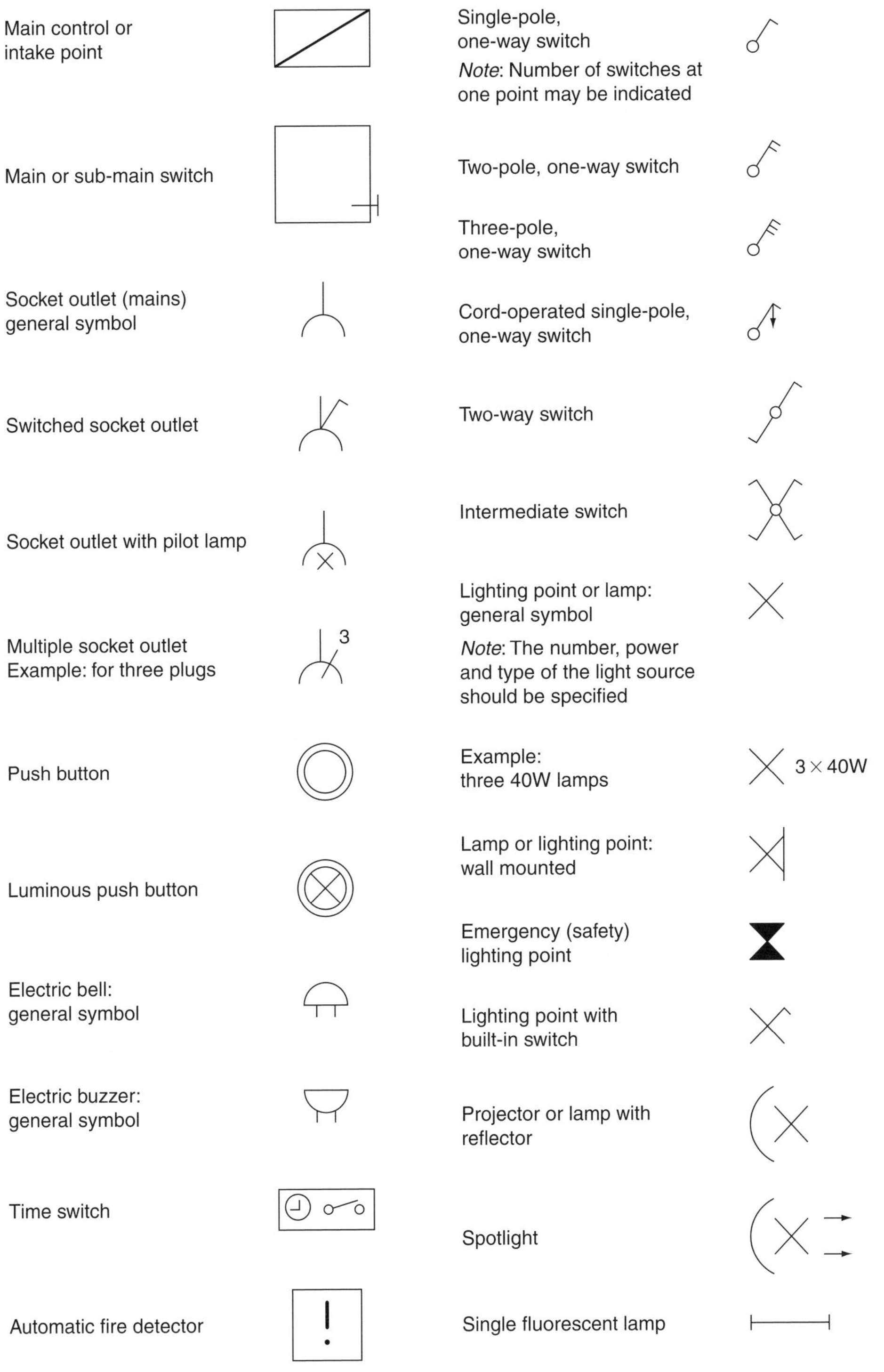

Figure 3.83 Some BS EN 60617 electrical installation symbols.

As-fitted drawings

When the installation is completed a set of drawings should be produced which indicate the final positions of all the electrical equipment. As the building and electrical installation progresses, it is sometimes necessary to modify the positions of equipment indicated on the layout drawing because, for example,

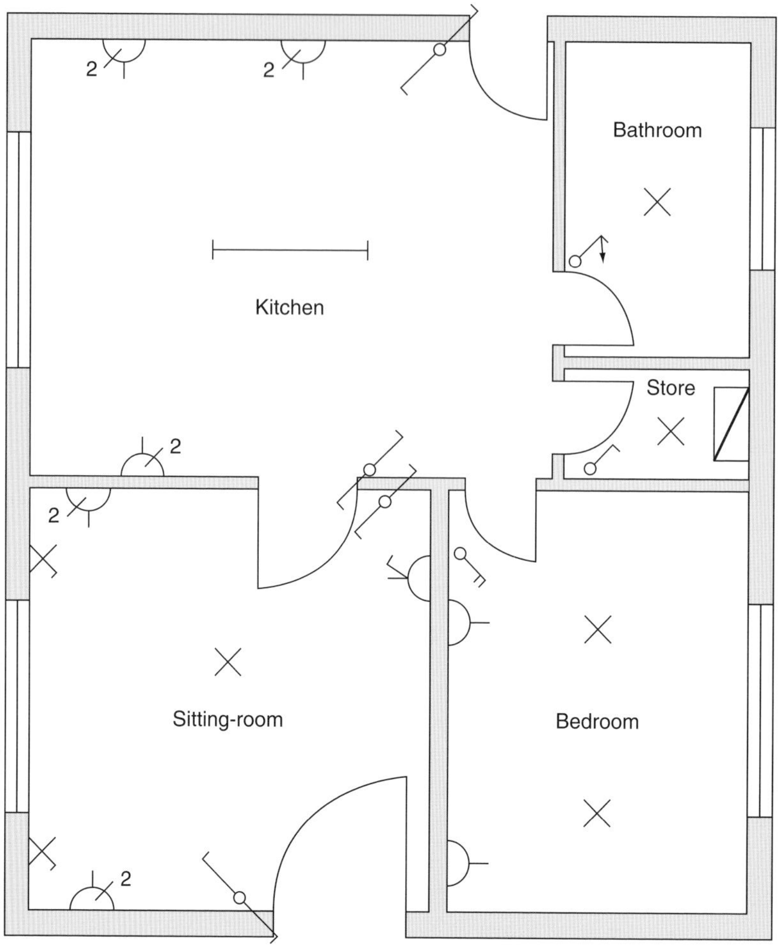

Figure 3.84 Layout drawing or site plan of a small electrical installation.

the position of a doorway has been changed. The layout drawings or site plans indicate the original intentions for the position of equipment, while the 'as-fitted' drawing indicates the actual positions of equipment upon completion of the contract.

Try this

Drawings

Take a moment to clarify the difference between:

- layout drawings; and
- as-fitted drawings.

Detail drawings and assembly drawings

These are additional drawings produced by the architect to clarify some point of detail. For example, a drawing might be produced to give a fuller description of a suspended ceiling arrangement or the assembly arrangements of the metalwork for the suspended ceiling.

Location drawings

Location drawings identify the place where something is located. It might be the position of the manhole covers giving access to the drains. It might be the position of all water stop taps or the position of the emergency lighting fittings. This type of information may be placed on a blank copy of the architect's site plan or on a supplementary drawing.

Distribution cable route plans

On large installations there may be more than one position for the electrical supplies. Distribution cables may radiate from the site of the electrical mains intake position to other sub-mains positions. The site of the sub-mains and the route taken by the distribution cables may be shown on a blank copy of the architect's site plan or on the electrician's 'as-fitted' drawings.

Block diagrams

A block diagram is a very simple diagram in which the various items or pieces of equipment are represented by a square or rectangular box. The purpose of the block diagram is to show how the components of the circuit relate to each other and, therefore, the individual circuit connections are not shown. Figure 3.85 shows the block diagram of a space-heating control system.

Wiring diagrams

A wiring diagram or connection diagram shows the detailed connections between components or items of equipment. They do not indicate how a piece of equipment or circuit works. The purpose of a wiring diagram is to help someone with the actual connection of the circuit conductors. Figure 3.86 shows the wiring diagram for a space-heating control system and Fig. 3.87 to 3.90 the wiring diagrams for a one-way, two-way and intermediate switch control of a light.

Circuit diagrams

A circuit diagram shows most clearly how a circuit works. All the essential parts and connections are represented by their graphical symbols. The purpose

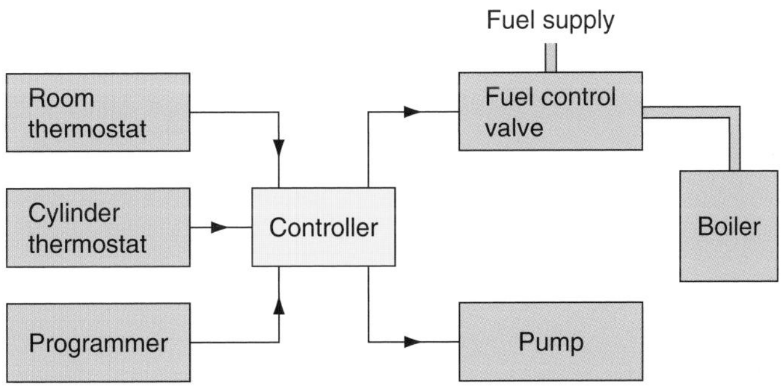

Figure 3.85 Block diagram – space-heating control system (Honeywell Y. Plan).

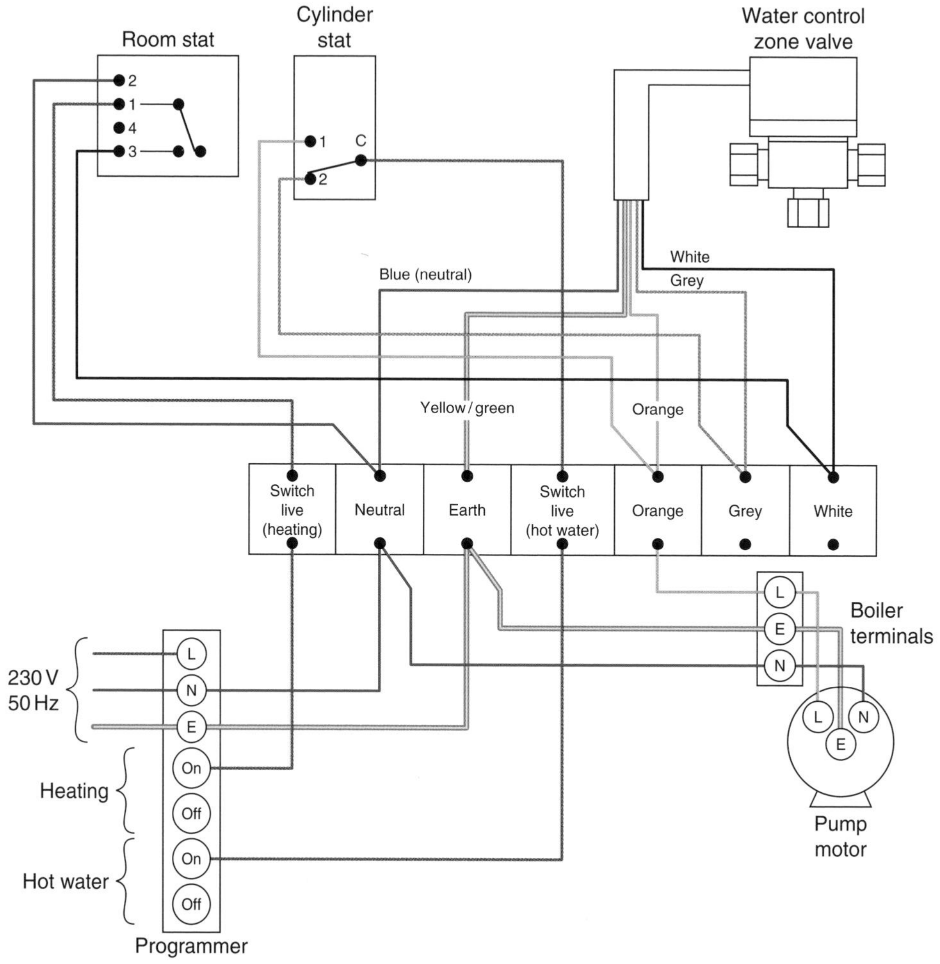

Figure 3.86 Wiring diagram – space-heating control system (Honeywell Y. Plan).

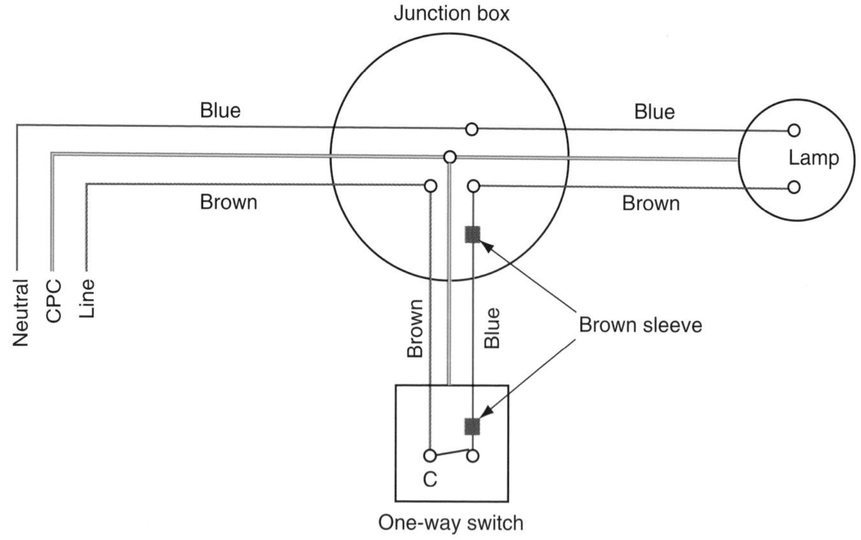

Figure 3.87 Wiring diagram of one-way switch control.

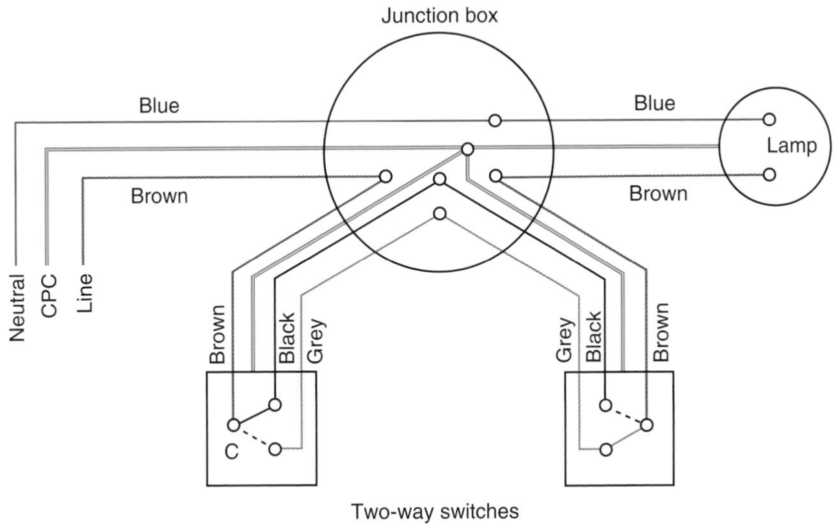

Figure 3.88 Wiring diagram of one-way converted to two-way switch control.

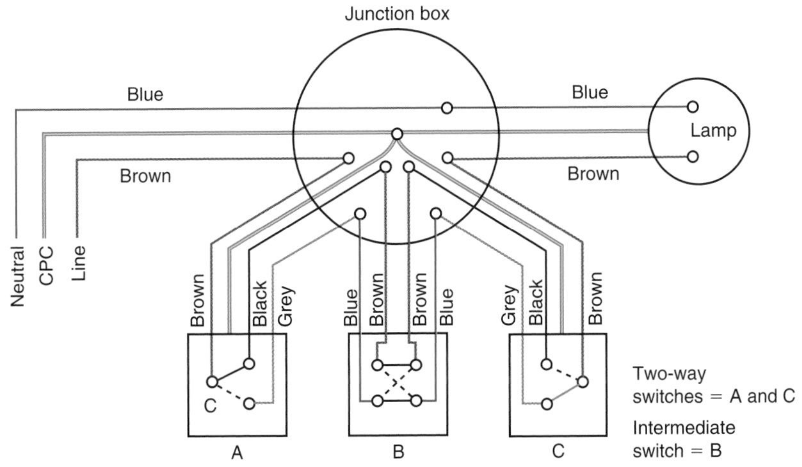

Figure 3.89 Wiring diagram of intermediate switch control.

of a circuit diagram is to help our understanding of the circuit. It will be laid out as clearly as possible, without regard to the physical layout of the actual components and, therefore, it may not indicate the most convenient way to wire the circuit. Figure 3.91 shows the circuit diagram of our same space-heating control system.

Schematic diagrams

A schematic diagram is a diagram in outline of, for example, a motor starter circuit. It uses graphical symbols to indicate the interrelationship of the electrical elements in a circuit. These help us to understand the working operation of the circuit but are not helpful in showing us how to wire the components. An electrical schematic diagram looks very like a circuit diagram. Figure 3.92 shows a schematic diagram.

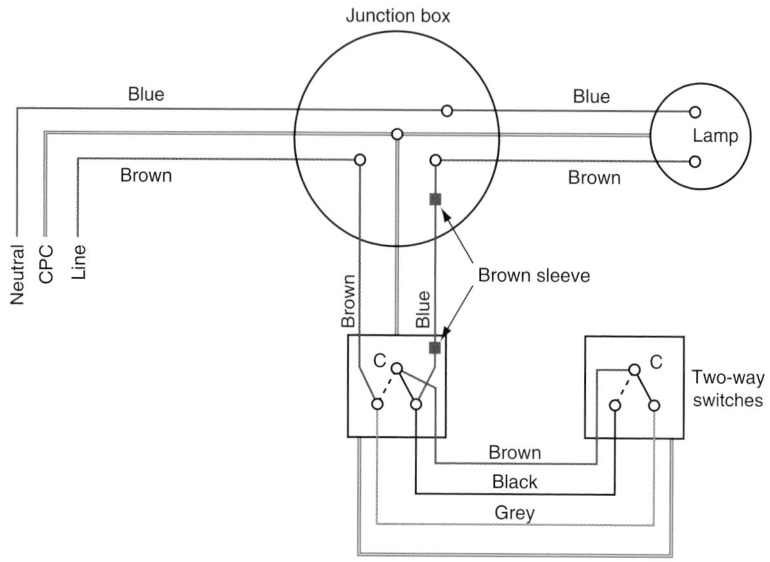

Figure 3.90 Wiring diagram of one-way converted to two-way switch control.

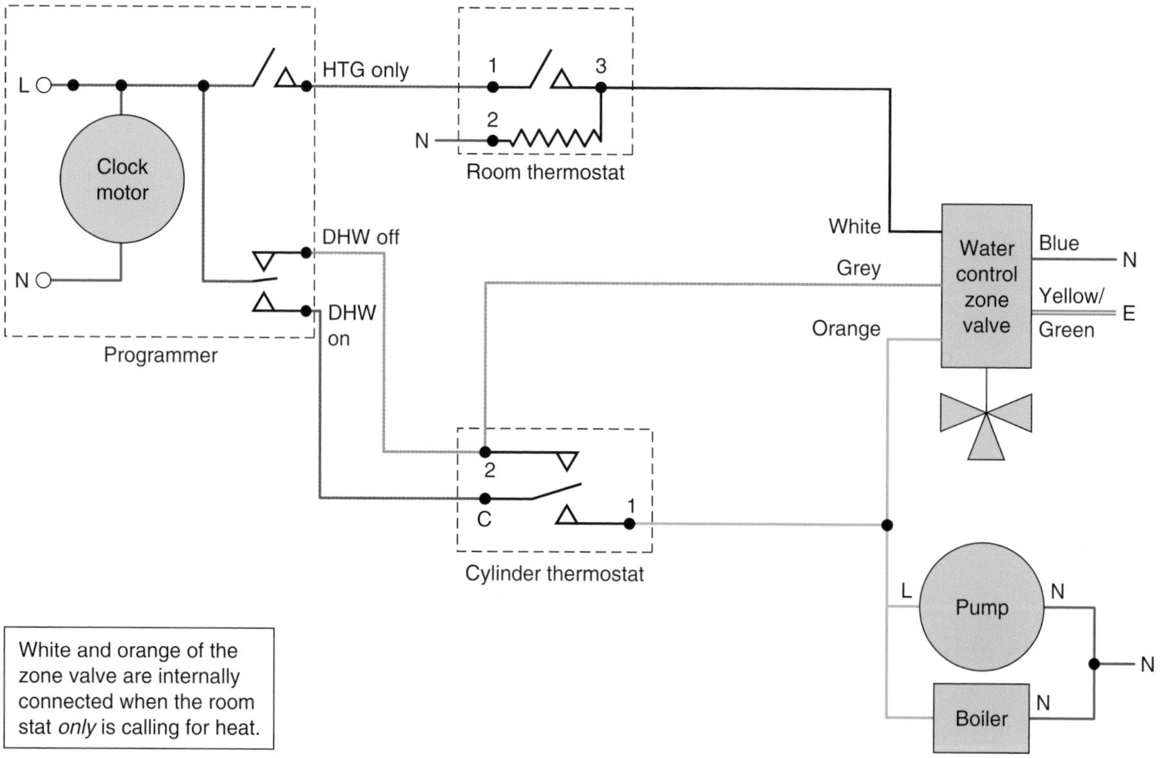

Figure 3.91 Circuit diagram – space-heating control system (Honey Well Y. Plan).

Freehand working diagrams

Freehand working drawings or sketches are another important way in which we communicate our ideas. The drawings of the spring toggle bolt in Fig. 3.67 were done from freehand sketches. A freehand sketch may be done as an initial draft of an idea before a full working drawing is made. It is often much easier to

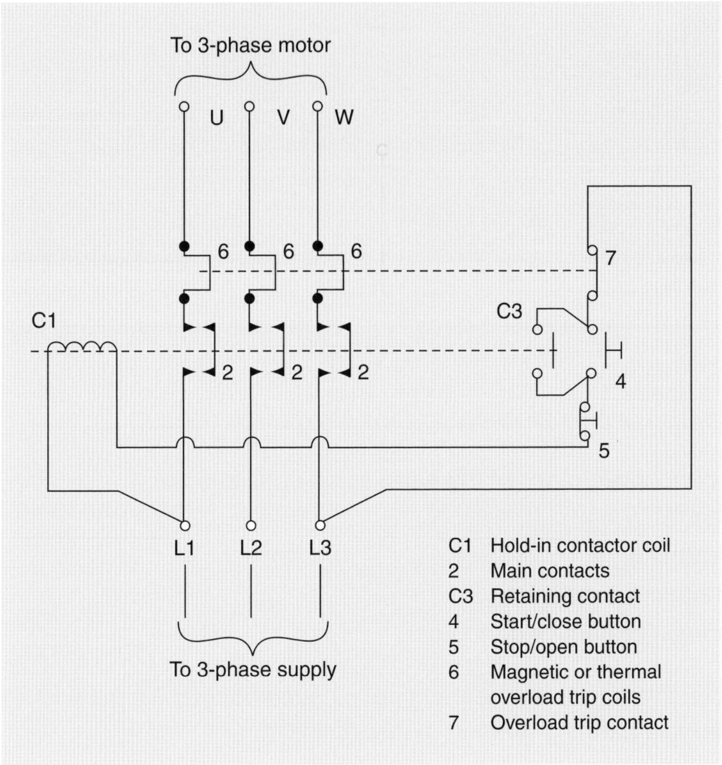

Figure 3.92 Schematic diagram – DOL motor starter.

produce a sketch of your ideas or intentions than to describe them or produce a list of instructions.

To convey the message or information clearly it is better to make your sketch large rather than too small. It should also contain all the dimensions necessary to indicate clearly the size of the finished object depicted by the sketch.

All drawings and communications must be aimed at satisfying the client's wishes for the project. It is the client who will pay the final bill which, in turn, pays your wages. The detailed arrangements of what must be done to meet the client's wishes are contained in the client's specification documents and all your company's efforts must be directed at meeting the whole specification, but no more.

Check your understanding

When you have completed the questions, check out the answers at the back of the book.

Note: more than one multiple-choice answer may be correct.

1 'The conductive mass of the earth' is one definition of:
   a. earth
   b. earthing
   c. bonding conductor
   d. circuit protective conductor.

2 A protective conductor connecting exposed conductive parts of equipment to the main earthing terminal is one definition of:
   a. earth
   b. earthing
   c. bonding conductor
   d. circuit protective conductor.

3 A protective conductor connecting exposed and extraneous parts together is one definition of:
   a. earth
   b. earthing
   c. bonding conductor
   d. circuit protective conductor.

4 The act of connecting the exposed conductive parts of the installation to the main earthing terminal of the installation is called:
   a. earth
   b. earthing
   c. bonding conductor
   d. circuit protective conductor.

5 The act of linking together the exposed and extraneous metal parts is called:
   a. earthing
   b. bonding
   c. basic protection
   d. fault protection.

6 The protection provided by insulating live parts is called:
   a. extraneous conductive parts
   b. basic protection

c. exposed conductive parts
d. fault protection.

7. The metalwork of the electrical installation is called:
    a. extraneous conductive parts
    b. basic protection
    c. exposed conductive parts
    d. fault protection.

8 The metalwork of the building and other service pipes is called:
    a. extraneous conductive parts
    b. basic protection
    c. exposed conductive parts
    d. fault protection.

9 The protection provided by equipotential bonding and automatic disconnection of the supply is called:
    a. extraneous conductive parts
    b. basic protection
    c. exposed conductive parts
    d. fault protection.

10 Cutting off the electrical supply in order to ensure the safety of those working on the equipment is one definition of:
    a. basic protection
    b. fault protection
    c. protective bonding
    d. isolation switching.

11 A current which exceeds the rated current in an otherwise healthy circuit is one definition of:
    a. fault protection
    b. an over-current
    c. an overload current
    d. a short-circuit current.

12 The weakest link in the circuit designed to melt when an over-current flows is one definition of:
    a. fault protection
    b. a circuit protective conductor
    c. a fuse
    d. a consumer unit.

13 According to IET Regulation 411.3.2.2 all final circuits not exceeding 32 A in a building supplied with a 230 V TN supply shall have a maximum disconnection time not exceeding:
    a. 0.2 s
    b. 0.4 s
    c. 5.0 s
    d. unlimited.

14 To ensure the effective operation of the over-current protective devices, the earth fault loop path must have:
a. a 230V supply
b. a very low resistance
c. fuses or MCBs in the live conductor
d. a very high resistance.

15 Electricity is generated in large commercial power-stations at:
a. 230V
b. 400V
c. 25kV
d. 132kV.

16 Transmission of electricity on the National Grid network takes place at:
a. 230V
b. 400V
c. 25kV
d. 132kV.

17 Electricity is transmitted at very high voltages because for a given power the:
a. current is reduced
b. current is increased
c. losses are reduced
d. losses are increased.

18 Electricity is distributed from the local substations by underground cables at:
a. 230V
b. 400V
c. 25kV
d. 132kV.

19 A load connected to the three phases of a star-connected three-phase four-wire supply system from the local substation would have a voltage of:
a. 230V
b. 400V
c. 25kV
d. 132kV.

20 A load connected to phase and neutral of a star-connected three-phase four-wire supply system from the local substation would have a voltage of:
a. 230V
b. 400V
c. 25kV
d. 132kV.

21 The phase voltage of a star-connected load is 100V. The line voltage will be:
a. 57.73V
b. 100V
c. 173.2V
d. 230V.

22 The phase voltage of a delta-connected load is 100 V. The line voltage will be:

a. 57.73 V
b. 100 V
c. 173.2 V
d. 230 V.

23 The phase current of a star-connected load is 100 A. The line current will be:

a. 57.73 A
b. 100 A
c. 173.2 A
d. 230 A.

24 The phase current of a delta-connected load is 100 A. The line current will be:

a. 57.73 A
b. 100 A
c. 173.2 A
d. 230 A.

25 An electrical cable is made up of three parts which are:

a. conduction, convection and radiation
b. conductor, insulation and outer sheath
c. heating, magnetic and chemical
d. conductors and insulators.

26 An appropriate wiring method for a domestic installation would be a:

a. metal conduit installation
b. trunking and tray installation
c. PVC cables
d. PVC/SWA cables.

27 An appropriate wiring method for an underground feed to a remote building would be a:

a. metal conduit installation
b. trunking and tray installation
c. PVC cables
d. PVC/SWA cables.

28 An appropriate wiring method for a high-temperature installation in a boiler house is:

a. metal conduit installation
b. trunking and tray installation
c. FP 200 cables
d. MI cables.

29 The cables suspended from the transmission towers of the National Grid network are made from:

a. copper and brass
b. copper with PVC insulation
c. aluminium and steel
d. aluminium and porcelain.

30 An appropriate wiring system for a three-phase industrial installation would be:
   a. PVC cables
   b. PVC conduit
   c. one which meets the requirements of Part 2 of the IET Regulations
   d. one which meets the requirements of Part 5 of the IET Regulations.

31 A current which exceeds the rated value in an otherwise healthy circuit is one definition of:
   a. earthing
   b. bonding
   c. overload
   d. short circuit.

32 An over-current resulting from a fault of negligible impedance is one definition of:
   a. earthing
   b. bonding
   c. overload
   d. short circuit.

33 The connection of the exposed conductive parts of an installation to the main protective earthing terminal of the installation is one definition of:
   a. earthing
   b. protective equipotential bonding
   c. overload
   d. short circuit.

34 The linking together of the exposed or extraneous conductive parts of an installation for the purpose of safety is one definition of:
   a. earthing
   b. protective equipotential bonding
   c. exposed conductive parts
   d. extraneous conductive parts.

35 The conduit and trunking parts of the electrical installation are:
   a. earth conductors
   b. bonding conductors
   c. exposed conductive parts
   d. extraneous conductive parts.

36 The gas, water and central heating pipes of the building not forming a part of the electrical installation are called:
   a. earthing conductors
   b. bonding conductors
   c. exposed conductive parts
   d. extraneous conductive parts.

37 The protection provided by insulating the live parts of the electrical installation is called:
   a. overload protection
   b. short-circuit protection

c. basic protection
d. fault protection.

38 The protection provided by main protective bonding and automatic disconnection of the supply is called:
a. overload protection
b. short-circuit protection
c. basic protection
d. fault protection.

39 A PVC insulated and sheathed wiring system would be suitable for the following type of installation:
a. commercial
b. domestic
c. horticultural
d. industrial.

40 A PVC conduit installation would be suitable for the following type of installation:
a. commercial
b. domestic
c. horticultural
d. industrial.

41 A steel conduit installation would be suitable for the following type of installation:
a. commercial
b. domestic
c. horticultural
d. industrial.

42 A steel trunking installation would be suitable for the following type of installation:
a. commercial
b. domestic
c. horticultural
d. industrial.

43 Which of the following fixing methods would be suitable for holding a lightweight load on a plasterboard partition?
a. cable clip
b. Rawlbolt
c. screw fixing to plastic plug
d. spring toggle.

44 Which of the following fixing methods would be suitable for holding a medium-weight load onto a brick wall?
a. cable clip
b. Rawlbolt
c. screw fixing to plastic plug
d. spring toggle.

45 Which of the following fixing methods would be suitable for holding a PVC insulated and sheathed cable onto a wood surface such as a ceiling joist?
a. cable clip
b. Rawlbolt
c. screw fixing to plastic plug
d. spring toggle.

46 Which of the following fixing methods would be suitable for securing a heavy electric motor to a concrete bed?
a. cable clip
b. Rawlbolt
c. screw fixing to plastic plug
d. spring toggle.

47 A scale drawing which shows the original intention for the position of electrical equipment is called a:
a. wiring diagram
b. detail assembly drawing
c. site plan or layout drawing
d. as-fitted drawing.

48 A scale drawing which shows the actual position of the electrical equipment upon completion of the contract is called a:
a. wiring diagram
b. detail assembly drawing
c. site plan or layout drawing
d. as-fitted drawing.

49 A scale drawing showing the position of equipment by graphical symbols is a description of a:
a. block diagram
b. site plan or layout diagram
c. wiring diagram
d. circuit diagram.

50 A diagram which shows the detailed connections between individual items of equipment is a description of a:
a. block diagram
b. site plan or layout diagram
c. wiring diagram
d. circuit diagram.

51 A diagram which shows most clearly how a circuit works, with all items represented by graphical symbols, is a description of a:
a. block diagram
b. site plan or layout diagram
c. wiring diagram
d. circuit diagram.

52 An electrical cable is made up of three parts which are:
a. conduction, convection and radiation
b. conductor, insulation and outer sheath

c. heating, magnetic and chemical
d. conductors and insulators.

53 An appropriate wiring method for a domestic installation would be a:
a. metal conduit installation
b. trunking and tray installation
c. PVC cables
d. PVC/SWA cables.

54 An appropriate wiring method for an underground feed to a remote building would be a:
a. metal conduit installation
b. trunking and tray installation
c. PVC cables
d. PVC/SWA cables.

55 An appropriate wiring method for a high-temperature installation in a boiler house is:
a. metal conduit installation
b. trunking and tray installation
c. FP200 cables
d. MI cables.

56 The cables suspended from the transmission towers of the National Grid network are made from:
a. copper and brass
b. copper with PVC insulation
c. aluminium and steel
d. aluminium and porcelain.

57 An appropriate wiring system for a three-phase industrial installation would be:
a. PVC cables
b. PVC conduit
c. one which meets the requirements of Part 2 of the IET Regulations
d. one which meets the requirements of Part 5 of the IET Regulations.

58 Briefly explain why an electrical installation needs protective devices.

59 List the four factors on which the selection of a protective device depends.

60 List the five essential requirements for a device designed to protect against over-current.

61 Briefly describe the action of a fuse under fault conditions.

62 State the meaning of 'discrimination' or 'selectivity' as applied to circuit protective devices.

63 Use a sketch to show how 'effective coordination' can be applied to a piece of equipment connected to a final circuit.

64 List typical 'exposed parts' of an installation.

65 List typical 'extraneous parts' of a building.

66 Use a sketch to show the path taken by an earth fault current.

67 Use bullet points and a simple sketch to briefly describe the operation of an RCD.

68 State the need for RCDs in an electrical installation:
   a. supplying socket outlets with a rated current not exceeding 20 A, and
   b. for use by mobile equipment out of doors as required by IET Regulation 411.3.3.

69 Briefly describe an application for RCBOs.

70 State the meaning of fault protection.

71 State the meaning of basic protection.

72 State the meaning of a 'polyphase supply system'.

73 State the standard voltages used for generation, transmission and distribution in the United Kingdom.

74 Environmentalists often say that steel transmission towers are a 'blot on the landscape'. Why do we continue to use steel towers for transmission on the National Grid network?

75 Why is the distribution from local sub-stations to end users, for the most part, by underground cables in the United Kingdom?

76 State the reasons for balancing single-phase loads across a three-phase supply.

77 Briefly describe how and why we generate a three-phase supply compared to a single-phase supply.

78 What are the advantages of connecting a three-phase supply:
   a. in delta
   b. in star?

79 Produce a quick coloured sketch of a PVC insulated and sheathed cable and name the parts.

80 Produce a quick coloured sketch of a PVC/SWA cable and name the parts.

81 Produce a quick sketch of an electric circuit and name the five component parts.

82 Give an example of a device or accessory for each component part. For example, the supply might be from the a.c. mains or a battery.

83 In your own words, state the meaning of circuit overload and short-circuit protection. What will provide this type of protection?

84 State the purpose of earthing and earth protection. What do we do to achieve it and why do we do it?

85 In your own words, state the meaning of exposed and extraneous conductive parts and give examples of each.

86 In your own words, state the meaning of earthing and bonding. What types of cables and equipment would an electrician use to achieve earthing and bonding on an electrical installation?

87 In your own words, state what we mean by 'basic protection' and how it is achieved.

88 In your own words, state what we mean by 'fault protection' and how it is achieved.

89 What methods could you use to find and store some information about:
- Health and safety at work
- British Standards
- Electrical accessories and equipment?

90 What are the advantages and disadvantages of having sources of technical information on:
a. some form of electronic storage system such as a CD, DVD or USB memory stick, or
b. hard copy such as a catalogue, drawings or *On Site Guide*?

Would it make a difference if you were at the office or on a construction site?

91 List at least two statutory regulations which have an impact upon all electrical activities and state your reasons why.

92 List at least one code of practice which has an impact upon all electrical activities and state your reason why.

93 List seven sections within the IET Regulations which have an impact upon electrical safety in buildings.

94 List six types of building or situation to which the IET Regulations alone will apply.

95 List two types of installation or situation to which the IET Regulations alone will not apply because the installation or situation is considered to be too dangerous.

96 Use bullet points to list the main stages involved in lifting a heavy box from the floor, carrying it across a room and placing it on a worktop, using a safe manual handling technique.

97 Describe a safe manual handling technique for moving a heavy electric motor out of the stores, across a yard and into the back of a van for delivery to site.

98 State the advantages of a conduit trunking and tray cable enclosure system for a commercial installation such as a shopping centre.

99 Compare PVC/SWA cables with MI cables and give their advantages, disadvantages and typical applications.

100 Use a sketch with notes of explanation to show how mini-trunking and skirting trunking could be used to contain all the electrical supplies in a school's computing classroom.

101 Explain the meaning of 'segregation' of circuits.

102 Use a sketch with notes of explanation to show how trunking and tray may be suspended from the girders of a building structure by appropriate brackets.

103 Sketch a site plan or layout drawing for the room which you normally use at college and indicate the position of all the electrical accessories in the room using BS EN 60617 symbols.

CHAPTER 4

C&G Unit 204

Installation of wiring systems and enclosures

Unit 204 of the City and Guilds 2365-02 syllabus

Learning outcomes – when you have completed this chapter you should:

- know safe working practice in the workplace;
- know the tools used for installing wiring systems;
- be able to install and terminate wiring systems;
- be able to bond main services to the main earthing terminal;
- know how to inspect an electrical installation;
- be able to test an electrical installation.

This chapter has free associated content, including animations and instructional videos, to support your learning.

When you see the logo, visit the companion website for more on this topic www.routledge.com/cw/linsley

Safe working practice in the workplace

In Chapter 1 we looked at some of the laws and regulations that affect our working environment. We looked at safety signs and personal protective equipment (PPE), and how to recognize and use different types of fire extinguishers.

If your career in the electrical industry is to be a long, happy and safe one, you must always wear appropriate PPE such as footwear and head protection and behave responsibly and sensibly in order to maintain a safe working environment. Before starting work, make a safety assessment; what is going to be hazardous, will you require PPE, do you need any special access equipment?

Construction sites can be hazardous because of the temporary nature of the construction process. The surroundings and systems are always changing as the construction process moves to its completion date when everything is finally in place.

Safe methods of working must be demonstrated by everyone at every stage. 'Employees have a duty of care to protect their own health and safety and that of others who might be affected by their work activities.'

Figure 4.1 Safe methods of working must be demonstrated by everyone at every stage.

To make the work area safe before starting work and during work activities, it may be necessary to:

- use barriers or tapes to screen off potential hazards;
- place warning signs as appropriate;
- inform those who may be affected by any potential hazard;
- use a safe isolation procedure before working on live equipment or circuits;
- obtain any necessary 'permits to work' before work begins.

Get into the habit of always working safely and being aware of the potential hazards around you when you are working.

Having chosen an appropriate wiring system which meets the intended use and structure of the building and satisfies the environmental conditions of the installation, you must install the system conductors, accessories and equipment in a safe and competent manner.

The structure of the building must be made good if it is damaged during the installation of the wiring system. For example, where conduits and trunking pass through walls and floors.

All connections in the wiring system must be both electrically and mechanically sound and accessible for testing and inspection (IET Regulations 131.14 and 526.3). All conductors must be chosen so that they will carry the design current under the installed conditions.

If the wiring system is damaged during installation it must be made good to prevent future corrosion. For example, where galvanized conduit trunking or tray is cut or damaged by pipe vices, it must be made good to prevent localized corrosion.

All tools must be used safely and sensibly. Cutting tools should be sharpened and screwdrivers ground to a sharp, square end on a grindstone.

It is particularly important to check that the plug top and cables of handheld electrically powered tools and extension leads are in good condition. Damaged plug tops and cables must be repaired before you use them. All electrical power tools of 110 and 230V must be tested with a portable appliance tester (PAT) in accordance with the company's health and safety procedures, but probably at least once each year.

Tools and equipment that are left lying about in the workplace can become damaged or stolen and may also be the cause of people slipping, tripping or falling. Tidy up regularly and put power tools back in their boxes (**called 'good housekeeping**'). You personally may have no control over the condition of the workplace in general, but keeping your own work area clean and tidy is the mark of a skilled and conscientious craftsman.

Section F (B2.3) of the *Electricians Guide to the Building Regulations* says, 'tidiness, wherever work is carried out, is the foundation of safety; good housekeeping will help to ensure a clean, tidy and safe place to work'.

Finally, when the job is finished, clean up and dispose of all waste material responsibly, as described in Chapter 1.

Every year thousands of people have accidents at their place of work despite the legal requirements laid down by the Health and Safety Executive. Many people recover quickly but an accident at work can result in permanent harm or even death.

At the very least, injuries hurt individuals. They may prevent you from doing the things you enjoy in your spare time and they cost a lot of money, to you in loss of earnings and to your employer in loss of production and possibly damage to equipment. Your place of work may look harmless but it can be dangerous.

If there are five or more people employed by your company then the company must have its own safety policy, as described in Chapter 1. This must spell out the organization and arrangements which have been put in place to ensure that you and your workmates are working in a safe place.

Your employer must also have carried out an assessment on the risks to your health and safety in the place where you are working. You should be told about the safety policy and risk assessment; for example, you may have been given a relevant leaflet when you started work. Risk assessment was discussed in some detail in Chapter 1.

You have a responsibility under the Health and Safety at Work Act to:

- learn how to work safely and to follow company procedures of work;
- obey all safety rules, notices and signs;
- not interfere with or misuse anything provided for safety;
- report anything that seems damaged, faulty or dangerous;
- behave sensibly, not play practical jokes and not distract other people at work;
- walk sensibly and not run around the workplace;
- use the prescribed walkways;
- drive only those vehicles for which you have been properly trained and passed the necessary test;
- not wear jewellery which could become caught in moving parts if you are using machinery at work;
- always wear appropriate clothing and PPE if necessary.

Common causes of accidents at work

Slips, trips and falls are still the major causes of accidents at work.

To help prevent them:

- keep work areas clean and tidy;
- keep walkways clear;

- do not leave objects, tools and equipment lying around blocking up walkways;
- clean up spills or wet patches on the floor straight away.

Manual handling, that is, moving objects by hand, may result in strains, sprains and trap injury pains.

To help prevent them:

- use a mechanical aid to move heavy objects, such as a sack truck or flat bed truck;
- only lift and carry what you can manage easily;
- wear gloves to avoid rough or sharp edges;
- use a good manual lifting technique as discussed in Chapter 1.

When using equipment, machines and certain tools, such as angle grinders which are difficult to use.

To help prevent injuries:

- wear goggles;
- wear appropriate PPE;
- make sure tools and equipment are in good condition and carry an 'in date' PAT test label.

Badly stored equipment can become unstable and fall onto someone.

To help prevent objects from falling:

- stack equipment sensibly and securely;
- stack heavy objects low down;
- stack objects so that they can be reached without stretching or reaching over.

Fire safety was discussed earlier in Chapter 1.

Electricity and its safe use is what the electrical industry and the regulations are all about.

To prevent electrical accidents always use the 'safe isolation procedure' before any work begins, as described in Chapter 1.

Carrying out the bullet-pointed activities listed above, which help to prevent the causes of accidents, can all be considered to be 'good housekeeping' because they contribute individually to a safer work environment.

Try to get into the habit of doing a risk assessment every time you arrive in the workplace. Not a formal written risk assessment like the one described in Chapter 1, but an informal one for yourself, assessing the potential hazards around your area of work. You can then reduce any potential hazards by appropriate safety measures which will make your working environment safer.

Definition

A hazard is something with the potential to cause harm, for example, chemicals, electricity, working above ground.

Definition

A risk is the likelihood or chance of harm actually being done by the hazard.

Definition

PPE is defined as all equipment designed to be worn, or held, to protect against a risk to health and safety.

Safe working procedures to prevent injury

Where a particular **hazard** exists in the working environment, an employer must carry out a risk assessment and establish procedures which will reduce or eliminate the risk. When the **risk** cannot be completely removed, an employer must provide personal protective equipment (PPE) to protect his or her employees from a risk to health and safety.

Personal protective equipment

PPE is defined as all equipment designed to be worn, or held, to protect against a risk to health and safety. This includes most types of protective clothing, and

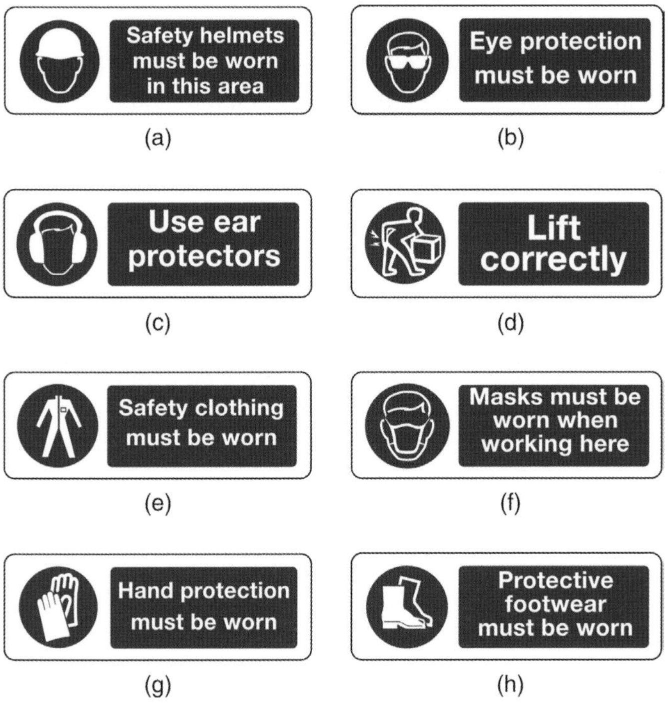

Figure 4.2 Safety signs showing types of PPE to be worn.

equipment such as eye, foot and head protection, safety harnesses, life-jackets and high-visibility clothing.

Under the Health and Safety at Work Act, employers must provide free of charge any PPE and employees must make full and proper use of it. Safety signs such as those shown in Fig. 4.2 are useful reminders of the type of PPE to be used in a particular area. The vulnerable parts of the body which may need protection are the head, eyes, ears, lungs, torso, hands and feet and, in addition, protection from falls may need to be considered. Objects falling from a height present the major hazard against which head protection is provided. Other hazards include striking the head against projections and hair becoming entangled in machinery. Typical methods of protection include helmets, light-duty scalp protectors called 'bump caps' and hair-nets.

The eyes are very vulnerable to liquid splashes, flying particles and light emissions such as ultraviolet light, electric arcs and lasers. Types of eye protectors include safety spectacles, safety goggles and face shields. Screen-based workstations are being used increasingly in industrial and commercial locations by all types of personnel. Working with VDUs (visual display units) can cause eye strain and fatigue and, therefore, every display screen operator is entitled to a free eye test.

Noise is accepted as a problem in most industries and may be defined as any disagreeable or undesirable sound or sounds, generally of a random nature, which do not have clearly defined frequencies. The usual basis for measuring noise or sound level is the decibel scale. Whether noise of a particular level is harmful or not also depends upon the length of exposure to it. This is the basis of the widely accepted limit of 85 dB of continuous exposure to noise for eight hours per day.

Where individuals must be subjected to some noise at work it may be reduced by ear protectors. These may be disposable ear plugs, reusable ear plugs or ear

Figure 4.3 Breathing protection signs.

muffs. The chosen ear protector must be suited to the user and suitable for the type of noise, and individual personnel should be trained in its correct use.

Breathing reasonably clean air is the right of every individual, particularly at work. Some industrial processes produce dust which may present a potentially serious hazard. The lung disease asbestosis, caused by the inhalation of asbestos dust or particles, and the coal dust disease pneumoconiosis, suffered by many coal-miners, have made people aware of the dangers of breathing in contaminated air.

Some people may prove to be allergic to quite innocent products, such as flour dust in the food industry or wood dust in the construction industry. The main effect of inhaling dust is a measurable impairment of lung function. This can be avoided by wearing an appropriate mask, respirator or breathing apparatus, as recommended by the company's health and safety policy and indicated by local safety signs, such as those shown in Fig. 4.3.

A worker's body may need protection against heat or cold, bad weather, chemical or metal splash, impact or penetration and contaminated dust. Alternatively, there may be a risk of the worker's own clothes causing contamination of the product, as in the food industry. Appropriate clothing will be recommended in the company's health and safety policy. Ordinary working clothes and clothing provided for food hygiene purposes are not included in the PPE at Work Regulations.

Hands and feet may need protection from abrasion, temperature extremes, cuts and punctures, impact or skin infection. Gloves or gauntlets provide protection from most industrial processes but should not be worn when operating machinery because they may become entangled in it. Care in selecting the appropriate protective device is required; for example, barrier creams provide only a limited protection against infection.

Boots or shoes with in-built toe-caps can give protection against impact or falling objects and, when fitted with a mild steel sole plate, can also provide protection from sharp objects penetrating through the sole. Special slip-resistant soles can also be provided for employees working in wet areas.

Whatever the hazard to health and safety at work, the employer must be able to demonstrate that he or she has carried out a risk assessment, made recommendations which will reduce that risk and communicated these recommendations to the workforce. Where there is a need for PPE to protect against personal injury and to create a safe working environment, the employer must provide that equipment and any necessary training which might be required and the employee must make full and proper use of such equipment and training.

Working at height regulations

Working above ground level creates added dangers and slows down the work rate of the electrician. New Work at Height Regulations came into force on 6 April 2005. Every precaution should be taken to ensure that the working platform is appropriate for the purpose and in good condition.

We looked at this topic as it applies to electrical personnel in Chapter 1 under the subheading 'Working above ground'.

The Work at Height Regulations 2005 tell us that a person is at height if that person could be injured by falling from it. The regulations require that:

- we should avoid working at height if at all possible;
- no work should be done at height which can be done on the ground. For example, equipment can be assembled on the ground then taken up to height, perhaps for fixing;
- ensure the work at height is properly planned;
- take account of any risk assessments carried out under Regulation 3 of the Management of Health and Safety at Work Regulations.

Tools used to install wiring systems

A craftsman earns a living by hiring out their skills or selling products made using their skills and expertise. They shape their environment, mostly for the better, improving the living standards for themselves and others.

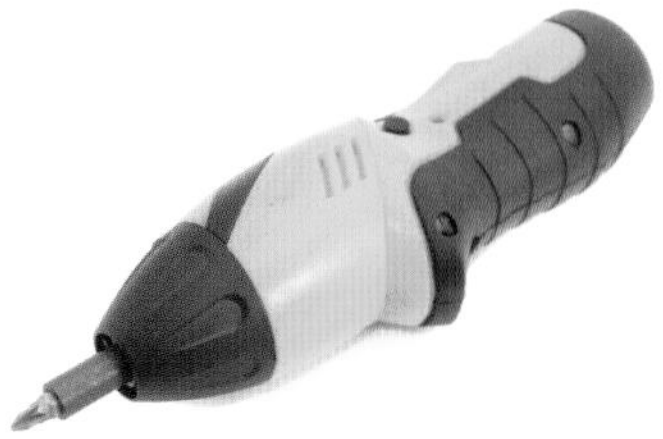

Figure 4.4 A torque screwdriver.

Tools extend the limited physical responses of the human body and therefore good-quality, sharp tools are important to a craftsman. An electrician is no less a craftsman than a wood carver. Both must work with a high degree of skill and expertise, and both must have sympathy and respect for the materials they use. Modern electrical installations using new materials are lasting longer than 50 years. Therefore they must be properly installed. **Good design, good workmanship by skilled (electrically) or instructed (electrically) persons and the use of proper materials are essential if the installation is to comply with the relevant regulations (IET Regulation 134.1.1) and reliably and safely meet the requirements of the customer for over half a century**.

An electrician must develop a number of basic craft skills particular to their own trade, but they also require some of the skills used in many other trades. An electrician's toolkit will reflect both the specific and general nature of the work.

The basic tools required by an electrician are those used in the stripping and connecting of conductors.

These are pliers, side cutters, a knife and an assortment of screwdrivers, as shown in Fig. 4.5.

The torque screwdriver is becoming increasingly popular in the electrical industry. **A torque screwdriver** is a screwdriver with an in-built mechanism to ensure that the screw can be tightened to a specified, pre-set torque or turning force. This ensures that tightening is sufficient to make a safe electrical connection, but not so excessive as to cause damage.

The tools required in addition to these basic tools will depend upon the type of installation work being undertaken. When wiring new houses or rewiring old ones, additional tools are required. Examples are shown in Fig. 4.6.

Safety first

Hand Tools

- keep them clean;
- keep them sharp;
- keep them safe.

Figure 4.5 The tools of an electrician.

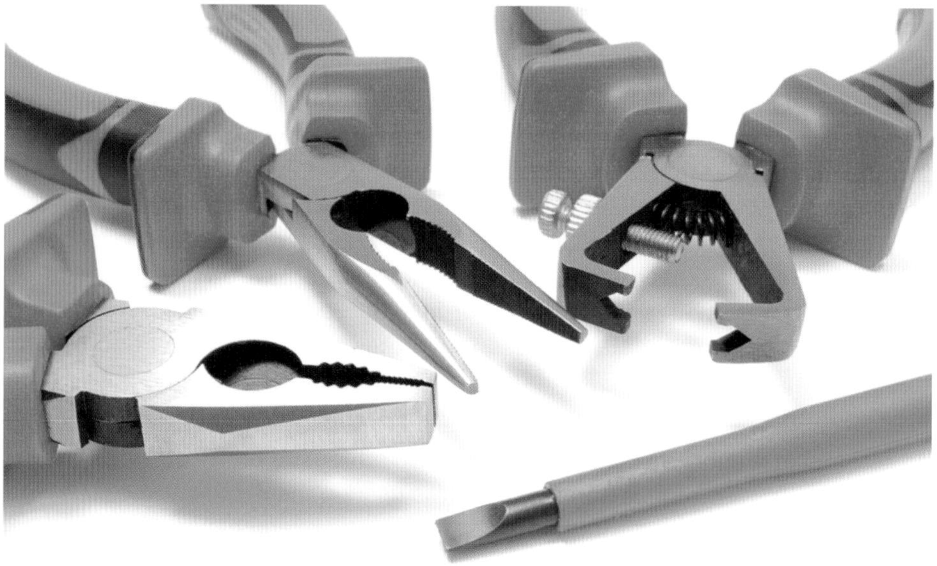

Figure 4.6 Some of the essentials.

When working on industrial installations, installing conduit and trunking, the additional tools required by an electrician would more normally be those associated with a fitter or sheet-metal fabricator.

Where special tools are required, for example, those required to terminate mineral insulated (MI) cables or the bending and cutting tools for conduit and cable trays, as shown in Fig. 4.7, they will often be provided by an employer, but most hand tools are provided by the electrician.

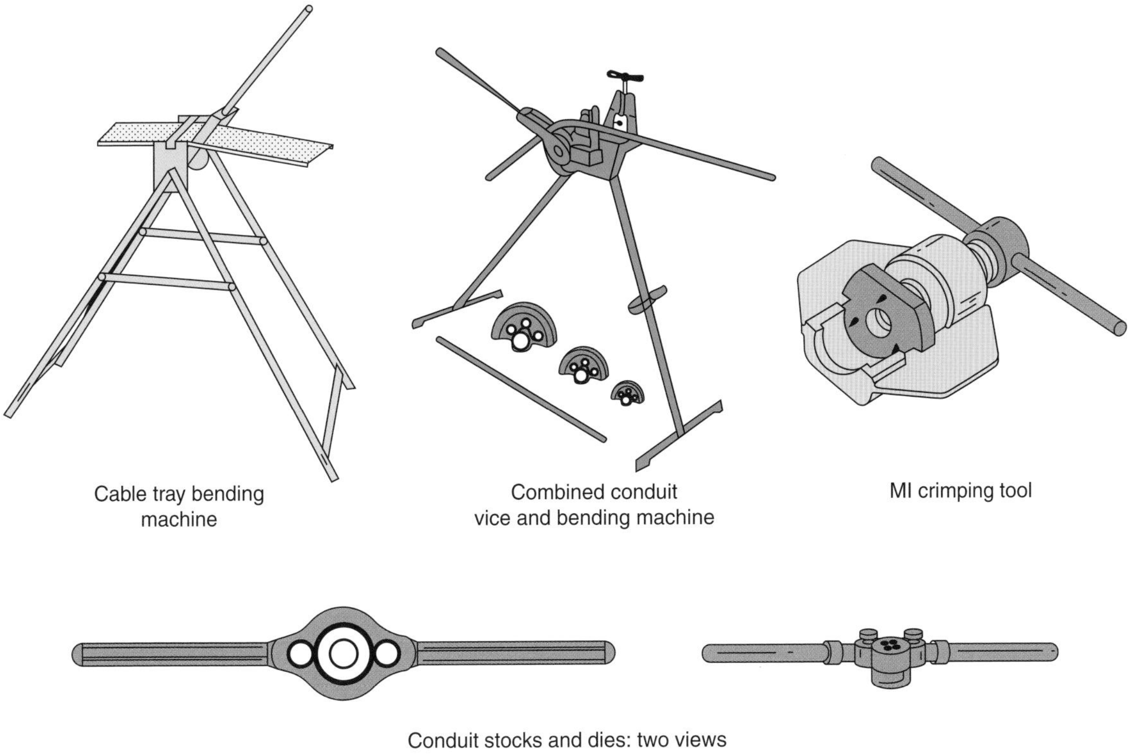

Figure 4.7 Some special tools required by an electrician engaged in industrial installations.

In general, good-quality tools last longer and stay sharper than those of inferior quality, but tools are very expensive to buy. A good set of tools can be assembled over the training period if the basic tools are bought first and the extended toolkit acquired one tool at a time.

Another name for an installation electrician is a 'journeyman' electrician and, as the name implies, an electrician must be mobile and prepared to carry his or her tools from one job to another. Therefore, a good toolbox is an essential early investment, so that the right tools for the job can be easily transported.

Tools should be cared for and maintained in good condition if they are to be used efficiently and remain serviceable. Screwdrivers should have a flat, squared-off end and wood chisels should be very sharp. Access to a grindstone will help an electrician to maintain their tools in first-class condition. In addition, wood chisels will require sharpening on an oilstone to give them a very sharp edge.

Electrical power tools

Portable electrical tools can reduce much of the hard work for any tradesperson and increase his or her productivity. Electrical tools should be maintained in a good condition and be appropriate for the purpose for which they are used. Many construction sites now insist on low-voltage or battery tools being used which can further increase safety without any loss of productivity. Some useful electrical tools are shown in Fig. 4.8.

Electric drills are probably used most frequently of all electrical tools. They may be used to drill metal or wood. Wire brushes are made which fit into the drill

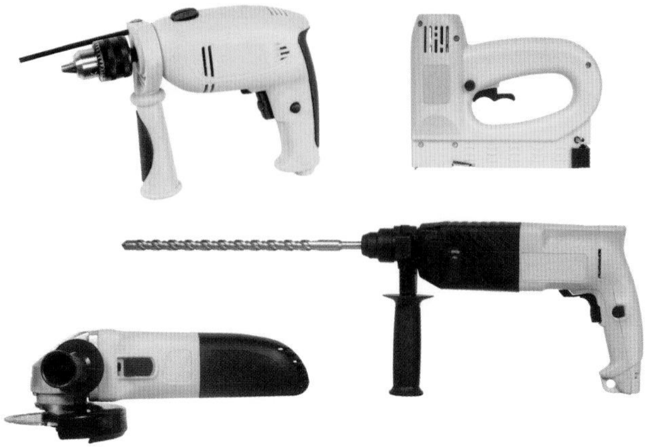

Figure 4.8 Electrical hand tools.

chuck for cleaning metal. Variable-speed electric drills, which incorporate a vibrator, will also drill brick and concrete as easily as wood when fitted with a masonry drill bit.

Hammer drills give between 2,000 to 3,000 impacts per minute and are used for drilling concrete walls and floors.

Cordless electric drills are also available which incorporate a rechargeable battery, usually in the handle. They offer the convenience of electric drilling when an electrical supply is not available or if an extension cable is impractical.

Angle grinders are useful for cutting chases in brick or concrete. The discs are interchangeable. Silicon carbide discs are suitable for cutting slate, marble, tiles, brick and concrete, and aluminium oxide discs for cutting iron and steel such as conduit and trucking.

Jigsaws can be fitted with wood- or metal-cutting blades. With a wood-cutting blade fitted they are useful for cutting across floorboards and skirting boards or any other application where a padsaw would be used. With a metal-cutting blade fitted, they may be used to cut trunking.

When a lot of trunking work is to be undertaken, an electric nibbler is a worthwhile investment. This nibbles out the sheet metal, is easily controllable and is one alternative to the jigsaw or hacksaw.

All tools must be used safely and sensibly. Cutting tools should be sharpened and screwdrivers ground to a sharp square end on a grindstone.

It is particularly important to check that the plug tops and cables of handheld electrically powered tools and extension leads are in good condition. Damaged plug tops and cables must be repaired before you use them. All electrical power tools of 110 V and 230 V must be inspected and tested with a portable appliance tester (PAT) in accordance with the company's procedures, but probably at least once each year. PAT testing tests the quality of the insulation resistance and the earth continuity. Inspection checks the condition of the plug top, fuse and lead.

Tools and equipment that are left lying about in the workplace can become damaged or stolen, and may also be the cause of people slipping, tripping or falling. Tidy up regularly and put power tools back in their boxes. You personally may have no control over the condition of the workplace in general, but keeping your own work area clean and tidy is the mark of a skilled and conscientious craftsman.

Top tip

Drilling brick or stone

When using masonry drills, withdraw several times while drilling to remove dust and reduce the chance of the drill jamming in the hole.

Top tip

Drilling tiles

When drilling glazed surfaces or tiles, place masking tape on the surface. This allows you to mark the hole centre and prevents the drill bit from skidding.

Top tip

Drill speed

Remember, the larger the drill bit, the slower the drill speed.

Safety first

Power tools

Many construction sites now only allow:

- low-voltage tools;
- PAT tested tools;
- battery-powered tools.

Try this

Power tools

Look at the power tools at work.

- Do they have a PAT Test label?
- If so, is the PAT Test label 'in date'?

Marking out the work on-site

When the electrician arrives on-site to begin the work, that is, installing the trunking, conduit or cables and boxes to the final fixing positions in the building, they will need to know what is being installed and exactly where it is being installed.

On a larger job they will probably be given a site plan or layout drawing. This is a scale drawing based upon the architect's site plan of the building and will show the position of the electrical equipment to be installed.

The electrical equipment will be identified by graphical symbols on the plan, as described in Chapter 5 and shown in Fig. 5.5. The drawing will be to scale, and if we say the scale was 1:100, then 10 mm or 1 cm on the site plan will represent 1 m in the building. The electrician can take measurements from the drawing and transfer these to the building, probably using a retractable steel rule. They can then mark on the walls of the building with a piece of chalk or a marker the final positions of the electrical equipment which is to be installed.

If the job is a small one the electrician's supervisor or foreman may walk them around the site showing them what is to be installed and where. On a very small job it might be the customer who describes what is to be done by the electrician.

When the electrician knows what is required, they can begin to plan out cable routes and mark out where they will install their trunking and conduits, cables or tray.

Trunking, conduits, cables and tray must be installed securely, but most of all they must be installed horizontally or vertically for a professional finish. A long spirit-level is very useful in determining what is horizontal and vertical, although a plumb-line can also be used to determine the vertical. A plumb-line is a piece of string with a steel weight attached to the bottom. Holding the string up against a wall, the steel weight at the bottom will initially swing like a pendulum, but quickly come to rest in a true vertical line. If a little chalk is first rubbed onto the line, it becomes a chalk line. If an assistant holds the steel weight in its rest position so that the line presses against the wall, and then flicks the string so that it twangs against the wall, the chalk will transfer to the wall, leaving a true vertical line on the wall.

When the measuring and marking out is completed, the electrician can begin the process of installing the trunking, conduits, cables or tray, using the hand tools and power tools described previously in this chapter.

Wiring systems and enclosures

The final choice of a wiring system must rest with those designing the installation and those ordering the work, but whatever system is employed, good

Safety first

Site safety

The Health and Safety at Work Act makes *you* responsible for:

- following the site safety rules;
- demonstrating safe working practice;
- wearing or using appropriate PPE.

Top tip

Cutting

Before cutting anything, trunking, tray or lengths of cable, **measure twice and cut once** to reduce errors.

workmanship by competent persons and the use of proper materials is essential for compliance with the Regulations (IET Regulation 134.1.1). When installing wiring systems and enclosures the safe working practices and procedures discussed earlier in this chapter must be followed. The necessary skills can be acquired by an electrical trainee who has the correct attitude and dedication to his or her craft.

Table A1 in Appendix A of the *On Site Guide* deals with the assumed current demand of points, and states that for lighting outlets we should assume a current equivalent to a minimum of 100 W per lamp holder. This means that for a domestic lighting circuit rated at 5 A, a maximum of 11 lighting outlets could be connected to each circuit. In practice, it is usual to divide the fixed lighting outlets into a convenient number of circuits of seven or eight outlets each. In this way the whole installation is not plunged into darkness if one lighting circuit fuses and complies with Regulation 314.1 which tells us to **'divide into circuits to minimize inconvenience and avoid danger'**.

Lighting circuits are usually wired in 1.0 or 1.5 mm cable using either a loop-in or joint-box method of installation. The loop-in method is universally employed with conduit installations or when access from above or below is prohibited after installation, as is the case with some industrial installations or blocks of flats. In this method the only joints are at the switches or lighting points, the live conductors being looped from switch to switch and the neutrals from one lighting point to another.

The use of junction boxes with fixed brass terminals is the method often adopted in domestic installations, since the joint boxes can be made accessible but are out of sight in the loft area and under floorboards. However, every connection must remain accessible for inspection, testing and maintenance (IET Regulation 526.3).

The live conductors must be broken at the switch position in order to comply with the polarity regulations 612.7. A ceiling rose may only be connected to installations operating at 250 V maximum and must only accommodate one flexible cord unless it is specially designed to take more than one (IET Regulations 559.6.1.2 and 559.6.1.3). Lamp holders suspended from flexible cords must be capable of suspending the mass of the luminaire fixed to the lamp holder (IET Regulation 559.6.1.5).

Safety first

Fire

- Mini-spots get hot in use.
- Cutting a hole in a ceiling may compromise the fire integrity of the room.
- Fix spots away from combustible material.
- Follow manufacturers' fitting instructions.
- Consider installing fire hoods.

The method of fixing must be capable of carrying a mass of not less than 5 kg. Suspended ceilings are considered to be 'stable' or firmly fixed and may, therefore, support luminaires.

A luminaire, that is, a light fitting or small spotlight, must be fixed at an adequate distance from combustible material, or as recommended by the manufacturer, or be enclosed in non-flammable material (IET Regulations 422.3.1, 422.3.8, 422.4.100 and 559.5.1).

The type of circuit and wiring system used will depend upon the installation conditions and the customer's requirements.

PVC insulated and sheathed cable installations

PVC insulated and sheathed wiring systems are used extensively for lighting and socket installations in domestic dwellings. Mechanical damage to the cable caused by impact, abrasion, penetration, compression or tension must be minimized during installation (Regulation 522.6.1). The cables are generally fixed using plastic clips incorporating a masonry nail, which means the cables can be fixed to wood, plaster or brick with almost equal ease. Cables should be run horizontally or vertically, not diagonally, down a wall. All kinks should be removed so that the cable is run straight and neatly between clips fixed at equal distances providing adequate support for the cable so that it does not become damaged by its own weight (Regulation 522.8.4 and Table D1 of the *On Site Guide*). Table D1 of the IET *On Site Guide* is shown in Table 4.1. Where cables are bent, the radius of the bend should not cause the conductors to be damaged (Regulation 522.8.3 and Table D5 of the *On Site Guide*).

The 3rd Amendment to the 17th Edition of the IET Regulations introduced a new Regulation 521.200.1, requiring robust support of wiring systems in escape routes so that they will not become compromised or collapse in the event of a fire.

Terminations or joints in the cable may be made in ceiling roses, junction boxes, or behind sockets or switches, provided that they are enclosed in a non-ignitable material, are properly insulated and are mechanically and electrically secure (IET Regulation 526). All joints must be accessible for inspection testing and maintenance when the installation is completed (IET Regulation 526.3).

Where PVC insulated and sheathed cables are concealed in walls, floors or partitions, they must be provided with a box incorporating an earth terminal at each outlet position. PVC cables do not react chemically with plaster, as do some cables, and consequently PVC cables may be buried under plaster. Further protection by channel or conduit is only necessary if mechanical protection from nails or screws is required or to protect them from the plasterer's trowel. However, Regulation 522.6.201 now tells us that where PVC cables are to be embedded in a wall or partition at a depth of less than 50 mm they should be run along one of the permitted routes shown in Fig. 4.10. Figure 4.9 shows a typical PVC installation. To identify the most probable cable routes, Regulation 522.6.201 tells us that outside a zone formed by a 150 mm border all around a wall edge, cables can only be run horizontally or vertically to a point or accessory unless they are contained in a substantial earthed enclosure, such as a conduit, which can withstand nail penetration, as shown in Fig. 4.10.

Where the accessory or cable is fixed to a wall which is less than 100 mm thick, protection must also be extended to the reverse side of the wall if a position can be determined.

Where none of this protection can be complied with then the cable must be given additional protection with a 30 mA RCD (IET Regulation 522.6.201).

Where cables pass through walls, floors and ceilings the hole should be made good with incombustible material such as mortar or plaster to prevent the spread of fire (Regulations 527.1.2 and 527.2.1). Cables passing through metal boxes should be bushed with a rubber grommet to prevent abrasion of the cable. Holes drilled in floor joists through which cables are run should be 50 mm below the top or 50 mm above the bottom of the joist to prevent damage to the cable by nail penetration (Regulation 522.6.200), as shown in Fig. 4.11. PVC cables should not be installed when the surrounding temperature is below 0°C or when the cable temperature has been below 0°C for the previous 24 hours because the insulation becomes brittle at low temperatures and may be damaged during installation.

Other types of wiring circuits and wiring enclosures are discussed in detail in Chapter 3.

Table 4.1 Spacing of cable supports

Spacings of supports for cables in accessible positions								
Overall diameter of cable* (mm)	**Maximum spacings of clips**							
	Non-armoured thermosetting, thermoplastic or lead sheathed cables				**Armoured cables**		**Mineral insulated copper sheathed or aluminium sheathed cables**	
	Generally		**In caravans**					
	Horizontal (mm)**	**Vertical** (mm)**	**Horizontal** (mm)**	**Vertical** (mm)**	**Horizontal** (mm)**	**Vertical** (mm)**	**Horizontal** (mm)**	**Vertical** (mm)**
1	**2**	**3**	**4**	**5**	**6**	**7**	**8**	**9**
Not exceeding 9	250	400			–	–	600	800
Exceeding 9 and not exceeding 15	300	400	250 (for all sizes)	400 (for all sizes)	350	450	900	1200
Exceeding 15 and not exceeding 20	350	450			400	550	1500	2000
Exceeding 20 and not exceeding 40	400	550			450	600	–	–

Note: For the spacing of supports for cables having an overall diameter exceeding 40 mm, and for single-core cables having conductors of cross-sectional area 300 mm^2 and larger, the manufacturer's recommendations should be observed.
*For flat cables taken as the dimension of the major axis.
**The spacings stated for horizontal runs may be applied also to runs at an angle of more than 30° from the vertical. For runs at an angle of 30° or less from the vertical, the vertical spacings are applicable.
Adapted from the IET *On Site Guide* by kind permission of the Institution of Engineering and Technology.

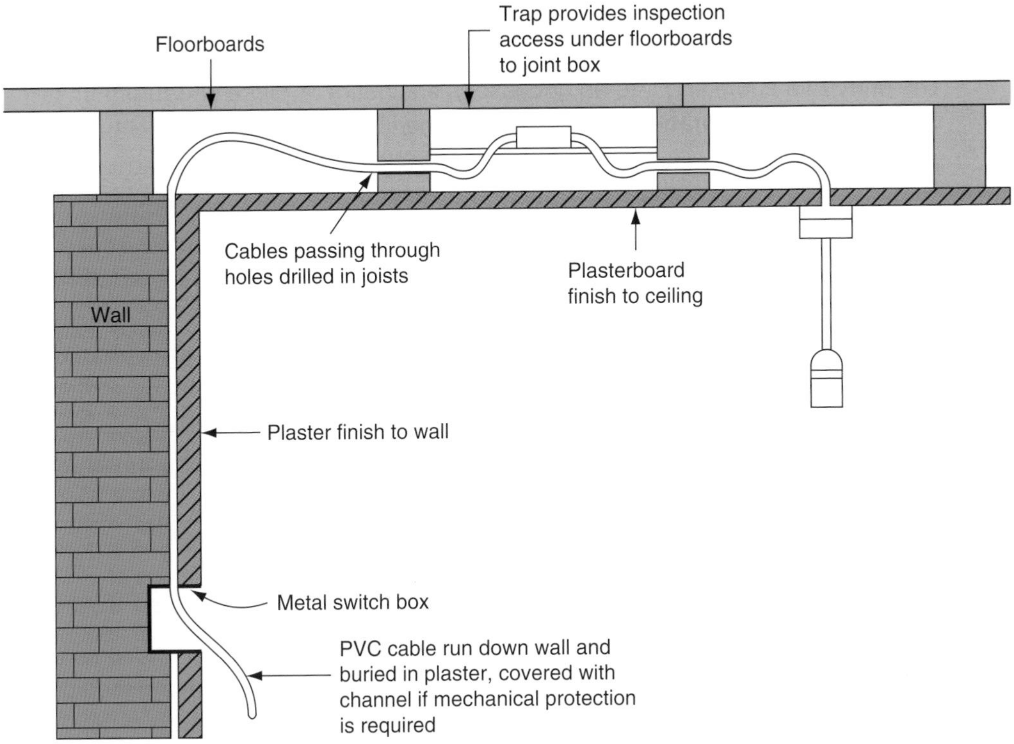

Figure 4.9 A concealed PVC sheathed wiring system.

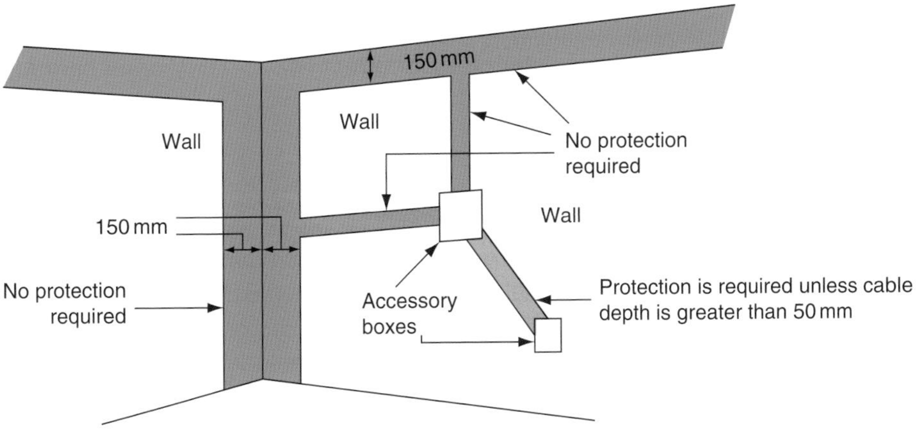

Figure 4.10 Permitted cable routes.

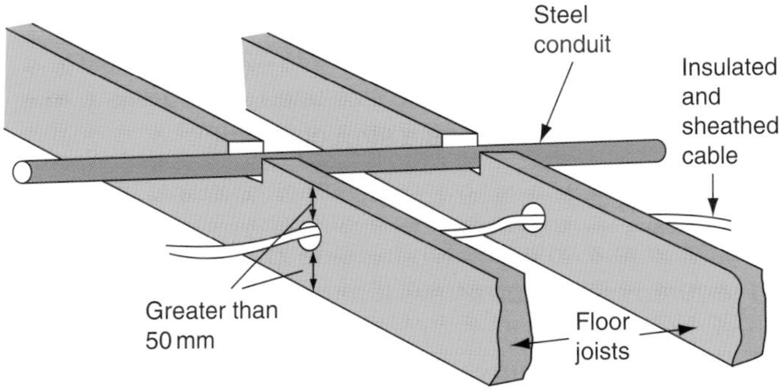

Notes:

1. Maximum diameter of hole should be 0.25 × joist depth.
2. Holes on centre line in a zone between 0.25 and 0.4 × span.
3. Maximum depth of notch should be 0.125 × joist depth.
4. Notches on top in a zone between 0.1 and 0.25 × span.
5. Holes in the same joist should be at least 3 diameters apart.

Figure 4.11 Correct installation of cables in floor joists.

Terminating wiring systems

Definition

The entry of a cable end into an accessory, enclosure or piece of equipment is what we call a *termination*.

The entry of a cable end into an accessory, enclosure or piece of equipment is what we call a **termination**. Section 526 of the IET Regulations tells us that:

1. Every connection between conductors and equipment shall be durable, provide electrical continuity and mechanical strength and protection.
2. Every termination and joint in a live conductor shall be made within a suitable accessory, piece of equipment or enclosure that complies with the appropriate product standard.
3. Every connection shall be accessible for inspection, testing and maintenance.
4. The means of connection shall take account of the number and shape of the wires forming the conductor.
5. The connection shall take account of the cross-section of the conductor and the number of conductors to be connected.
6. The means of connection shall take account of the temperature attained in normal service.
7. There must be no mechanical strain on the conductor connections.

There is a wide range of suitable means of connecting conductors and we shall look at these in a moment. Whatever method is used to connect live conductors, the connection must be contained in an enclosed compartment such as an accessory; for example, a switch or socket box or a junction box. Alternatively, an equipment enclosure may be used; for example, a motor enclosure or an enclosure partly formed by non-combustible building material (IET Regulation 526.5). This is because faulty joints and terminations in live conductors can attain very high temperatures due to the effects of resistive heating. They might also emit arcs, sparks or hot particles with the consequent risk of fire or other harmful thermal effects to adjacent materials.

Types of terminal connection

Junction boxes

Junction boxes are probably the most popular method of making connections in domestic properties. Brass terminals are fixed inside a bakelite container. The two important factors to consider when choosing a junction box are the number of terminals required and the current rating. Socket outlet junction boxes have larger brass terminals than lighting junction boxes.

Key fact

Junction boxes are probably the most popular method of making connections in domestic properties.

Strip connectors

Strip connectors or a chocolate block is a very common method of connecting conductors. The connectors are mounted in a moulded plastic block in strips of 10 or 12. The conductors are inserted into the block and secured with the grub-screw. In order that the conductors do not become damaged, the screw connection must be firm but not over-tightened. The size used should relate to the current rating of the circuit. Figure 4.12 shows a strip connector.

Key fact

Strip connectors or a chocolate block is a very common method of connecting conductors.

Pillar terminal

A pillar terminal is a brass pillar with a hole through the side into which the conductor is inserted and secured with a set-screw. If the conductor is small in relation to the hole it should be doubled back. In order that the conductor does not become damaged, the screw connection should be tight but not over-tightened. Figure 4.12 shows a pillar terminal.

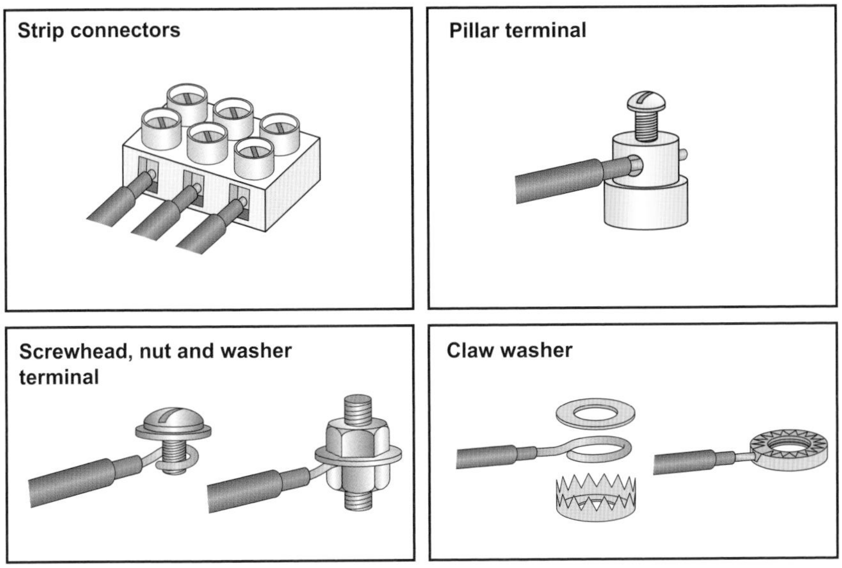

Figure 4.12 Types of terminal.

Screwhead, nut and washer terminals

The conductor being terminated is formed into an eye, as shown in Fig. 4.12. The eye should be slightly larger than the screw shank but smaller than the outside diameter of the screwhead, nut or washer. The eye should be placed on the screw shank in such a way that the rotation of the screwhead or nut will tend to close the joint in the eye.

Claw washers

In order to avoid inappropriate separation or spreading of individual wires of multiwire, claw washers are used to obtain a good sound connection. The looped conductor is laid in the pressing, as shown in Fig. 4.12, a plain washer is placed on top of the conductor and the metal points folded over the washer.

Crimp terminals

Crimp terminals are made of tinned sheet copper. The chosen crimp terminal is slipped over the end of the conductor and crimped with the special crimping tool. This type of connection is very effective for connecting equipotential bonding conductors to approved earth clamps.

Soldered joints or compression joints

Although the soldering of large underground cables is still common today, joints up to about 100 A are now usually joined with a compression joint. This uses the same principle as for the crimp termination above; it is just a little larger.

Whatever method is used to make the connection in conductors, the connection must be both electrically and mechanically sound if we are to avoid high-resistance joints, corrosion and erosion at the point of termination.

When terminating very fine multiwire conductors, the IET Regulation 526.9 gives us the following advice:

1. To avoid separation or spreading of individual wires, suitable terminals must be used or the conductor ends treated. For example, by enclosing the individual wires of multiwire in a brass ferrule or claw washer as shown in Fig 4.12.

Key fact

Whatever method is used to make the connection in conductors, the connection must be both electrically and mechanically sound if we are to avoid high-resistance joints, corrosion and erosion at the point of termination.

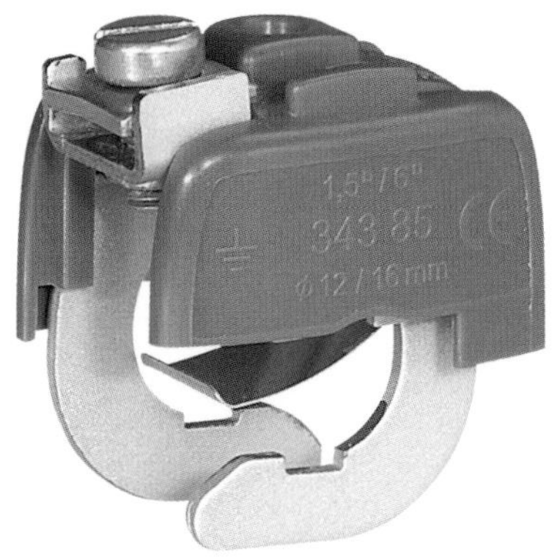

Figure 4.13 Rapid earth clamp.

Definition

Protective bonding is bonding for the purpose of safety.

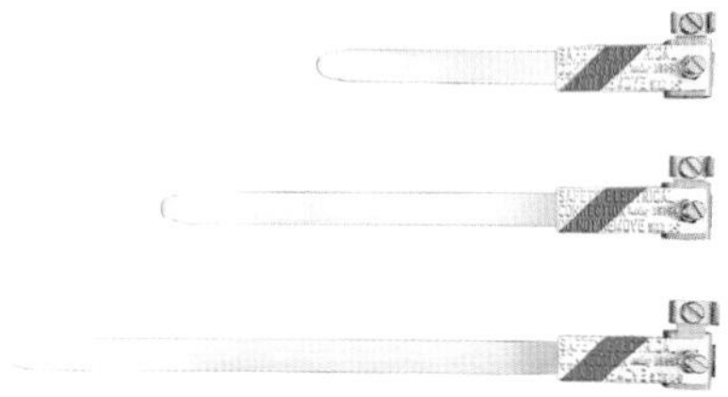

Figure 4.14 Adjustable bonding clamps.

2. Soldering or tinning of the whole conductor end of multiwire is not permitted if screw terminals are used.
3. Soldered or tinned conductor ends are not permissible at connection and junction points which may be subject in service to relative movement or vibration.

Protective bonding of main services – water and gas

Where earthed electrical equipment may come into contact with the metalwork of other services, they too must be effectively connected to the main earthing terminal of the installation (IET Regulations 411.3.1.2 and 544).

Other services are described as:

- main water pipes;
- main gas pipes;
- other service pipes and ducting;
- central heating and air-conditioning systems;
- exposed metal parts of the building structure;
- lightning protective conductors.

The main protective bonding conductors should be made to gas and water services at their point of entry into the building, as shown in Fig. 4.15, using insulated bonding conductors of not less than half the cross-section of the incoming main earthing conductor. The minimum permitted size is 6 mm^2 but the cross-section need not exceed 25 mm^2 (IET Regulation 544.1.1). The bonding clamp must be fitted on the consumer's side of the gas meter between the outlet union, before any branch pipework but within 600 mm of the meter (IET Regulation 544.1.2).

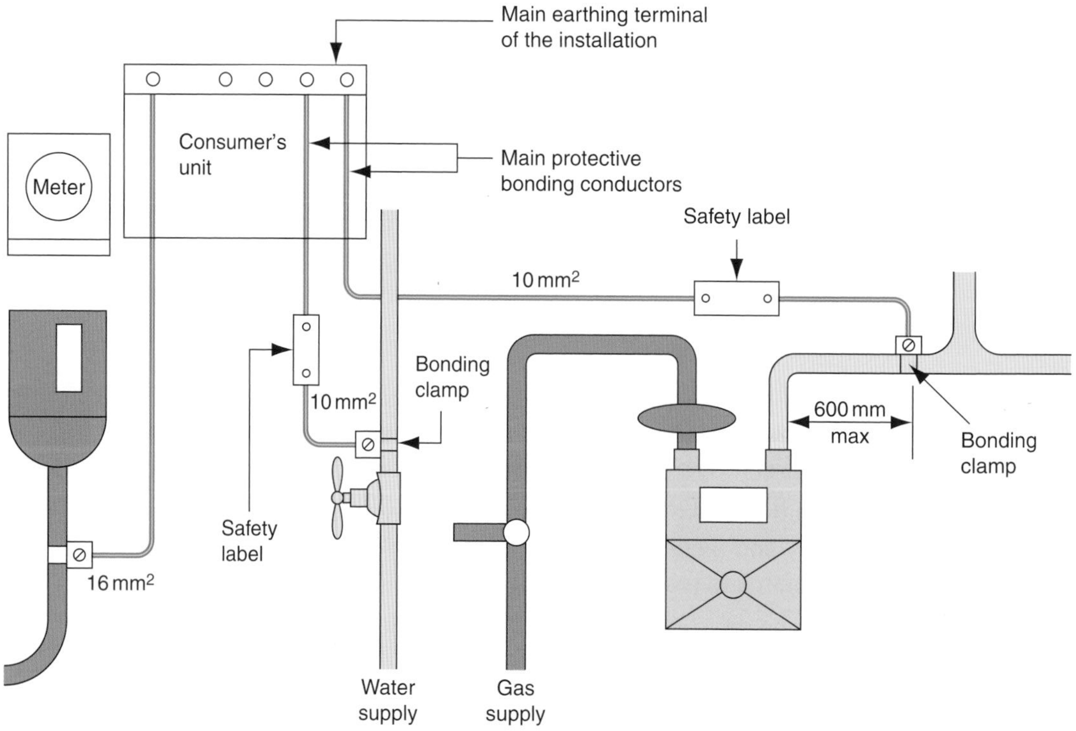

Figure 4.15 Main protective bonding of gas and water supplies.

A permanent label must also be fixed in a visible position at or near the point of connection of the bonding conductor with the words 'Safety Electrical Connection – Do Not Remove' (IET Regulation 514.13.1). Supplementary bonding is described in Chapter 3 and shown at Fig. 3.74.

Electrical installation testing

The electrical contractor is charged with a responsibility to carry out a number of tests on an electrical installation and electrical equipment. The individual tests are dealt with in Part 6 of the IET Regulations and described later in this chapter.

The reasons for testing the installation are:

- to ensure that the installation complies with the IET Regulations;
- to ensure that the installation meets the specification;
- to ensure that the installation is safe to use.

Those who are to carry out the electrical tests must first consider the following safety factors:

- an assessment of safe working practice must be made before testing begins;
- all safety precautions must be put in place before testing begins;
- everyone must be notified that the test process is about to take place, for example, the client and other workers who may be affected by the tests;
- 'permits-to-work' must be obtained where relevant;
- all sources of information relevant to the tests have been obtained;
- the relevant circuits and equipment have been identified;
- safe isolation procedures have been carried out – care must be exercised here, in occupied premises, not to switch off computer systems without first obtaining permission;
- those who are to carry out the tests are electrically skilled to do so.

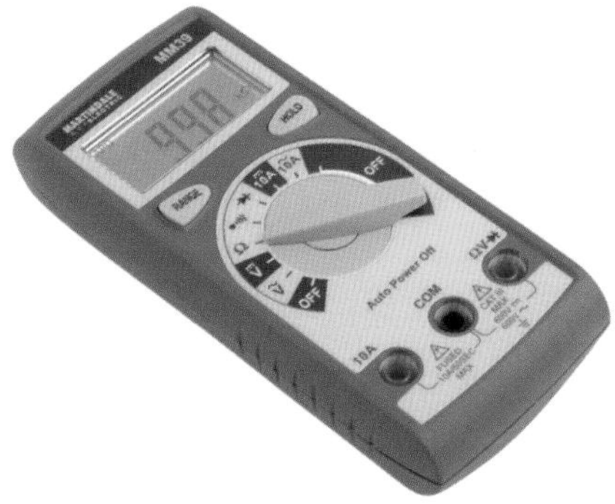

Figure 4.16 Martindale mm39 standard digital multimeter – auto ranging.

The electrical contractor is required by the IET Regulations to test all new installations and major extensions during erection and upon completion before being put into service. The contractor may also be called upon to test installations and equipment in order to identify and remove faults. These requirements imply the use of appropriate test instruments, and in order to take accurate readings consideration should be given to the following points:

- Is the instrument suitable for this test?
- Has the correct scale been selected?
- Is the test instrument correctly connected to the circuit?

Many commercial instruments are capable of making more than one test or have a range of scales to choose from. A range selector switch is usually used to choose the appropriate scale. A scale range should be chosen which suits the range of the current, voltage or resistance being measured. For example, when taking a reading in the 8 or 9 V or RV range, the obvious scale choice would be one giving 10 or 12V full scale deflection. To make this reading on an instrument with 100 V full-scale deflection would lead to errors, because the deflection is too small.

Ammeters must be connected in series with the load, and voltmeters in parallel across the load, as shown in Fig. 4.17. The power in a resistive load may be calculated from the readings of voltage and current since $P = VI$. This will give accurate calculations on both a.c. and d.c. supplies, but when measuring the power of an a.c. circuit which contains inductance or capacitance, a wattmeter must be used because the voltage and current will be out of phase.

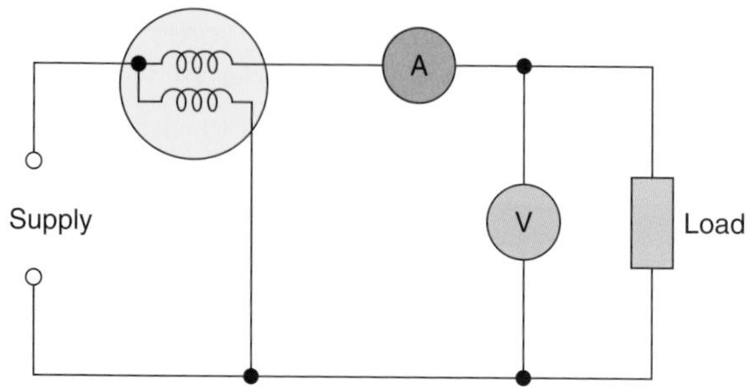

Figure 4.17 Wattmeter, ammeter and voltmeter correctly connected to a load.

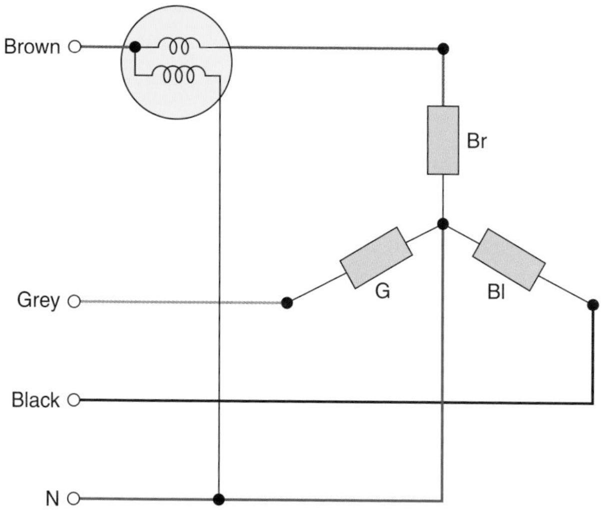

Figure 4.18 One-wattmeter measurement of power.

Measurement of power in a three-phase circuit

One-wattmeter method

When three-phase loads are balanced, for example, in motor circuits, one wattmeter may be connected into any phase, as shown in Fig. 4.18. This wattmeter will indicate the power in that phase and, since the load is balanced, the total power in the three-phase circuit will be given by:

$$\text{Total power} = 3 \times \text{wattmeter reading}$$

Two-wattmeter method

This is the most commonly used method for measuring power in a three-phase, three-wire system since it can be used for both balanced and unbalanced loads connected in either star or delta. The current coils are connected to any two of the lines, and the voltage coils are connected to the other line, the one without a current coil connection, as shown in Fig. 4.19. Then,

$$\text{Total power} = W_1 + W_2$$

This equation is true for any three-phase load, balanced or unbalanced, star or delta connection, provided there is no fourth wire in the system.

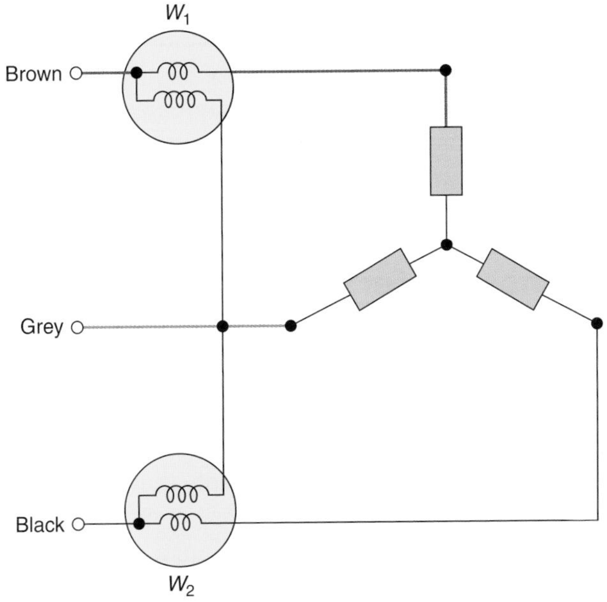

Figure 4.19 Two-wattmeter measurement of power.

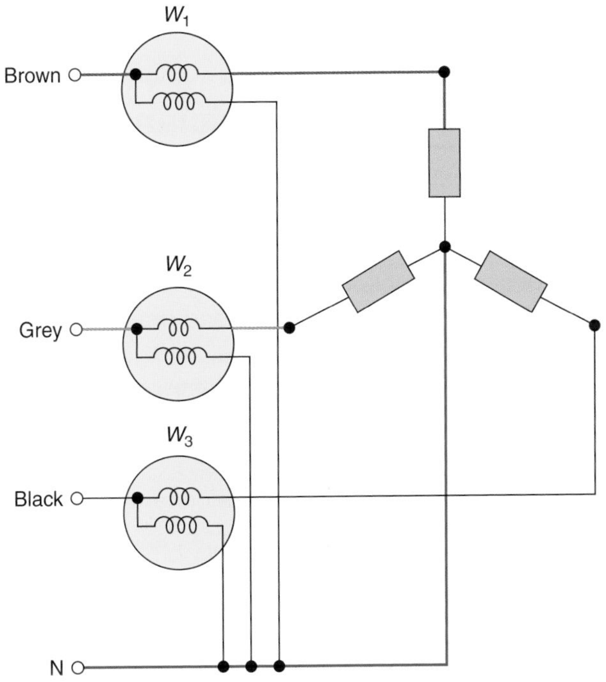

Figure 4.20 Three-wattmeter measurement of power.

Three-wattmeter method

If the installation is four-wire, and the load on each phase is unbalanced, then three-wattmeter readings are necessary, connected as shown in Fig. 4.20. Each wattmeter measures the power in one phase and the total power will be given by:

$$\text{Total power} = W_1 + W_2 + W_3$$

Figure 4.21 Tong tester or clip-on ammeter.

Tong tester

The tong tester, or clip-on ammeter, works on the same principle as the transformer. The laminated core of the transformer can be opened and passed over the busbar or single-core cable. In this way a measurement of the current being carried can be made without disconnection of the supply. The construction is shown in Fig. 4.21.

Phase sequence testers

Phase sequence is the order in which each phase of a three-phase supply reaches its maximum value. The normal phase sequence for a three-phase supply is brown–black–grey, which means that first brown, then black and finally the grey phase reaches its maximum value. IET Regulation 612.12 tells us that for polyphase circuits a test must be made to verify that the phase sequence is maintained.

Phase sequence has an important application in the connection of three-phase transformers. The secondary terminals of a three-phase transformer must not be connected in parallel until the phase sequence is the same.

A phase sequence tester can be an indicator which is, in effect, a miniature induction motor, with three clearly colour-coded connection leads. A rotating disc with a pointed arrow shows the normal rotation for phase sequence brown–black–grey. If the sequence is reversed, the disc rotates in the opposite direction to the arrow.

Test equipment used by electricians

The Health and Safety Executive (HSE) has published a guidance note (GS 38) which advises electricians and other electrically competent people on the selection of suitable test probes, voltage-indicating devices and measuring instruments. This is because they consider suitably constructed test equipment to be as vital for personal safety as the training and practical skills of the electrician. In the past, unsatisfactory test probes and voltage indicators have frequently been the cause of accidents, and therefore all test probes must now incorporate the following features:

1. The probes must have finger barriers or be shaped so that the hand or fingers cannot make contact with the live conductors under test.
2. The probe tip must not protrude more than 2 mm, and preferably only 1 mm, be spring-loaded and screened.
3. The lead must be adequately insulated and coloured so that one lead is readily distinguished from the other.
4. The lead must be flexible and sufficiently robust.
5. The lead must be long enough to serve its purpose but not too long.
6. The lead must not have accessible exposed conductors even if it becomes detached from the probe or from the instrument.
7. Where the leads are to be used in conjunction with a voltage detector they must be protected by a current limiting resistor.

A suitable probe and lead is shown in Fig. 4.22.

GS 38 also tells us that where the test is being made simply to establish the presence or absence of a voltage, the preferred method is to use a proprietary test lamp or voltage indicator which is suitable for the working voltage, rather than a multimeter. Accident history has shown that incorrectly set multimeters

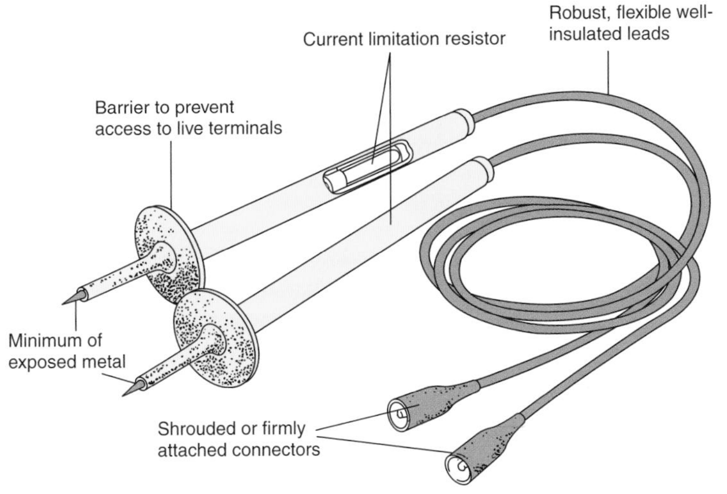

Figure 4.22 Recommended type of test probe and leads.

or makeshift devices for voltage detection have frequently caused accidents. Figure 4.23 shows a suitable voltage indicator. Test lamps and voltage indicators are not fail safe, and therefore GS 38 recommends that they should be regularly proved, preferably before and after use, as described in the flowchart for a safe isolation procedure shown in Fig 4.27.

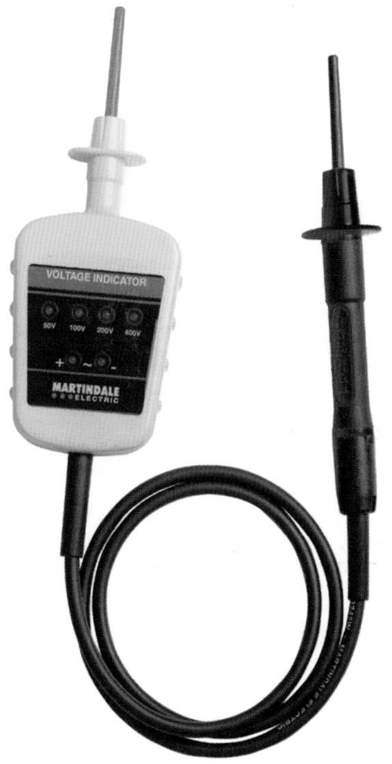

Figure 4.23 Typical voltage indicator.

Before testing begins

1. The circuits must be isolated using a 'safe isolation procedure', such as that described below, before beginning to test.
2. All test equipment must be 'approved' and connected to the test circuits by recommended test probes as described by GS 38. The test equipment used must also be 'proved' on a known supply or by means of a proving unit such as that shown in Fig. 1.26 in Chapter 1.
3. Isolation devices must be 'secured' in the 'off' position as shown in Fig. 4.25 and 4.26.
4. Warning notices must be posted.
5. All relevant safety and functional tests must be completed before restoring the supply.

Live testing

The **Electricity at Work Regulations** tell us that it is 'preferable' that supplies be made dead before work commences (Regulation 4(3)). However, it does acknowledge that some work, such as fault-finding and testing, may require the electrical equipment to remain energized. Therefore, if the fault-finding and testing can only be successfully carried out 'live', then the person carrying out the fault diagnosis must:

- be trained so that he or she understands the equipment and the potential hazards of working live and can, therefore, be deemed to be 'electrically skilled' to carry out the activity;
- only use approved test equipment;
- set up barriers and warning notices so that the work activity does not create a situation dangerous to others.

Definition

The *Electricity at Work Regulations* tell us that it is 'preferable' that supplies be made dead before work commences (Regulation 4(3)).

Safety first

Note that while live testing may be required in order to find the fault, live repair work must not be carried out. The individual circuit or item of equipment must first be isolated.

Figure 4.25 Always ensure that circuits are 'locked off' before working on them.

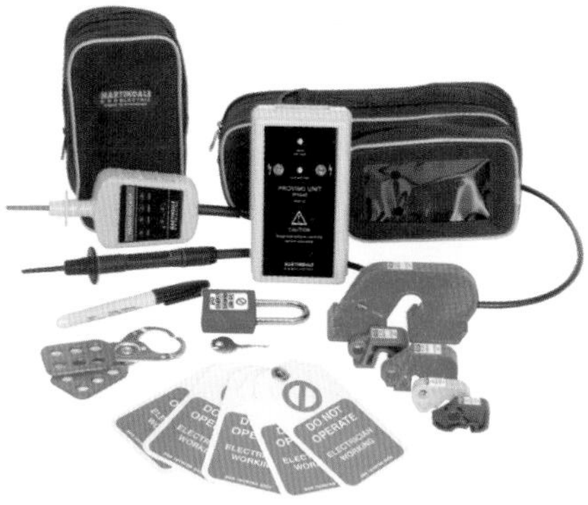

Figure 4.24 Comprehensive lock-out kit for use as part of a safe isolation procedure.

Note that while live testing may be required in order to find the fault, live repair work must not be carried out. The individual circuit or item of equipment must first be isolated.

Electrical circuits – safe isolation procedure

We looked at 'secure electrical isolation' in Chapter 1 as an essential part of reducing the risk of an electrical shock when working on-site.

In this chapter City & Guilds asks us once again to look at a 'safe isolation' procedure as a preparation for inspecting and testing an electrical installation.

If you are confident that you remember this topic from Chapter 1, go straight on to the next section, 'Inspection and testing an electrical installation'.

The Electricity at Work Regulations are very specific in describing the procedure to be used for isolation of the electrical supply. Regulation 12(1) tells us that *isolation* means the disconnection and separation of the electrical equipment from every source of electrical energy in such a way that this disconnection and separation is secure. Regulation 4(3) tells us that we must also prove the conductors dead before work commences and that the test instrument used for this purpose must itself be proved immediately before, and immediately after, testing the conductors. To isolate an individual circuit or item of equipment successfully, competently and safely we must follow a procedure such as that given by the flow diagram in Fig. 4.27. Start at the top and work your way down the flowchart. When you get to the heavy-outlined amber boxes, pause and ask yourself whether everything is satisfactory up to this point. If the answer is yes, move on. If no, go back as indicated by the diagram.

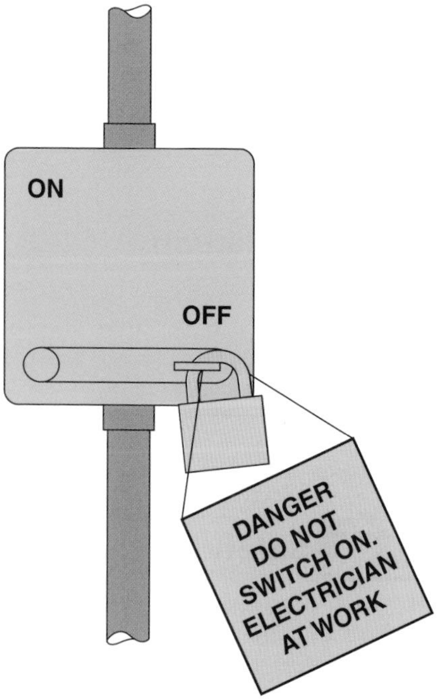

Figure 4.26 Secure isolation of a supply.

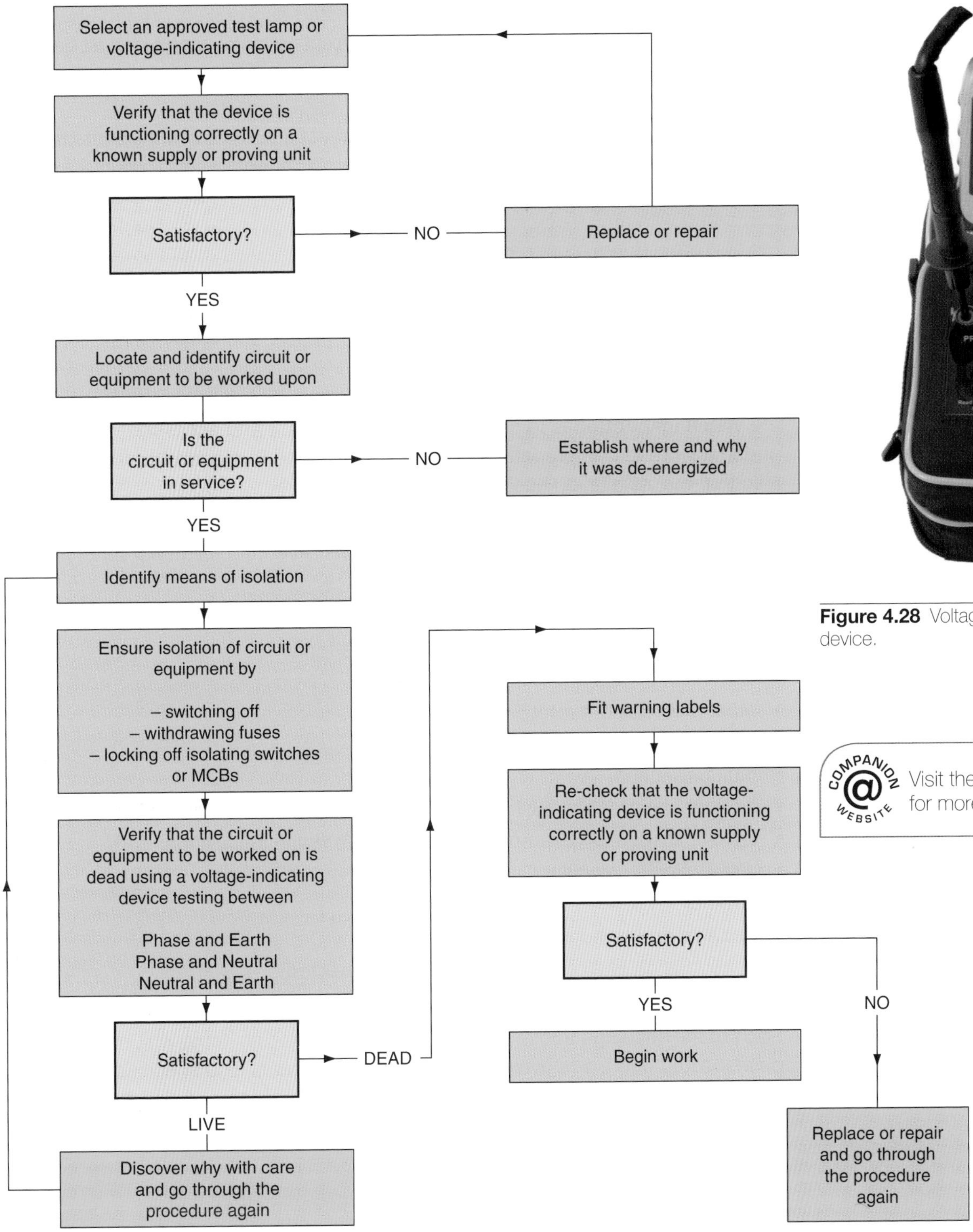

Figure 4.27 Method statement or flowchart for a secure isolation procedure.

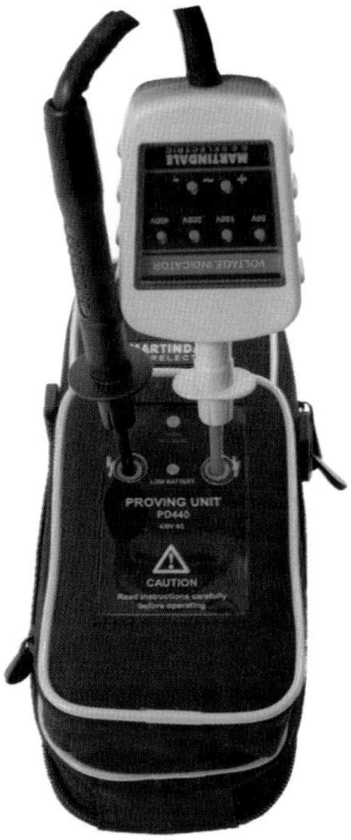

Figure 4.28 Voltage tester and proving device.

COMPANION WEBSITE @ Visit the companion website for more on this topic.

Inspection and testing an electrical installation

The testing of an installation implies the use of instruments to obtain readings. However, a test is unlikely to identify a cracked socket outlet, a chipped or loose switch plate or a missing conduit-box lid or saddle, so it is also necessary to make a visual inspection of the installation.

All new installations must be inspected and tested during erection and upon completion before being put into service. All existing installations should be periodically inspected and tested to ensure that they are safe and meet the IET Regulations (IET Regulations 610–634).

The method used to test an installation may inject a current into the system. This current must not cause danger to any person or equipment in contact with the installation, even if the circuit being tested is faulty. The test results must be compared with any relevant data, including the tables in the IET Regulations, and the test procedures must be followed carefully and in the correct sequence, as indicated by IET Regulation 612.1. This ensures that the protective conductors are correctly connected and secure before the circuit is energized.

Definition

The aim of the *visual inspection* is to confirm that all equipment and accessories are undamaged and comply with the relevant British and European Standards, and also that the installation has been securely and correctly erected.

Visual inspection

The installation must be visually inspected before testing begins. The aim of the **visual inspection** is to confirm that all equipment and accessories are undamaged and comply with the relevant British and European Standards, and also that the installation has been securely and correctly erected.

The 3rd Amendment to the IET Regulations in Appendix 6 gives a schedule or checklist of examples of items requiring inspection during the initial verification of the installation, including:

- electrical intake equipment;
- parallel or switched alternative sources of supply;
- automatic disconnection of supply;
- other methods of protection;
- basic protection;
- fault protection;
- additional protection such as RCD and bonding;
- specific inspection examples as appropriate to the installation;
- distribution equipment;
- **circuits**;
- isolation and switching;
- current using equipment (permanently connected);
- Part 7 Special installations and locations.

Key fact

Testing

- All new installations must be inspected and tested during erection and upon completion.
- All existing installations must be inspected and tested periodically.

IET Regulations 610–634

Each of the headings above is broken down into specific items which are to be inspected and the relevant IET Regulation number is also specified. So, for example, the heading **circuits** above is further subdivided as follows:

- identification of conductors (514.3.1);
- cables correctly supported throughout (522.8.5);
- examination of cables for damage during installation (522.6.2);
- concealed cables adequately protected against damage (522.6.101 and 522.6.103);
- provision of fire barriers and sealing arrangements to reduce spread of fire (Section 527);
- termination of cables at enclosures (Section 526);
- suitability of circuit accessories for external influences (512.2);
- single-pole switching devices in line conductor (132.14.1 and 522.8.5);
- circuit accessories not damaged during erection.
- the quality or adequacy of connections within accessories and equipment (Section 526).

Safety first

Live working

- **NEVER** work **LIVE**.
- Some 'live testing' is allowed by 'competent persons'.
- Otherwise, isolate and secure the isolation.
- Prove the supply dead before starting work.

See Appendix 6 of the 3rd Amendment to the 17th Edition of the IET Regulations for the complete list of items requiring inspection during initial verification.

It is thought that the new inspection schedule introduced by the 3rd Amendment to the 17th Edition for all new work will help electricians carry out the work in an organized and efficient manner. The schedule of inspections must be appended to the electrical installation certificate and given to the customer or client upon completion.

Following the inspection, those tests which are relevant to the installation must then be carried out in the sequence given in IET Regulation 612.1 for reasons of safety and accuracy. These tests are as follows:

Before the supply is connected:

1 Test for continuity of protective conductors, including protective bonding and supplementary bonding.
2 Test the continuity of all ring final circuit conductors.
3 Test for insulation resistance.
4 Test for polarity using the continuity method.
5 Test the earth electrode resistance.

With the supply connected:

6 Re-check polarity using a voltmeter or approved test lamp.
7 Test the earth fault loop impedance.
8 Check phase sequence.
9 Functional test – Carry out additional protection testing (e.g. operation of residual current devices, RCDs).

If any test fails to comply with the IET Regulations, then *all* the preceding tests must be repeated after the fault has been rectified. This is because the earlier test results may have been influenced by the fault (IET Regulation 612.1).

There is an increased use of electronic devices in electrical installation work; for example, in dimmer switches and ignitor circuits of discharge lamps. These devices should be temporarily disconnected so that they are not damaged by the test voltage of, for example, the insulation resistance test (IET Regulation 612.3).

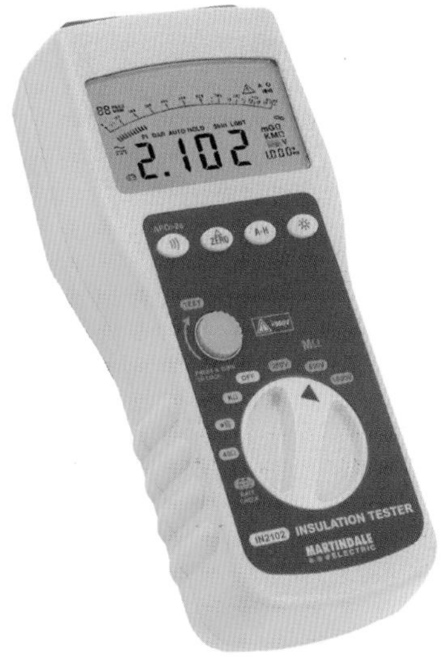

Figure 4.29 Multivoltage insulation and continuity tester.

Approved test instruments

The **test instruments and test leads** used by the electrician for testing an electrical installation must meet all the requirements of the relevant regulations. The HSE has published guidance note GS 38 for test equipment used by electricians. The IET Regulations (BS 7671) also specify the test voltage or current required to carry out particular tests satisfactorily. All test equipment must be chosen to comply with the relevant parts of BS EN 61557. All testing must, therefore, be carried out using an ‘approved’ test instrument if the test results are to be valid. The test instrument must also carry a calibration certificate, otherwise the recorded results may be void. **Calibration certificates** usually last for a year. Test instruments must, therefore, be tested and recalibrated each year by an approved supplier. This will maintain the accuracy of the instrument to an acceptable level, usually within 2% of the true value.

Definition

The *test instruments and test leads* used by the electrician for testing an electrical installation must meet all the requirements of the relevant regulations.

Definition

Calibration certificates usually last for a year. Test instruments must, therefore, be tested and recalibrated each year by an approved supplier.

Modern digital test instruments are reasonably robust, but to maintain them in good working order they must be treated with care. An approved test instrument costs as much as a good-quality camera; it should, therefore, receive the same care and consideration.

Let us now look at the requirements of four often-used test meters.

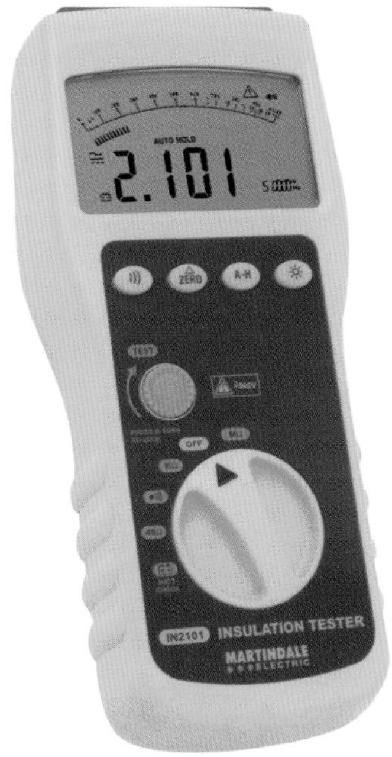

Figure 4.30 Insulation and continuity tester.

Continuity tester

To measure accurately the resistance of the conductors in an electrical installation, we must use an instrument which is capable of producing an open-circuit voltage of between 4 and 24 V a.c. or d.c., and delivering a short-circuit current of not less than 200 mA (IET Regulation 612.2.1). The functions of continuity testing and insulation resistance testing are usually combined in one test instrument.

Insulation resistance tester

The test instrument must be capable of detecting insulation leakage between live conductors and between live conductors and earth. To do this and comply with IET Regulation 612.3 the test instrument must be capable of producing a test voltage of 250, 500 or 1000 V and delivering an output current of not less than 1 mA at its normal voltage.

Earth fault-loop impedance tester

The test instrument must be capable of delivering fault currents as high as 25 A for up to 40 ms using the supply voltage. During the test, the instrument does an Ohm's law calculation and displays the test result as a resistance reading.

RCD tester

Where circuits are protected by an RCD we must carry out a test to ensure that the device will operate very quickly under fault conditions and within the time limits set by the IET Regulations. The instrument must, therefore, simulate a fault and measure the time taken for the RCD to operate. The instrument is, therefore, calibrated to give a reading measured in milliseconds to an in-service accuracy of 10%.

If you purchase good-quality 'approved' test instruments and leads from specialist manufacturers they will meet all the regulations and standards and therefore give valid test results. However, to carry out all the tests required by the IET Regulations will require a number of test instruments and this will represent a major capital investment in the region of £1,000.

Let us now consider the individual tests.

1 Testing for continuity of protective conductors, including main and supplementary protective bonding (Regulation 612.2.1)

The object of the test is to ensure that the circuit protective conductor (CPC) is correctly connected, is electrically sound and has a total resistance which is low enough to permit the over-current protective device to operate within the disconnection time requirements of IET Regulation 411.4.6, should an earth fault occur. Every protective conductor must be separately tested from the consumer's main protective earthing terminal to verify that it is electrically sound and correctly connected, including the main protective bonding and supplementary protective bonding conductors, as shown in Figs. 4.31 and 3.79. The IET Regulations describe the need to consider additional protection by supplementary protective bonding in situations where there is a high risk of electric shock such as kitchens and bathrooms (IET Regulation 415.2).

A d.c. test using an ohmmeter continuity tester is suitable where the protective conductors are of copper or aluminium up to 35 mm^2. The test is made with the supply disconnected, measuring from the consumer's main protective earthing terminal to the far end of each CPC, as shown in Fig. 4.32. The resistance of the

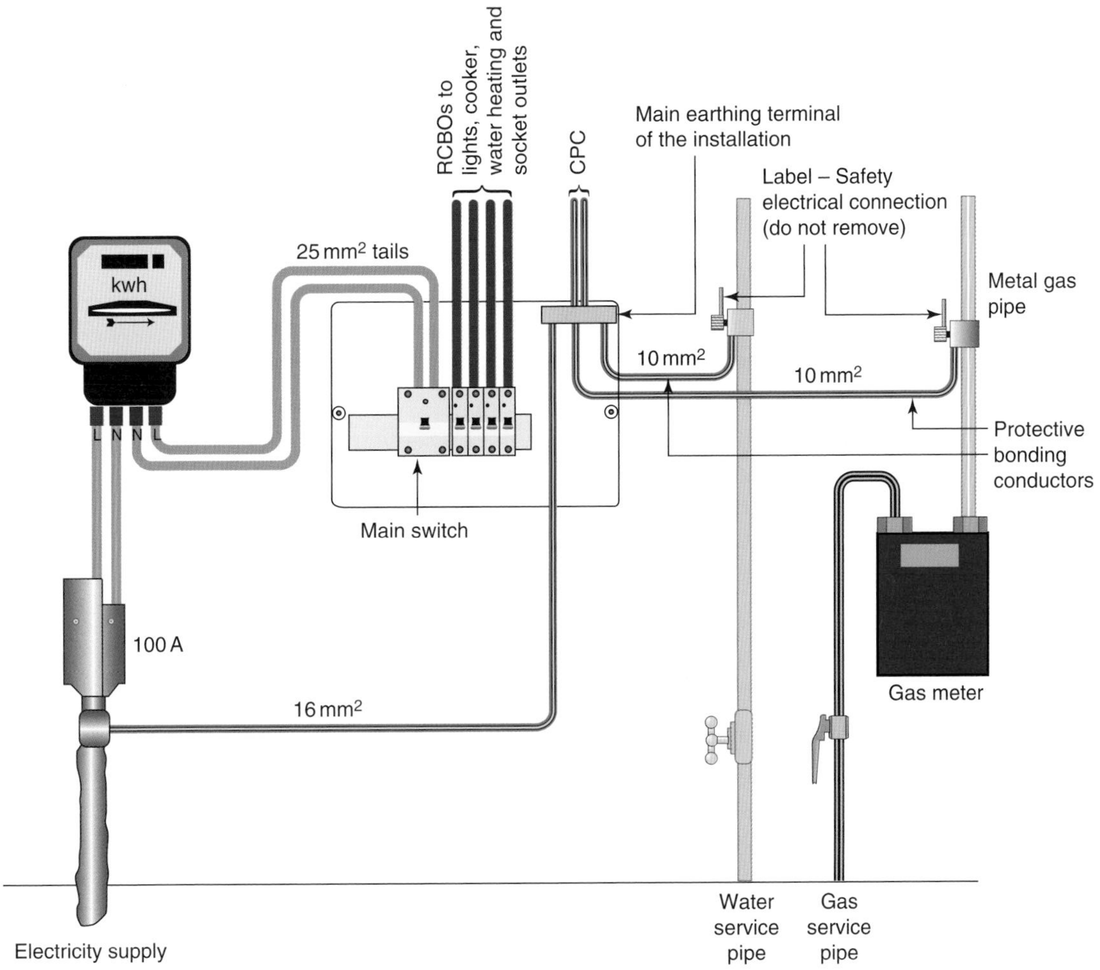

Figure 4.31 Cable sheath earth supplies (TN-S system): showing earthing and main protective bonding arrangements.

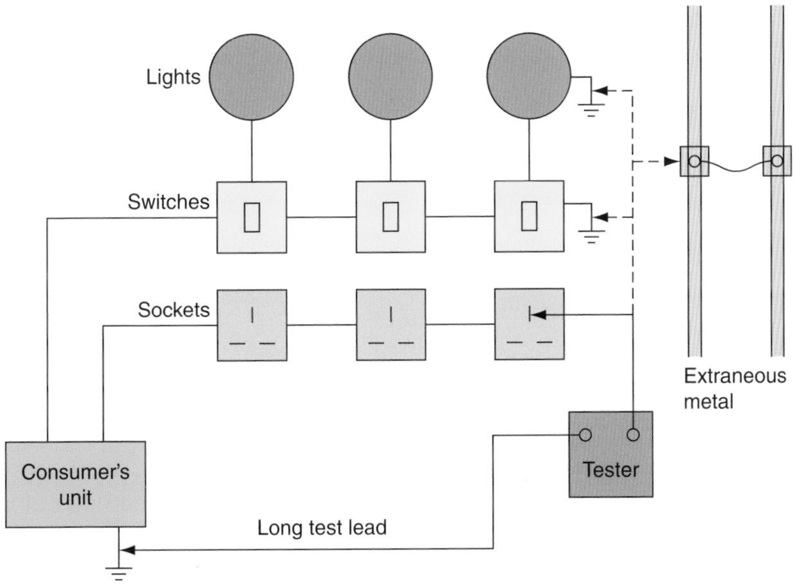

Figure 4.32 Testing continuity of protective conductors.

long test lead is subtracted from these readings to give the resistance value of the CPC. The result is recorded on an installation schedule such as that given in Appendix 6 of the IET Regulations.

A satisfactory test result for the bonding conductors will be in the order of 0.05 Ω or less (IET Guidance Note 3).

Where steel conduit or trunking forms the protective conductor, the standard test described above may be used, but additionally the enclosure must be visually checked along its length to verify the integrity of all the joints.

If the inspecting engineer has grounds to question the soundness and quality of these joints then the phase earth loop impedance test described later in this chapter should be carried out.

If, after carrying out this further test, the inspecting engineer still questions the quality and soundness of the protective conductor formed by the metallic conduit or trunking, then a further test can be done using an a.c. voltage not greater than 50 V at the frequency of the installation and a current approaching 1.5 times the design current of the circuit, but not greater than 25 A.

This test can be done using a low-voltage transformer and suitably connected ammeters and voltmeters, but a number of commercial instruments are available, such as the Clare tester, which give a direct reading in ohms.

Because fault currents will flow around the earth fault-loop path, the measured resistance values must be low enough to allow the over-current protective device to operate quickly. For a satisfactory test result, the resistance of the protective conductor should be consistent with those values calculated for a line conductor of similar length and cross-sectional area. Values of resistance per metre for copper and aluminium conductors are given in Table I1 of the *On Site Guide*. The resistances of some other metallic containers are given in Table 4.2.

Example

The CPC for a ring final circuit is formed by a $1.5\,mm^2$ copper conductor of 50 m approximate length. Determine a satisfactory continuity test value for the CPC using the value given in Table I1 of the *On Site Guide*.

Table I1 gives resistance/metre for a $1.5\,mm^2$ copper conductor $= 12.10\,m\Omega/m$

$$\text{Therefore, the resistance of 50m} = 50 \times 12.10 \times 10^{-3} = 0.605\,\Omega$$

The protective conductor resistance values calculated by this method can only be an approximation since the length of the CPC can only be estimated. Therefore, in this case, a satisfactory test result would be obtained if the resistance of the protective conductor was about 0.6 Ω. A more precise result is indicated by the earth fault-loop impedance test which is carried out later in the sequence of tests.

Table 4.2 Resistance values of some metallic containers

Metallic sheath	Size (mm)	Resistance at 20°C (mΩ/m)
Conduit	20	1.25
	25	1.14
	32	0.85
Trunking	50 × 50	0.949
	75 × 75	0.526
	100 × 100	0.337

2 Testing for continuity of ring final circuit conductors (Regulation 612.2.2)

The object of the test is to ensure that all ring circuit cables are continuous around the ring; that is, that there are no breaks and no interconnections in the ring, and that all connections are electrically and mechanically sound. This test also verifies the polarity of each socket outlet.

The test is made with the supply disconnected, using an ohmmeter as follows:

Disconnect and separate the conductors of both legs of the ring at the main fuse. There are three steps to this test:

Step 1

Measure the resistance of the line conductors (L_1 and L_2), the neutral conductors (N_1 and N_2) and the protective conductors (E_1 and E_2) at the mains position, as shown in Fig. 4.33. End-to-end live and neutral conductor readings should be approximately the same (i.e. within 0.05 Ω) if the ring is continuous. The protective conductor reading will be 1.67 times as great as these readings if 2.5/1.5 mm cable is used. Record the results on a table such as that shown in Table 4.3.

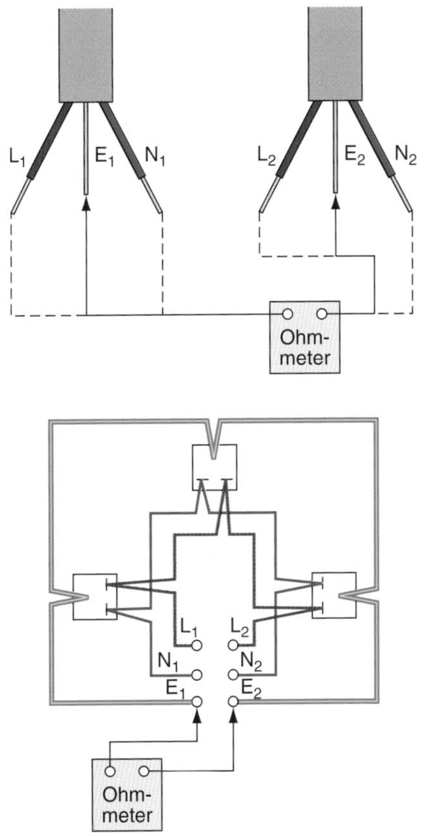

Figure 4.33 Step 1 test: measuring the resistance of phase, neutral and protective conductors.

Table 4.3 Table which may be used to record the readings taken when carrying out the continuity of ring final circuit conductors tests according to IET Regulation 612.2.2

Test	Ohmmeter connected to	Ohmmeter readings	This gives a value for
Step 1	L_1 and L_2		r_1
	N_1 and N_2		
	E_1 and E_2		r_2
Step 2	Live and neutral at each socket		
Step 3	Live and earth at each socket		$R_1 + R_2$
As a check ($R_1 + R_2$) value should equal ($r_1 + r_2$)/4.			

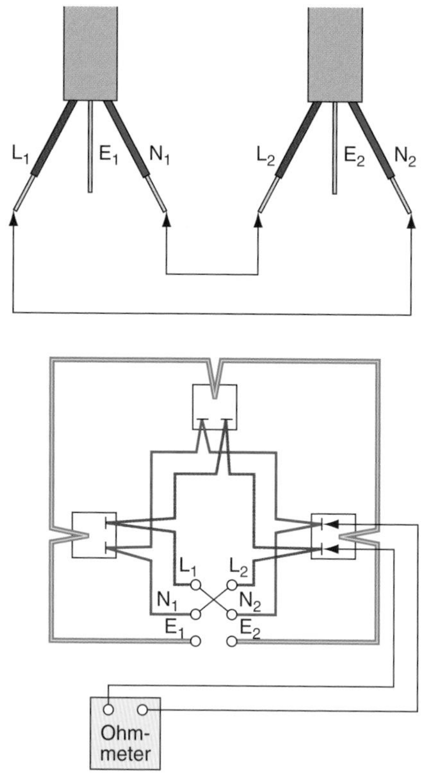

Figure 4.34 Step 2 test: connection of mains conductors and test circuit conditions.

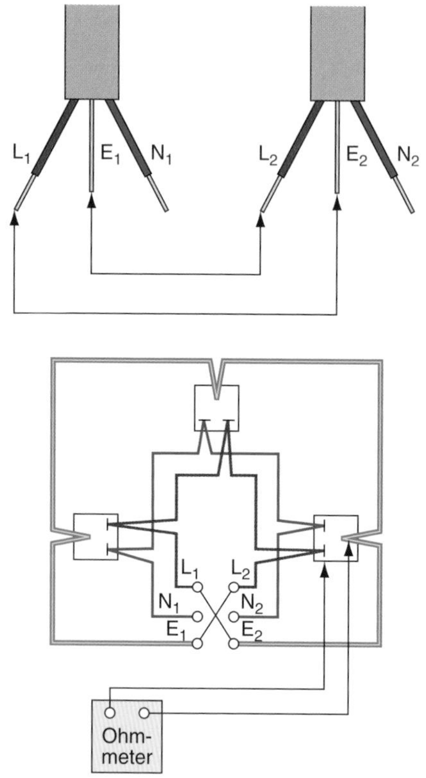

Figure 4.35 Step 3 test: connection of mains conductors and test circuit conditions.

Step 2

The live and neutral conductors should now be temporarily joined together, as shown in Fig. 4.34. An ohmmeter reading should then be taken between live and neutral at *every* socket outlet on the ring circuit. The readings obtained should be substantially the same, provided that there are no breaks or multiple loops in the ring. Each reading should have a value of approximately half the live and neutral ohmmeter readings measured in Step 1 of this test. Sockets connected as a spur will have a slightly higher value of resistance because they are fed by only one cable, while each socket on the ring is fed by two cables. Record the results on a table such as that shown in Table 4.3.

Step 3

Where the CPC is wired as a ring, for example, where twin and earth cables or plastic conduit is used to wire the ring, temporarily join the live and CPCs together, as shown in Fig. 4.35. An ohmmeter reading should then be taken between live and earth at *every* socket outlet on the ring. The readings obtained should be substantially the same provided that there are no breaks or multiple loops in the ring. This value is equal to $R_1 + R_2$ for the circuit. Record the results on an installation schedule such as that given in Appendix 6 of the IET Regulations or a table such as that shown in Table 4.3. The Step 3 value of $R_1 + R_2$ should be equal to $(r_1 + r_2)/4$, where r_1 and r_2 are the ohmmeter readings from Step 1 of this test (see Table 4.3).

3 Testing insulation resistance (Regulation 612.3)

The object of the test is to verify that the quality of the insulation is satisfactory and has not deteriorated or short-circuited. The test should be made at the consumer's unit with the mains switch off, all fuses in place and all switches closed. Neon lamps, capacitors and electronic circuits should be disconnected, since they will respectively glow, charge up and be damaged by the test.

There are two tests to be carried out using an insulation resistance tester which must have a test voltage of 500 V d.c. for 230 V and 400 V installations. These are line and neutral conductors to earth and between line conductors. The procedures are:

Line and neutral conductors to earth:

1 Remove all lamps.
2 Close all switches and circuit-breakers.
3 Disconnect appliances.
4 Test separately between the line conductor and earth *and* between the neutral conductor and earth, for *every* distribution circuit at the consumer's unit, as shown in Fig. 4.36 (a). Record the results on a schedule of test results such as that given in Appendix 6 of the IET Regulations.

Between line conductors:

1 Remove all lamps.
2 Close all switches and circuit-breakers.
3 Disconnect appliances.
4 Test between line and neutral conductors of *every* distribution circuit at the consumer's unit, as shown in Fig. 4.36 (b), and record the result.

The insulation resistance readings for each test must be not less than 1.0 MΩ for a satisfactory result (IET Regulation 612.3.2).

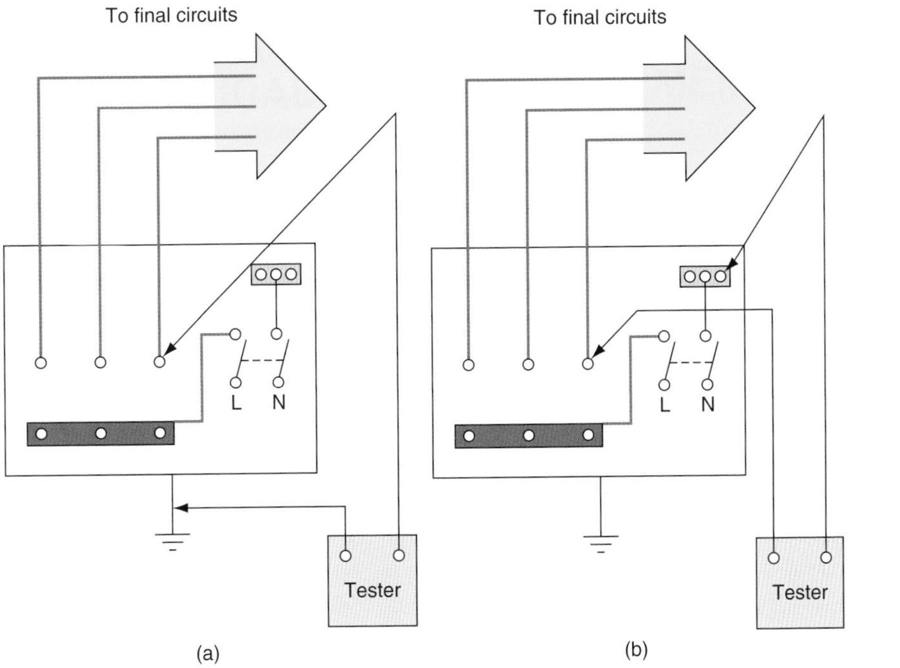

Figure 4.36 Insulation resistance test.

Where the circuit includes electronic equipment which might be damaged by the insulation resistance test, a measurement between all live conductors (i.e. live and neutral conductors connected together) and the earthing arrangements may be made. The insulation resistance of these tests should be not less than 1.0 MΩ (IET Regulation 612.3.3).

Although an insulation resistance reading of 1.0 MΩ complies with the regulations, the IET guidance notes tell us that much higher values than this can be expected and that a reading of less than 2 MΩ might indicate a latent, but not yet visible, fault in the installation. In these cases each circuit should be separately tested to obtain a reading greater than 2 MΩ.

4 Testing polarity (Regulation 612.6)

The object of this test is to verify that all fuses, circuit-breakers and switches are connected in the line or live conductor only, that all socket outlets are correctly wired and that Edison screw-type lamp holders have the centre contact connected to the live conductor. It is important to make a polarity test on the installation since a visual inspection will only indicate conductor identification.

The test is done with the supply disconnected using an ohmmeter or continuity tester as follows:

1. Switch off the supply at the main switch.
2. Remove all lamps and appliances.
3. Fix a temporary link between the line and earth connections on the consumer's side of the main switch.
4. Test between the 'common' terminal and earth at each switch position.
5. Test between the centre pin of any Edison screw lamp holders and any convenient earth connection.
6. Test between the live pin (i.e. the pin to the right of earth) and earth at each socket outlet, as shown in Fig. 4.37.

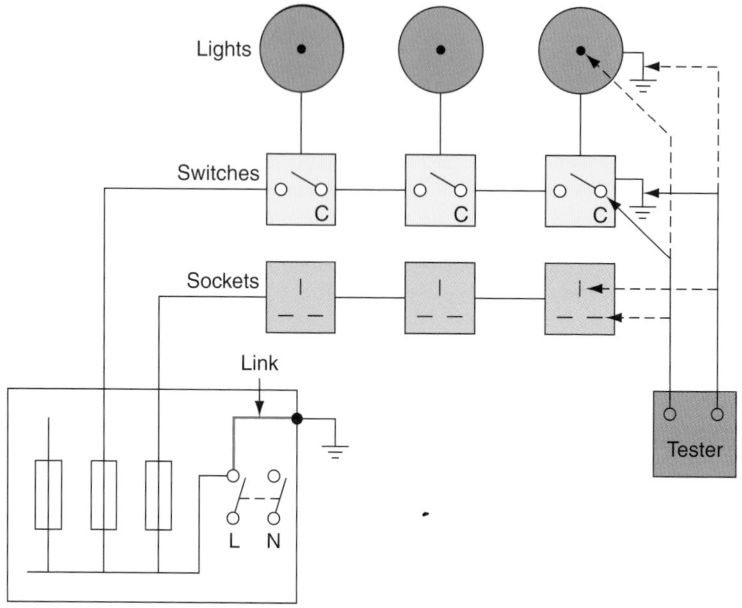

Figure 4.37 Polarity test.

For a satisfactory test result the ohmmeter or continuity meter should read very close to zero for each test.

Remove the test link and record the results on a schedule of test results such as that given in Appendix 6 of the IET Regulations.

5 Functional testing (Regulation 612.13)

The object of this test is to ensure that the accessories and equipment do what they are supposed to do within the installation. This is a hands-on test of all switching and control devices forming a part of the installation. For example:

- Do the switch toggles operate as they are supposed to operate?
- Do the electrical isolators such as that shown in Fig 4.26 switch on and off as they are supposed to do and are you confident that the device is robustly fixed and secure?
- Does the test button on any installed RCDs effectively operate the device?

Certification and reporting (Regulation 631)

Following the completion of all new electrical work or additional work to an existing installation, the installation must be inspected and tested and an installation certificate issued and signed by a competent person. From the introduction of Amendment 3 of the IET Regulations the 'competent' person is clarified as being a person who is skilled (electrically) or instructed (electrically). That person must have a sound knowledge of the type of work undertaken, be fully versed in the inspection and testing procedures contained in the IET Regulations (BS 7671) and employ adequate testing equipment.

A certificate of test results and a copy of the completed inspection schedule shall be issued to those ordering the work in the format given in Appendix 6 of the IET Regulations.

All installations must be periodically tested and inspected, and for this purpose an Electrical Installation Condition Report should be issued (IET Regulation 631.2).

The standard format is again shown in Appendix 6 of the IET Regulations. The 3rd Amendment of the IET Regulations now also requires an inspection within any accessible roof spaces where electrical equipment is present in that roof space.

In both cases the certificate must include the schedule of test results and the inspection schedule which verify that the installation complies with the IET Regulations at the time of testing.

Suggested frequency of periodic inspection intervals are given below:

- domestic installations – ten years;
- commercial installations – five years;
- industrial installations – three years;
- agricultural installations – three years;
- caravan site installations – one year;
- caravans – three years;
- temporary installations on construction sites – three months.

Safe working procedures when testing

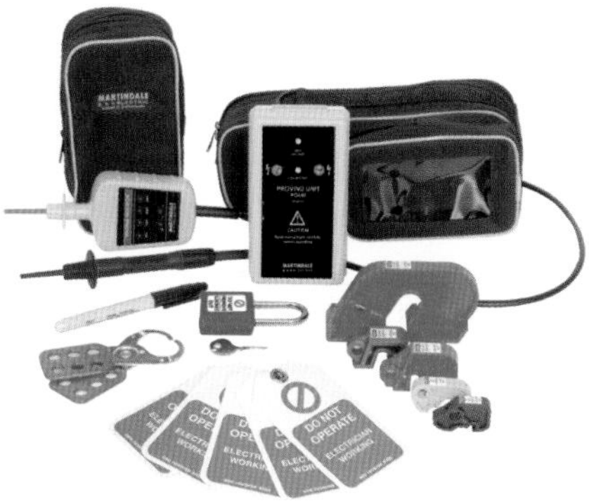

Figure 4.38 Comprehensive lock out kit for use as part of a safe isolation procedure

Whether you are carrying out the test procedure (i) as a part of a new installation, (ii) upon the completion of an extension to an existing installation, (iii) because you are trying to discover the cause of a fault on an installation or (iv) because you are carrying out an Electrical Installation Condition Report of a building, you must always be aware of your safety, the safety of others using the building and the possible damage which your testing might cause to other systems in the building.

For your own safety:

- always use 'approved' test instruments and probes;
- ensure that the test instrument carries a valid calibration certificate, otherwise the results may be invalid;
- secure all isolation devices in the 'off' position;
- put up warning notices so that other workers will know what is happening;
- notify everyone in the building that testing is about to start and for approximately how long it will continue;
- obtain a 'permit-to-work' if this is relevant;
- obtain approval to have systems shut down which might be damaged by your testing activities. For example, computer systems may 'crash' when supplies are switched off. Ventilation and fume extraction systems will stop working when you disconnect the supplies.

For the safety of other people:

- fix warning notices around your work area;
- use cones and highly visible warning tape to screen off your work area;
- make an effort to let everyone in the building know that testing is about to begin. You might be able to do this while you carry out the initial inspection of the installation;
- obtain verbal or written authorization to shut down information technology, emergency operation or stand-by circuits.

To safeguard other systems:

- computer systems can be severely damaged by a loss of supply or the injection of a high test voltage from, for example, an insulation resistance test. Computer systems would normally be disconnected during the test period but this will generally require some organization before the testing begins. Commercial organizations may be unable to continue to work without their computer systems and, in these circumstances, it may be necessary to test outside the normal working day;

- any resistance measurements made on electronic equipment or electronic circuits must be achieved with a battery-operated ohmmeter in order to avoid damaging the electronic circuits;
- farm animals are creatures of habit and may become very grumpy to find you testing their milking parlour equipment at milking time;
- hospitals and factories may have emergency stand-by generators which re-energize essential circuits in the event of a mains failure. Your isolation of the circuit for testing may cause the emergency systems to operate. Discuss any special systems with the person authorizing the work before testing begins.

Check your understanding

When you have completed these questions, check out the answers at the back of the book.

Note: more than one multiple-choice answer may be correct.

1 A tong test instrument can also correctly be called:
   a. a continuity tester
   b. a clip-on ammeter
   c. an insulation resistance tester
   d. a voltage indicator.

2 All electrical test probes and leads must comply with the standards set by the:
   a. BS EN 60898
   b. BS 7671
   c. HSE Guidance Note GS 38
   d. IET Regulations Part 2.

3 When making a test to determine the presence or absence of a voltage, the HSE recommends that for our own safety we should use:
   a. any old tester bought at a car-boot sale
   b. a multimeter set to the correct voltage
   c. a proprietary test lamp
   d. a voltage indicator.

4 For electrical test results to be valid the test instruments used:
   a. must be new
   b. must be of an approved type
   c. must have a calibration certificate
   d. must have a digital readout.

5 The test required by the IET Regulations to ascertain that the CPC is correctly connected is called:
   a. a basic protection
   b. continuity of ring final circuit conductors
   c. continuity of protective conductors
   d. earth electrode resistance.

6 One objective of the polarity test is to verify that:
   a. lamp holders are correctly earthed
   b. final circuits are correctly fused
   c. the CPC is continuous throughout the installation
   d. the protective devices are connected in the live conductor.

7 When testing a 230 V installation an insulation resistance tester must supply a voltage of:
   a. less than 50 V
   b. 500 V
   c. less than 500 V
   d. greater than twice the supply voltage but less than 1000 V.

8 The value of a satisfactory insulation resistance test on each final circuit of a 230 V installation must be:
   a. less than 1 Ω
   b. less than 0.5 MΩ
   c. not less than 0.5 MΩ
   d. not less than 1 MΩ.

9 Instrument calibration certificates are usually valid for a period of:
   a. Three months
   b. One year
   c. Three years
   d. Five years.

10 The maximum inspection and re-test period for a domestic electrical installation is:
   a. Three months
   b. Three years
   c. Five years
   d. Ten years.

11 A visual inspection of a new installation must be carried out:
   a. during the erection period
   b. during testing upon completion
   c. after testing upon completion
   d. before testing upon completion.

12 'To ensure that the accessories and equipment within a building work as they were intended to work' is one definition of the purpose of:
   a. testing electrical equipment
   b. inspecting electrical systems
   c. functional testing
   d. polarity testing.

13 The Electricity at Work Regulations tell us that 'we must ensure the disconnection and separation of electrical equipment from every source of supply and the separation must be secure'. A procedure to comply with this regulation is called:
   a. working at height
   b. a hazard risk assessment
   c. a safe isolation procedure
   d. a workstation risk assessment.

14 The Electricity at Work Regulations absolutely forbid the following work activity:
   a. working at height
   b. testing live electrical systems
   c. live repair work on electrical circuits
   d. working without the appropriate PPE.

15 'Good housekeeping' at work is about:
   a. cleaning up and putting waste in the skip
   b. working safely
   c. making the tea and collecting everyone's lunch
   d. putting tools and equipment away after use.

16 Which hand tools would you use for terminating conductors in a junction box?
   a. a pair of side cutters or knife
   b. a screwdriver
   c. a wood chisel and saw
   d. a tenon saw.

17 Which hand tools would you use for removing cable insulation?
   a. a pair of side cutters or knife
   b. a screwdriver
   c. a wood chisel and saw
   d. a tenon saw.

18 Which hand tools would you use to cut across a floorboard before lifting?
   a. a pair of side cutters or knife
   b. a screwdriver
   c. a wood chisel and saw
   d. a tenon saw.

19 Which hand tools would you use to cut and remove a notch in a floor joist?
   a. a pair of side cutters or knife
   b. a screwdriver
   c. a wood chisel and saw
   d. a tenon saw.

20 PAT testing is carried out on:
   a. hand tools
   b. domestic appliances only
   c. work electrical tools
   d. electrical equipment (e.g. 110 V transformers).

21 When PAT testing a 110 V electric drill we are testing the:
   a. efficiency of the drill
   b. earth continuity of the drill
   c. hammer action of the drill
   d. insulation resistance of the drill.

22 Under the Health and Safety at Work Act an employer is responsible for:
   a. maintaining plant and equipment
   b. providing PPE
   c. wearing PPE
   d. taking reasonable care to avoid injury.

23 Under the Health and Safety at Work Act an employee is responsible for:
   a. maintaining plant and equipment
   b. providing PPE
   c. wearing PPE
   d. taking reasonable care to avoid injury.

24 Before beginning work on a 'live' circuit or piece of equipment you should:
   a. only work 'live' if your supervisor is with you
   b. only work 'live' if you feel that you are 'competent' to do so
   c. isolate the circuit or equipment before work commences
   d. secure the isolation before work commences.

25 Use bullet points to state three reasons for testing a new electrical installation.

26 State five of the most important safety factors to be considered before electrical testing begins.

27 State the seven requirements of GS 38 when selecting probes, voltage indicators and measuring instruments.

28 Use bullet points to describe a safe isolation procedure of a final circuit fed from an MCB in a distribution board.

29 IET Regulation 611.3 gives a checklist of about 20 items to be considered in the initial visual inspection of an electrical installation. Make a list of ten of the most important items to be considered in the visual inspection process (perhaps by joining together similar items).

30 State three reasons why electricians must only use 'approved' test instruments.

31 State the first five tests to be carried out on a new electrical installation following a satisfactory 'inspection'. For each test:
   i. state the object (reason for) the test
   ii. state a satisfactory test result.

32 State the certification process for a:
   i. new electrical installation, and
   ii. an electrical installation that is being re-tested as part of an Electrical Installation Condition Report
   iii. what will be indicated on the test certificates
   iv. who will receive the test certificates, and
   v. who will issue the certificates
   vi. Finally, what type of person will carry out the actual testing.

33 State three safe working procedures relevant to your own safety when carrying out electrical testing.

34 State three safe working procedures relevant to the safety of other people when carrying out electrical testing.

35 State three safe working procedures relevant to the safety of other electrical systems when carrying out electrical testing.

36 State the two important tests that a PAT tester carries out on a portable appliance.

37 Use bullet points to state the reasons for commissioning a new building upon its completion.

38 Describe 'good practice' when handling and storing hand and power tools.

39 State five points that a PAT inspection and test checks on each appliance.

40 State what you would do with electrical tools on-site when you have finished using them in order to make sure:
   a. they remain in good condition
   b. they are available the next time you want to use them.

41 Briefly describe that we mean by 'good housekeeping' on-site.

42 State some of the actions you could take at work that would make the work environment safer and that could be considered 'good housekeeping'.

43 Slips, trips and falls are the most common causes of accidents at work. What can you do at work to reduce the possibility of an accident being caused by a slip, trip or fall?

44 Define what is meant by PPE.

45 State five pieces of PPE which a trainee could be expected to wear at work and the protection given by each piece.

46 What is a proving unit used for?

47 The HSE Guidance Note GS 38 tells us about suitable test probe leads. Use a sketch to identify the main recommendations.

48 Use a sketch to describe the main protective bonding arrangements for the gas, water and electricity supplies in a domestic property.

CHAPTER 5

C&G Unit 210

Communicating with others in building services engineering

Unit 210 of the City and Guilds 2365-02 syllabus

Learning outcomes – when you have completed this chapter you should:

- identify the members of the construction team;
- identify the members of the electrical team;
- identify statutory legislation and guidance information used in the electrical industry;
- know how to communicate with others in the workplace;
- identify technical information used in the workplace;
- identify the purpose of on site information;
- state company policies.

Basic Electrical Installation Work. 978-1-138-84884-9

This chapter has free associated content, including animations and instructional videos, to support your learning.

When you see the logo, visit the companion website for more on this topic www.routledge.com/cw/linsley

Organizations having electrotechnical activities in building services engineering

When we talk about the electrotechnical industry we are referring to all those different organizations or companies which provide an electrical service of some kind.

Electrical contractors install equipment and systems in new buildings. Once a building is fully operational, the electrical contractor may provide a maintenance service to that client or customer or alternatively the client may employ an 'in-house' electrician to maintain the installed electrical equipment. It all depends on the amount of work to be done and the complexity of the customers' systems.

The City and Guilds Syllabus directs us to look at *twelve* different organizations having electrotechnical activities and *ten* services provided by the electrotechnical industry, so here goes.

1. Electrical contractors
   - electrical contractors provide a design and installation service for all types of buildings and construction projects
   - the focus of this type of organization is on all types of electrotechnical activities in and around buildings
   - they install electrical equipment
   - they install electrical wiring systems
   - they carry out their installation work in domestic, commercial, industrial, agricultural and horticultural buildings.
2. Factories
   - factories contain lots of electrical plant and equipment
   - the wheels of all types of industry are driven by electromechanical devices and electrotechnical activities, such as motors and control systems.
3. Process plants
   - whether they process food or nuclear fuels, the prime mover for all processes is electrical plant, control and instrumentation equipment and machine drives.
4. Local councils
   - they are responsible for many different types of community buildings from town halls to swimming pools
   - the buildings all have electrical systems which require installation, maintenance and repair.
5. Commercial buildings and complexes
   - the 'office-type' activities carried out in these buildings require that electrical communication and data transmission systems are installed, maintained and repaired.
6. Leisure centres
   - these types of buildings contain lots of equipment driven by human sweat but which is also controlled and monitored by electrical and electronic systems

- leisure centres might contain a swimming pool or 'hot-air' sauna. Both types of electrical installation are considered 'Special Installations' by the IEE Regulations BS: 7671.

7. Panel builders
   - build specialist control, protection and isolation main switchgear systems for commerce and industry
   - the panel incorporates the isolation and protection systems required by the electrical installation.
8. Motor rewind and repair
   - electrical motors and their drives usually form an integral part of the industrial system or process
   - electrical motors and transformers sometimes break down or burn out
   - an exact new replacement can often be quickly installed
   - alternatively the existing motor can be rewound and reconditioned by a specialist company if time permits.
9. Railways
   - the prime mover for a modern inter-city-type electric train is an electric motor
   - electric trains require an infrastructure of electrical transmission lines throughout the network
   - all rail movements require signal and control systems
   - railway station buildings contain electrical and electronic installations.
10. The armed forces
    - operate in harsh, hostile and unpredictable environments
    - they need to adapt, modify and repair electrical and electronic systems in a war situation away from their home base and a comfortable, well-equipped workshop
    - a modern warship can contain as many people as an English village. They need electrotechnical systems to support them and to keep them safe 24 hours per day, 7 days per week.
11. Hospitals
    - contain a great deal of high-technology equipment
    - this equipment requires power and electronic systems
    - life monitoring equipment must continue to operate in a power failure
    - standby electrical supplies are, therefore, often an important part of a hospital's electrical installations.
12. Equipment and machine manufacturers
    - white goods, brown goods, computer hardware, motors and transformers are manufactured to meet the increasing demands of the domestic, commercial and industrial markets
    - they are manufactured to very high standards and often contain very sophisticated electrical and electronic circuits and systems
    - they manufacture to British and European Standards.

Try this

Information

- Which type of electrotechnical organization do you belong to?
- What types of work have you carried out so far?

Services provided by the electrotechnical industry

1. Lighting and power installations ensure that the building in which they are installed
   - is illuminated to an appropriate level
   - is heated to a comfortable level
   - has the power circuits to drive the electrical and electronic equipment required by those who will use the buildings.
2. Emergency lighting and security systems
   - ensure that the building is safe to use in unforeseen or adverse situations, and
   - is secure from unwanted intruders.
3. Building management and control systems
   - provide a controlled environment for the people who use commercial buildings
   - provide a pleasant environment so that people can work effectively and efficiently.
4. Instrumentation
   - electrical instrumentation allows us to monitor industrial processes and systems often at a safe distance.
5. Electrical maintenance
   - a programme of planned maintenance allows us to maintain the efficiency of all installed systems.
6. Live cable jointing
   - making connections to 'live' cables provides a means of connecting new installations and services to existing live supply cables without inconvenience to existing supplies caused by electrical shutdown. This work requires special training.
7. Highway electrical systems
   - illuminated motorways, roads and traffic control systems make our roads and pavements safe for vehicles and pedestrians.
8. Electrical panel building
   - main electrical panels provide a means of electrical isolation and protection
   - they also provide a means of monitoring and measuring electrical systems in our commercial and industrial buildings.
9. Electrical machine drive installations
   - electrical machine drives drive everything that makes our modern life comfortable, from
   - trains and trams to
   - lifts and air-conditioning units
   - refrigerators, freezers and all types of domestic appliances.
10. Consumer and commercial electronics
    - give us data processing and number crunching
    - electronic mail and access to information on the world wide web
    - access to high-quality audio and video systems.

Try this

Information

Which services does the company you work for provide within the electrotechnical industry?

Roles and responsibilities of workers in the electrotechnical industry

Any electrotechnical organization is made up of a group of individuals with various duties, all working together for their own good, the good of their employer and their customers.

There is often no clear distinction between the duties of the individual employees; each do some of the others' work activities.

Responsibilities vary, even by people holding the same job title, and some individuals hold more than one job title. However, let us look at some of the roles and responsibilities of those working in the electrotechnical industry.

1. Design engineer
   - will normally meet with clients and other trade professionals to interpret the customers' requirements
   - will produce the design specification which enables the cost of the project to be estimated.
2. Estimator/cost engineer
   - measures the quantities of labour and material necessary to complete the electrical project using the plans and specifications for the project
   - from these calculations and the company's fixed costs, a project cost can be agreed.
3. Contracts manager
   - may oversee a number of electrical contracts on different sites
   - will monitor progress in consultation with the project manager on behalf of the electrical companies
   - cost out variations to the initial contract
   - may have health and safety responsibilities because he or she has an overview of all company employees and contracts in progress.
4. Project manager
   - is responsible for the day-to-day management of one specific contract
   - will have overall responsibility on that site for the whole electrical installation
   - attends site meetings with other trades as the representative of the electrical contractor.
5. Service manager
   - monitors the quality of the service delivered under the terms of the contractor
   - checks that the contract targets are being met
   - checks that the customer is satisfied with all aspects of the project
   - the service manager's focus is customer specific while the project manager's focus is job specific.

6. Technician
   - will be more office based than site based
   - will carry out surveys of electrical systems
   - update electrical drawings
   - obtain quotations from suppliers
   - maintain records such as ISO 9000 quality systems
   - carry out testing inspections and commissioning of electrical installations
   - trouble shoot.
7. Supervisor/foreman
   - will probably be a mature electrician
   - have responsibility for small contracts
   - have responsibility for a small part of a large contract
   - be the leader of a small team (e.g. electrician and trainee) installing electrical systems.
8. Operative or skilled operative
   - will carry out the electrical work under the direction and guidance of a supervisor
   - will demonstrate a high degree of skill and competence in electrical work
   - will have, or be working towards, a recognized electrical qualification and status as an electrician, approved electrician or electrical technician.
9. Mechanic/fitter
   - an operative who usually has a 'core skill' or 'basic skill' and qualification in mechanical rather than electrical engineering
   - in production or process work, he or she would have responsibility for the engineering and fitting aspects of the contract, while the electrician and instrumentation technician would take care of the electrical and instrumentation aspects
   - all three operatives must work closely in production and process work
   - 'additional skilling' or 'multi-skilling' training produces a more flexible operative for production and process plant operations.
10. Maintenance manager/engineer
    - is responsible for keeping the installed electrotechnical plant and equipment working efficiently
    - takes over from the builders and contractors the responsibility of maintaining all plant equipment and systems under his or her control
    - might be responsible for a hospital or a commercial building, a university or college complex
    - will set up routine and preventative maintenance programmes to reduce possible future breakdowns
    - when faults or breakdowns do occur he or she will be responsible for the repair using the company's maintenance staff.

Try this

Information

- Where do you fit into your company's workforce?
- What is your job title – apprentice/trainee?
- What is your supervisor's job title?
- What is the name of the foreman?

Communications

When we talk about good communications we are talking about transferring information from one person to another both quickly and accurately. We do this by talking to other people, looking at drawings and plans, and discussing these with colleagues from the same company and with other professionals who have an interest in the same project. The technical information used within our industry comes from many sources. The IET Regulations (BS 7671) are the 'electrician's bible' and form the basis of all our electrical design calculations and installation methods. British Standards, European Harmonized Standards and Codes of Practice provide detailed information for every sector of the electrical industry, influencing all design-and-build considerations.

Members of the construction team and their role within the industry

An electrician working for an electrical contracting company works as a part of the broader construction industry. This is a multi-million-pound industry carrying out all types of building work, from basic housing to hotels, factories, schools, shops, offices and airports. The construction industry is one of the UK's biggest employers, and carries out contracts to the value of about 10% of the UK's gross national product.

Although a major employer, the construction industry is also very fragmented. Firms vary widely in size, from the local builder employing two or three people to the big national companies employing thousands. Of the total workforce of the construction industry, 92% are employed in small firms of fewer than 25 people.

The yearly turnover of the construction industry is about £35 billion. Of this total sum, about 60% is spent on new building projects and the remaining 40% on maintenance, renovation or restoration of mostly housing.

In all these various construction projects the electrical industries play an important role, supplying essential electrical services to meet the needs of those who will use the completed building.

The building team

The construction of a new building is a complex process which requires a team of professionals working together to produce the desired results. We can call this team of professionals the building team, and their interrelationship can be expressed as in Fig. 5.1.

The client is the person or group of people with the actual need for the building, such as a new house, office or factory. The client is responsible for financing all the work and, therefore, in effect, employs the entire building team.

The architect is the client's agent and is considered to be the leader of the building team. The architect must interpret the client's requirements and produce working drawings. During the building process the architect will supervise all aspects of the work until the building is handed over to the client.

The quantity surveyor measures the quantities of labour and materials necessary to complete the building work from drawings supplied by the architect.

Specialist engineers advise the architect during the design stage. They will prepare drawings and calculations on specialist areas of work, such as calculating the dimensions of structural supports.

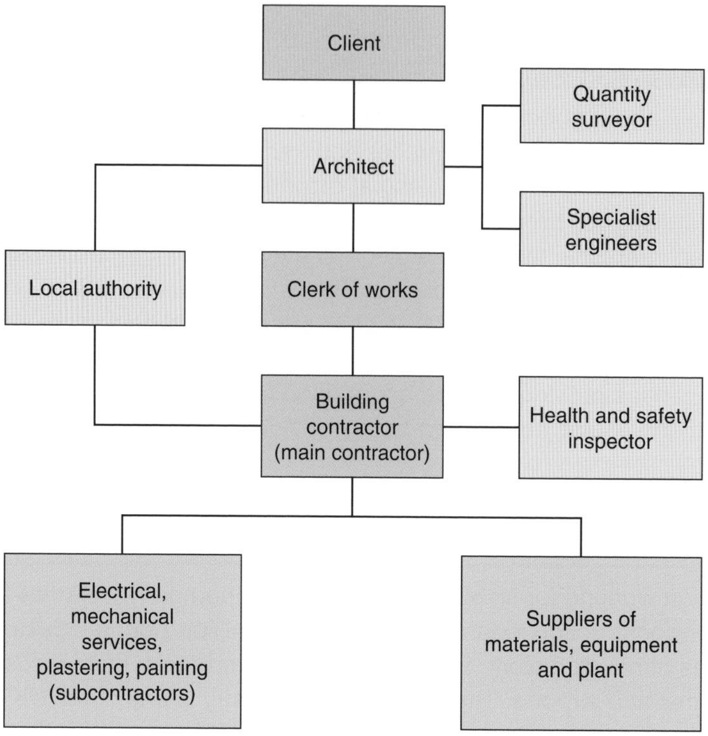

Figure 5.1 The building team.

The clerk of works is the architect's 'on-site' representative. He or she will make sure that the contractors carry out the work in accordance with the drawings and other contract documents. They can also agree general matters directly with the building contractor as the architect's representative.

The building control inspector from the local council will ensure that the proposed building conforms to the relevant planning and building legislation.

The health and safety inspectors will ensure that the government's legislation concerning health and safety is fully implemented by the building contractor.

The building contractor will enter into a contract with the client to carry out the construction work in accordance with contract documents. The building contractor is usually the main contractor and may engage subcontractors to carry out specialist services such as electrical installation, mechanical services, plastering and painting.

The electrical team

The electrical contractor is the subcontractor responsible for the installation of electrical equipment within the building.

Electrical installation activities include:

- installing electrical equipment and systems in new sites or locations;
- installing electrical equipment and systems in buildings that are being refurbished because of change of use;
- installing electrical equipment and systems in buildings that are being extended or updated;
- replacement, repairs and maintenance of existing electrical equipment and systems.

Try this

My team

Sketch a block diagram, similar to those shown in Figs 5.1 and 5.2, that represents the team in which you work.

An electrical contracting firm is made up of a group of individuals with varying duties and responsibilities. There is often no clear distinction between the duties of the individuals, and the responsibilities carried by an employee will vary from one employer to another. If the firm is to be successful, the individuals must work together to meet the requirements of their customers. Good customer relationships are important for the success of the firm and the continuing employment of the employee.

The customer or his or her representatives will probably see more of the electrician and the electrical trainee than the managing director of the firm and, therefore, the image presented by them is very important. They should always be polite and seen to be capable and in command of the situation. This gives a customer confidence in the firm's ability to meet his or her needs. The electrician and his trainee should be appropriately dressed for the job in hand, which probably means an overall of some kind. Footwear is also important, but is sometimes a difficult consideration for a journeyman electrician. For example, if working in a factory, the safety regulations may insist that protective footwear be worn, but rubber boots with toe protection may be most appropriate for a building site. However, neither of these would be the most suitable footwear for an electrician fixing a new light fitting in the home of the managing director!

The electrical installation in a building is often carried out alongside other trades. It makes sound sense to help other trades where possible and to develop good working relationships with other employees.

The employer has the responsibility of finding sufficient work for his employees, paying government taxes and meeting the requirements of the Health and Safety at Work Act described in Chapter 1. The rates of pay and conditions

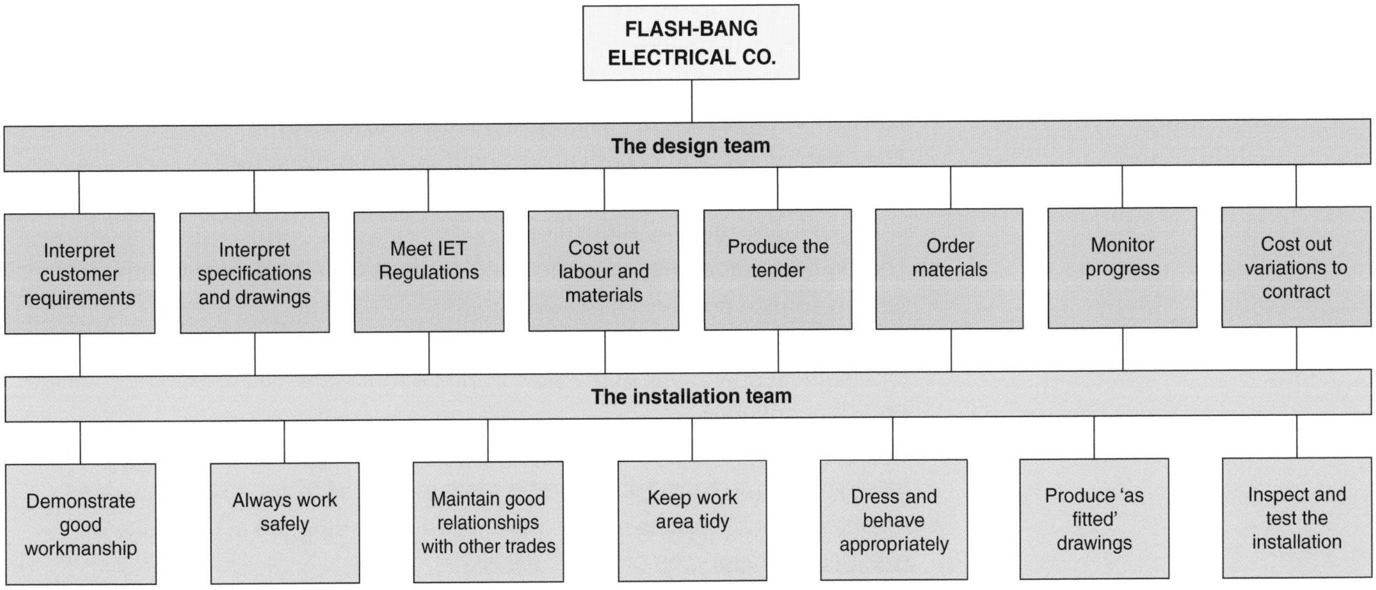

Figure 5.2 The electrical team.

for electricians and trainees are determined by negotiation between the Joint Industry Board and the Trades Unions, which will also represent their members in any disputes. Electricians are usually paid at a rate agreed for their grade; movements through the grades are determined by a combination of academic achievement and practical experience.

The electrical team will consist of a group of professionals and their interrelationship can be expressed as shown in Fig. 5.2.

Professional bodies supporting electrotechnical organizations

If you are reading this book I would guess that you are an electrical trainee working in one sector of the electrotechnical industry. You hope to eventually pass the City & Guilds Parts 2 and 3 qualifications, take your AM2 Practical Assessment and become a qualified electrician. Believe me, I do wish you well, because you are the future of the electrotechnical industry.

As a trainee, you are probably employed by an electrical company and attend your local college on either a 'day release' or 'block release' scheme. The combination of work and college will provide you with the skills you will need to become 'fully qualified'!

So, although you are doing all the work yourself, you are being sponsored or supported by the company that you work for, the JTL (JIB Training Ltd), EAL (EMTA Awards Ltd) and the City & Guilds of London Institute to become professionally qualified as an electrician.

It is in this same way that the professional bodies support the electrotechnical industry. They provide a structure of help, support and guidance to the individual companies that make up the electrotechnical industry.

So let us look at some of the professional bodies which support the electrotechnical organizations like the company you work for.

The Institution of Engineering and Technology (IET)

- The IET was formed in spring 2006 by bringing together the Institution of Electrical Engineers (IEE) and the Institution of Incorporated Engineers (IIE)
- The IET is Europe's largest professional society for engineers
- The IET produces the IET Wiring Regulations to BS 7671
- They also produce many other publications and provide training courses to help electricians, managers and supervisors to keep up to date with the changes in the relevant regulations
- The *On Site Guide* describes the 'requirements for electrical installations'
- Eight guidance notebooks are available
- The Electricians Guide to the Building Regulations clarifies the requirements for electrical operatives of the new Part P Regulations which came into effect on 1 January 2005
- *Wiring Matters* is a quarterly magazine published by the IET covering many of the topics which may trouble some of us in the electrotechnical industries
- All of these publications can be purchased by visiting the IET website at www.iet.org/resources or email: sales@iet.org.uk

The Electrical Contractors Association (ECA)

- The ECA was founded over 100 years ago and is a trade association representing electrotechnical companies
- Membership is made up of electrical contracting companies both large and small
- Customers employing an electrical contractor who has ECA membership are guaranteed that the work undertaken will meet all relevant regulations. If the work undertaken fails to meet the relevant standards, the ECA will arrange for the work to be rectified at no cost to the customer
- The work of the ECA member is regularly assessed by the Association's UKAS accredited inspection body
- Those electrotechnical companies which are members of the ECA are permitted to display the ECA logo on their company vehicles and stationery
- Further information can be found on the ECA website at www.eca.co.uk.

The National Inspection Council for Electrical Installation Contracting (NICEIC)

- The NICEIC is an independent consumer safety organization, set up to protect users of electricity against the hazards of unsafe electrical installations
- It is the electrical industry's safety regulatory body
- The NICEIC publishes a list of approved contractors whose standard of work is regularly assessed by local area engineers
- Customers employing an electrical contractor who has NICEIC membership can be assured that the work carried out will meet all relevant standards. If the work undertaken fails to meet all relevant standards, the name of the electrical contractor will be removed from the 'NICEIC Approved List'
- Some work, such as local authority work, is only available to NICEIC approved contractors
- Further information can be found at www.niceic.org.uk.

Trade unions

- Trade unions have a long history of representing workers in industry and commerce
- The relevant unions negotiate with employer organizations the pay and working conditions of their members
- The trade union which represents employees in the electrotechnical industry in the new millennium is called Unite
- Through a network of local area offices the union offers advice and support for its members. They will also provide legal advice and representation if a member has a serious accident as a result of a health and safety issue or has a dispute with an employer
- Further information can be found at www.unitetheunion.org.

Try this

Trade organizations

- Does the company you work for belong to a trade organization?
- Why do they belong, and what are the advantages?
- Do you belong to a trade union – if so, which one?
- If not, why? Trade union membership is often free while you are training.

Designing an electrical installation

The designer of an electrical installation must ensure that the design meets the requirements of the IET Wiring Regulations for electrical installations and any other regulations which may be relevant to a particular installation. The designer may be a professional technician or engineer whose job is to design electrical installations for a large contracting firm. In a smaller firm, the designer may also be the electrician who will carry out the installation to the customer's requirements. The **designer** of any electrical installation is the person who interprets the electrical requirements of the customer within the regulations, identifies the appropriate types of installation, the most suitable methods of protection and control and the size of cables to be used.

Definition

The *designer* of any electrical installation is the person who interprets the electrical requirements of the customer within the regulations.

A large electrical installation may require many meetings with the customer and his or her professional representatives in order to identify a specification of what is required. The designer can then identify the general characteristics of the electrical installation and its compatibility with other services and equipment, as indicated in Part 3 of the Regulations. The protection and safety of the installation, and of those who will use it, must be considered with due regard to Part 4 of the Regulations. An assessment of the frequency and quality of the maintenance to be expected will give an indication of the type of installation which is most appropriate.

The size and quantity of all the materials, cables, control equipment and accessories can then be determined. This is called a '**bill of quantities**'.

Definition

The size and quantity of all the materials, cables, control equipment and accessories can then be determined. This is called a '*bill of quantities*'.

It is a common practice to ask a number of electrical contractors to tender or submit a price for work specified by the bill of quantities. A tender is a formal offer to supply goods or carry out work at a stated price. The contractor must cost all the materials, assess the labour cost required to install the materials and add on profit and overhead costs in order to arrive at a final estimate for the work. This will be the 'tender' price. The contractor tendering the lowest cost is usually, but not always, awarded the contract.

To complete the contract in the specified time the electrical contractor must use the management skills required by any business to ensure that men and materials are on site as and when they are required. If alterations or modifications are made to the electrical installation as the work proceeds which are outside the original specification, then a **variation order** must be issued so that the electrical contractor can be paid for the additional work.

Definition

A *variation order* is one way of recording work which is additional to the original contract.

The specification for the chosen wiring system will be largely determined by the building construction and the activities to be carried out in the completed building.

An industrial building, for example, will require an electrical installation which incorporates flexibility and mechanical protection. This can be achieved by a conduit, tray or trunking installation.

In a block of purpose-built flats, all the electrical connections must be accessible from one flat without intruding upon the surrounding flats. A loop-in conduit system, in which the only connections are at the light switch and outlet positions, would meet this requirement.

For a domestic electrical installation an appropriate lighting scheme and multiple socket outlets for the connection of domestic appliances, all at a reasonable cost, are important factors which can usually be met by a PVC insulated and sheathed wiring system.

The final choice of a wiring system must rest with those designing the installation and those ordering the work, but whatever system is employed, good workmanship by competent persons is essential for compliance with all of the regulations. HSE Regulation 16 states that persons must be competent to prevent danger, so that the persons themselves or others are not placed at risk due to a lack of skill when dealing with electrical equipment.

A COMPETENT PERSON is therefore one who has the ability to perform a particular task safely and properly. Generally speaking an electrician will have the necessary skills to perform a wide range of electrical tasks competently.

The 3rd Amendment to the 17th Edition of the IET Regulations at 134.1.1 makes the following comment: 'good workmanship by skilled (electrically) or instructed (electrically) persons and proper materials shall be used in the erection of the electrical installation.'

So a SKILLED PERSON (electrically) is therefore one who has relevant education and experience to enable him or her to perceive risks and to avoid the hazards which electricity can create.

An INSTRUCTED PERSON (electrically) is a person adequately advised or supervised by electrically skilled persons to enable him or her to perceive risks and to avoid the hazards which electricity can create.

An ORDINARY PERSON is a person who is neither a skilled person nor an instructed person.

Try this

- People may be described as ordinary persons, skilled persons or instructed persons. Place the people you know into each category, for example yourself, your parents and relatives, and your work colleagues.
- The necessary skills can be acquired by an electrical trainee who has the correct attitude and dedication to the industry. NVQ 3 is 'skilled craft level' or the level required to be considered 'competent' or 'skilled'.

Legal contracts

Before work commences, some form of legal contract should be agreed between the two parties, that is, those providing the work (e.g. the subcontracting electrical company) and those asking for the work to be carried out (e.g. the main building company).

A contract is a formal document which sets out the terms of agreement between the two parties. A standard form of building contract typically contains four sections:

1 *The articles of agreement* – this names the parties, the proposed building and the date of the contract period.

2 *The contractual conditions* – this states the rights and obligations of the parties concerned, for example, whether there will be interim payments for work completed, or a penalty if work is not completed on time.
3 *The appendix* – this contains details of costings, for example, the rate to be paid for extras as daywork, who will be responsible for defects, how much of the contract tender will be retained upon completion and for how long.
4 *The supplementary agreement* – this allows the electrical contractor to recoup any value-added tax paid on materials at interim periods.

In signing the contract, the electrical contractor has agreed to carry out the work to the appropriate standards in the time stated and for the agreed cost. The other party, say, the main building contractor, is agreeing to pay the price stated for that work upon completion of the installation.

If a dispute arises the contract provides written evidence of what was agreed and will form the basis for a solution.

For smaller electrical jobs a verbal contract may be agreed, but if a dispute arises there is no written evidence of what was agreed and it then becomes a matter of one person's word against another's.

Statutory legislation and guidance information

Statutory legislation is the law of the land. Non-compliance may lead to prosecution by the courts and possible imprisonment. The statutory laws which apply to the construction and electrical industries must be complied with by all those who work in the industry (see Chapter 1 for statutory laws).

Examples of statutory legislation are:

- data protection;
- equal opportunities;
- health and safety;
- employment rights.

The health and safety regulations make the 'main contractor' responsible for the health, safety and welfare of all workers on the construction site. The main contractor must put in place a health and safety policy for each construction site and provide guardrails, working platforms, lighting, toilets, hand-washing facilities, first aid and site accommodation. There is also a duty to wear head protection on-site, including electricians.

Guidance information and codes of practice help us to comply with the statutory regulations.

Examples of guidance information and codes of practice are:

- IET Regulations;
- British Standards;
- Electricians Guide to the Building Regulations;
- *On Site Guide*;
- Manufacturer's installation instructions;
- Service and maintenance instructions.

The electrical subcontractor will be responsible for the electrical design and installation. If the design and installation meets the requirements of the IET Regulations then it will also comply with the statutory regulations such as the Electricity at Work Act.

Technical information used in the workplace

Technical information is communicated to electrical personnel in lots of different ways. It comes in the form of:

- *Specifications* – these are details of the client's requirements, usually drawn up by an architect. For example, the specification may give information about the type of wiring system to be employed or detail the types of luminaires or other equipment to be used.
- *Manufacturer's data* – if certain equipment is specified, let us say a particular type of luminaire or other piece of equipment, then the manufacturer's data sheet will give specific instructions for its assembly and fixing requirements. It is always good practice to read the data sheet before fitting the equipment. A copy of the data sheet should also be placed in the job file for the client to receive when the job is completed.
- *Reports and schedules* – a report is the written detail of something that has happened or the answer to a particular question asked by another professional person or the client. It might be the details of some problem on-site.

If the report is internal to the organization, a handwritten report is acceptable, but if the final report will go outside the organization, then it must be more formal and typed.

A **schedule** gives information about a programme or timetable of work. It might be a list or a chart giving details of when certain events will take place, for example, when the electricians will start to do the 'first fix' and how many days it will take. A simple **bar chart** is an easy-to-understand schedule of work that shows how different activities interact on a project. Figure 5.3 shows a bar chart or schedule of work where activity A takes two days to complete and activity B starts at the same time as activity A but carries on for eight days, etc.

User instructions – give information about the operation of a piece of equipment. Manufacturers of equipment provide 'user instructions' and a copy should be placed in the job file for the client to receive when the project is handed over.

Definition

A *schedule* is a list or programme of planned work.

Definition

Bar chart – the object of any bar chart is to establish the sequence and timing of the various activities involved in the whole job.

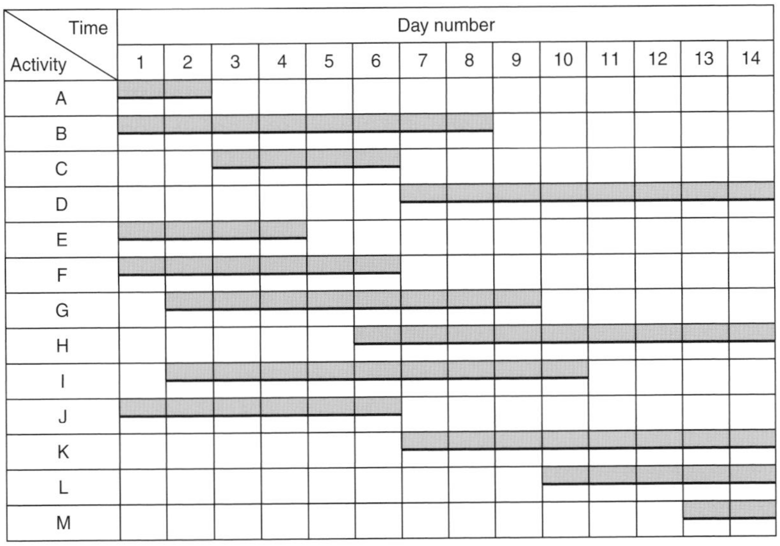

Figure 5.3 A simple bar chart or schedule of work.

Those who need or use technical information

Technical information is required by many of the professionals involved in any electrical activity, so who are the key people?

- **The operative** – in our case this will be the skilled electricians actually on-site, doing the job for the electrical company.
- **The supervisor** – he or she may have overall responsibility for a number of electricians on-site and will need the 'big picture'.
- **The contractor** – the main contractor takes on the responsibility of the whole project for the client. The main contractor may take on a subcontractor to carry out some part of the whole project. On a large construction site the electrical contractor is usually the subcontractor.
- **The site agent** – he or she will be responsible for the smooth running of the whole project and for bringing the contract to a conclusion on schedule and within budget. The site agent may be nominated by the architect.
- **The customer or client** – they are also the people ordering the work to be done. They will pay the final bill that pays everyone's wages.

Sources of technical information

Equipment and accessories available to use in a specific situation can often be found in the very comprehensive manufacturers' catalogues and the catalogues of the major wholesalers that service the electrical industries.

All of this technical information may be distributed and retrieved by using:

- conventional drawings and diagrams which we will look at in more detail below;
- sketch drawings to illustrate an idea or the shape of, say, a bracket to hold a piece of electrical equipment;
- the Internet can be used to download British Standards and Codes of Practice;
- the Internet can also be used to download health and safety information from the Health and Safety Executive at: www.hse.gov.uk or www.opsi.gov.uk;
- CDs, DVDs, USB memory sticks and email can be used to communicate and store information electronically;
- the facsimile (Fax) machine can be used to communicate with other busy professionals, say, information about a project you are working on together.

If you are working at your company office with access to online computers, then technical information is only a fingertip or mouse click away. However, a construction site is a hostile environment for a laptop and so a hard copy of any data is preferable on-site.

Let us now look at the types of drawings and diagrams which we use within our industry to communicate technical information between colleagues and other professionals. The type of diagram to be used in any particular situation is the one which most clearly communicates the desired information.

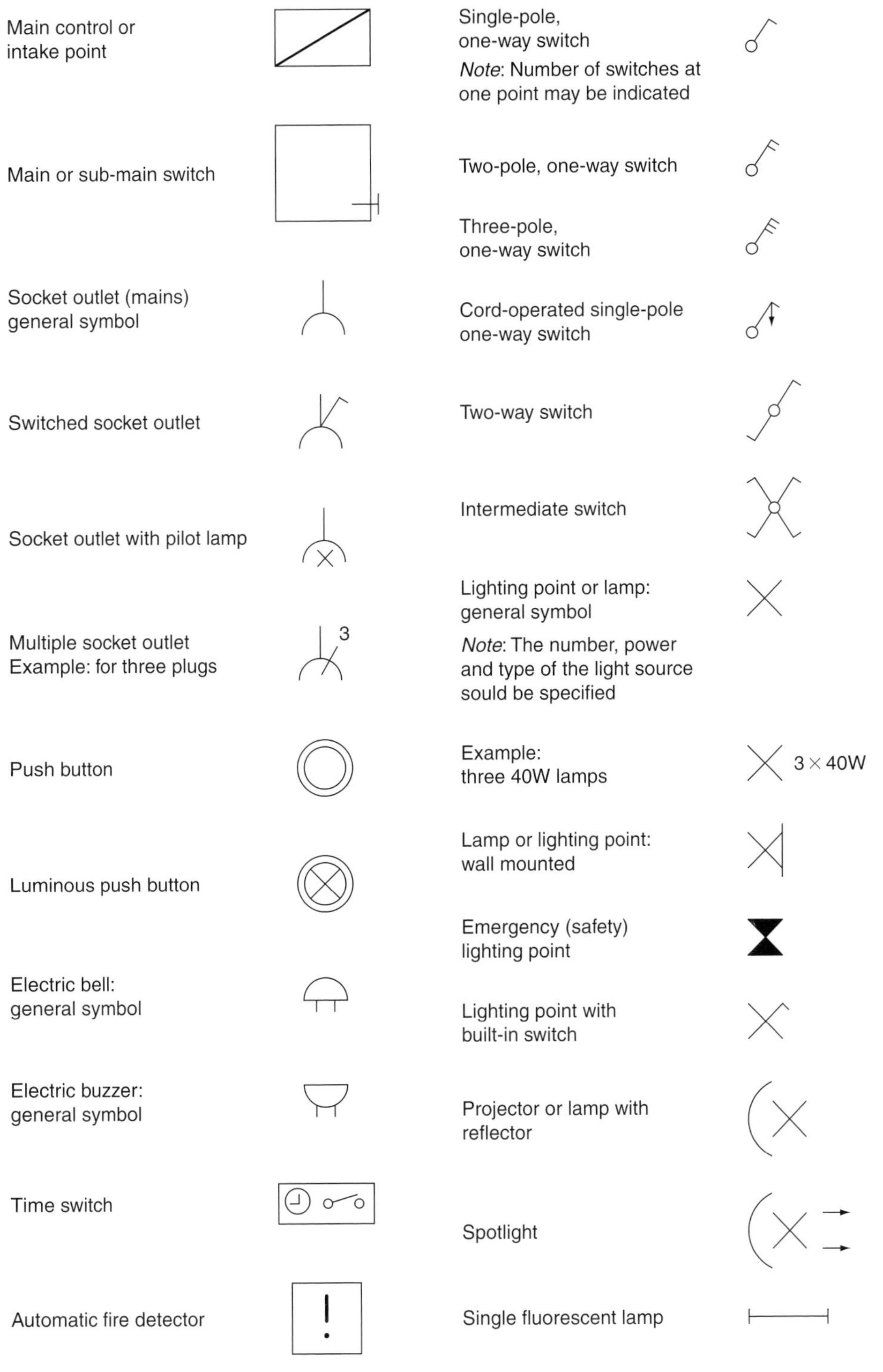

Figure 5.4 Some BS EN 60617 electrical installation symbols.

Site plans or layout drawings

These are scale drawings based upon the architect's site plan of the building and show the position of the electrical equipment which is to be installed. The electrical equipment is identified by a graphical symbol. The standard symbols used by the electrical contracting industry are those recommended by the British Standard EN 60617, *Graphical Symbols for Electrical Power, Telecommunications*

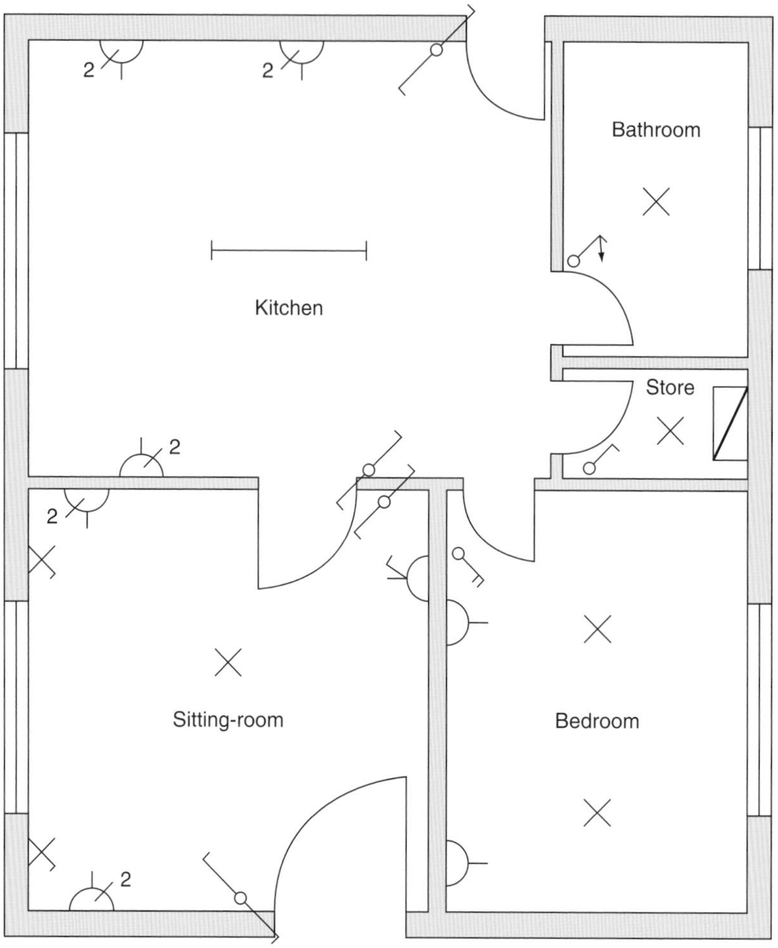

Figure 5.5 Layout drawing or site plan of a small electrical installation.

and Electronic Diagrams. Some of the more common electrical installation symbols are given in Fig. 5.4.

The site plan or layout drawing will be drawn to a scale smaller than the actual size of the building, so to find the actual measurement you must measure the distance on the drawing and multiply by the scale.

For example, if the site plan is drawn to a scale of 1:100, then 10 mm on the site plan represents 1 m measured in the building.

The layout drawing or site plan of a small domestic extension is shown in Fig. 5.5. It can be seen that the mains intake position, probably a consumer unit, is situated in the store-room which also contains one light controlled by a switch at the door. The bathroom contains one lighting point controlled by a one-way pull switch at the door. The kitchen has two doors and a switch is installed at each door to control the fluorescent luminaire. There are also three double sockets situated around the kitchen. The sitting room has a two-way switch at each door controlling the centre lighting point. Two wall lights with built-in switches are to be wired, one at each side of the window. Two double sockets and one switched socket are also to be installed in the sitting-room. The bedroom has two lighting points controlled independently by two one-way switches at the door. The wiring diagrams and installation procedures for all these circuits can be found in Chapter 3.

Try this

Drawings

The next time you are on-site ask your supervisor to show you the site plans. Ask them:

- How does the scale work?
- To help you put names to the equipment represented by British Standard symbols.

As-fitted drawings

When the installation is completed a set of drawings should be produced which indicate the final positions of all the electrical equipment. As the building and electrical installation progresses, it is sometimes necessary to modify the positions of equipment indicated on the layout drawing because, for example, the position of a doorway has been changed. The layout drawings or site plans indicate the original intentions for the position of equipment, while the 'as-fitted' drawings indicate the actual positions of equipment upon completion of the contract.

Try this

Drawings

Take a moment to clarify the difference between:

- layout drawings, and
- as-fitted drawings.

Detail drawings and assembly drawings

These are additional drawings produced by the architect to clarify some point of detail. For example, a drawing might be produced to give a fuller description of a suspended ceiling arrangement or the assembly arrangements of the metalwork for the suspended ceiling.

Location drawings

Location drawings identify the place where something is located. It might be the position of the manhole covers giving access to the drains. It might be the position of all water stop taps or the position of the emergency lighting fittings. This type of information may be placed on a blank copy of the architect's site plan or on a supplementary drawing.

Distribution cable route plans

On large installations there may be more than one position for the electrical supplies. Distribution cables may radiate from the site of the electrical mains intake position to other sub-mains positions. The site of the sub-mains and the route taken by the distribution cables may be shown on a blank copy of the architect's site plan or on the electrician's 'as-fitted' drawings.

Block diagrams

A block diagram is a very simple diagram in which the various items or pieces of equipment are represented by a square or rectangular box. The purpose of the block diagram is to show how the components of the circuit relate to each other and, therefore, the individual circuit connections are not shown. Figure 5.6 shows the block diagram of a space-heating control system.

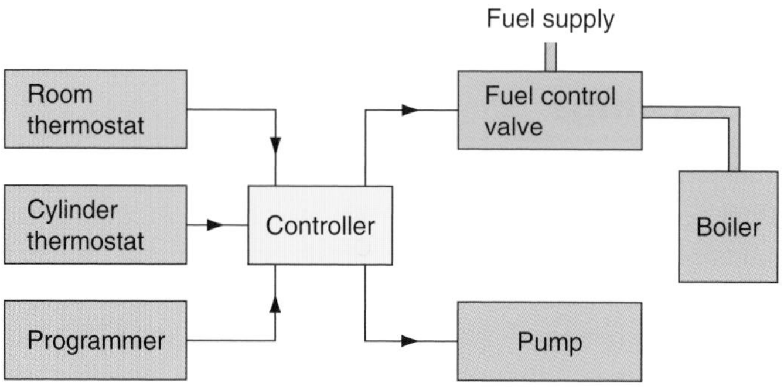

Figure 5.6 Block diagram – space-heating control system (Honeywell Y. Plan).

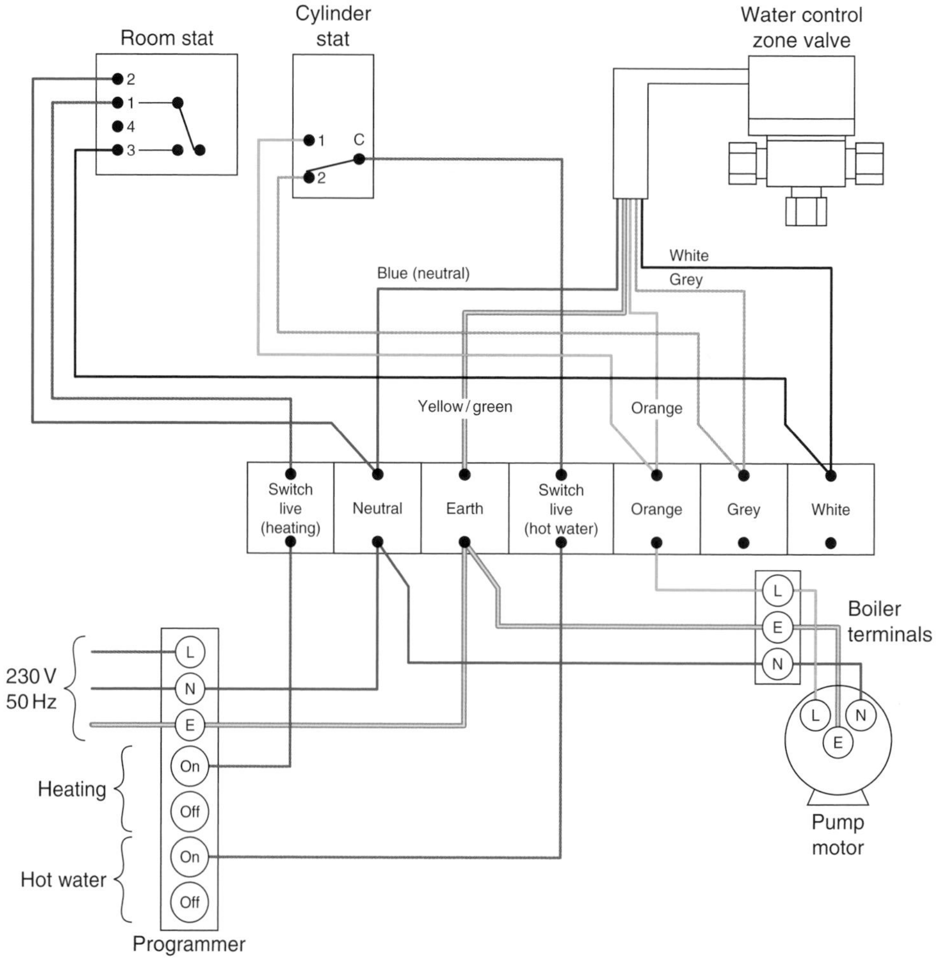

Figure 5.7 Wiring diagram – space-heating control system (Honeywell Y. Plan).

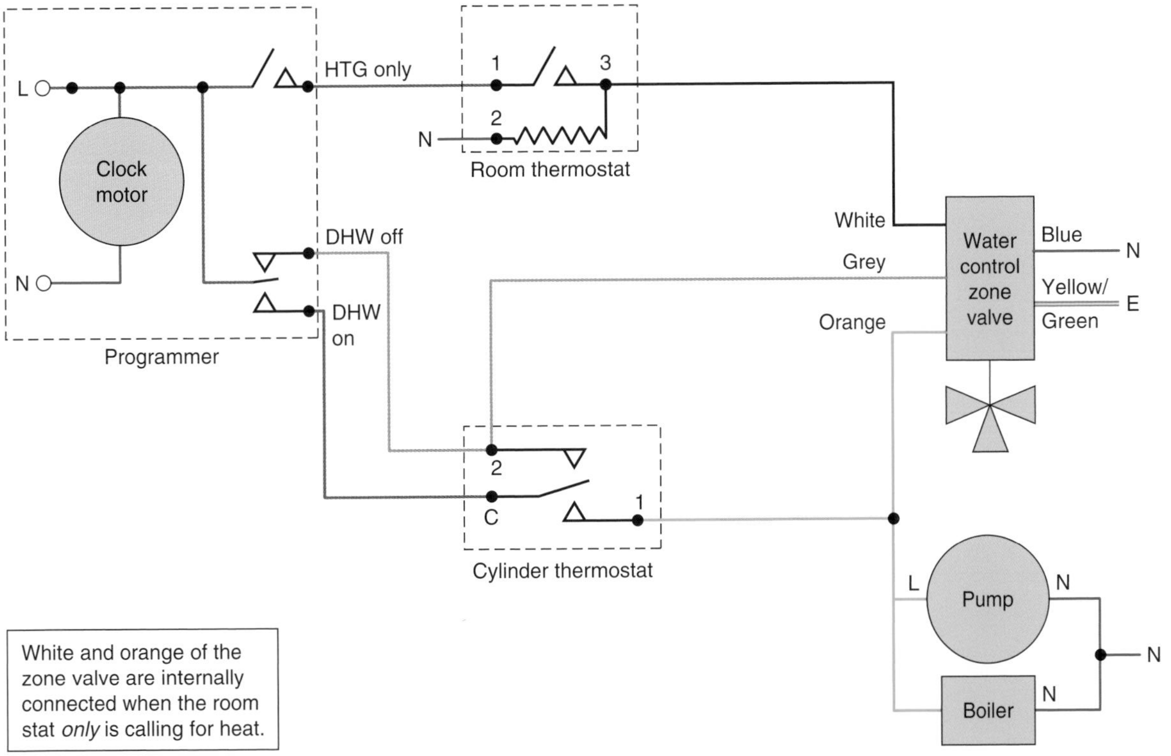

Figure 5.8 Circuit diagram – space-heating control system (Honeywell Y. plan).

Wiring diagrams

A wiring diagram or connection diagram shows the detailed connections between components or items of equipment. It does not indicate how a piece of equipment or circuit works. The purpose of a wiring diagram is to help someone with the actual wiring of the circuit. Figure 5.7 shows the wiring diagram for a space-heating control system.

Circuit diagrams

A circuit diagram shows most clearly how a circuit works. All the essential parts and connections are represented by their graphical symbols. The purpose of a circuit diagram is to help our understanding of the circuit. It will be laid out as clearly as possible, without regard to the physical layout of the actual components and, therefore, it may not indicate the most convenient way to wire the circuit. Figure 5.8 shows the circuit diagram of our same space-heating control system shown in the wiring diagram of Fig. 5.7.

Schematic diagrams

A schematic diagram is a diagram in outline of, for example, a motor starter circuit. It uses graphical symbols to indicate the interrelationship of the electrical elements in a circuit. These help us to understand the working operation of the circuit but are not helpful in showing us how to wire the components. An electrical schematic diagram looks very like a circuit diagram. Figure 5.9 shows a schematic diagram.

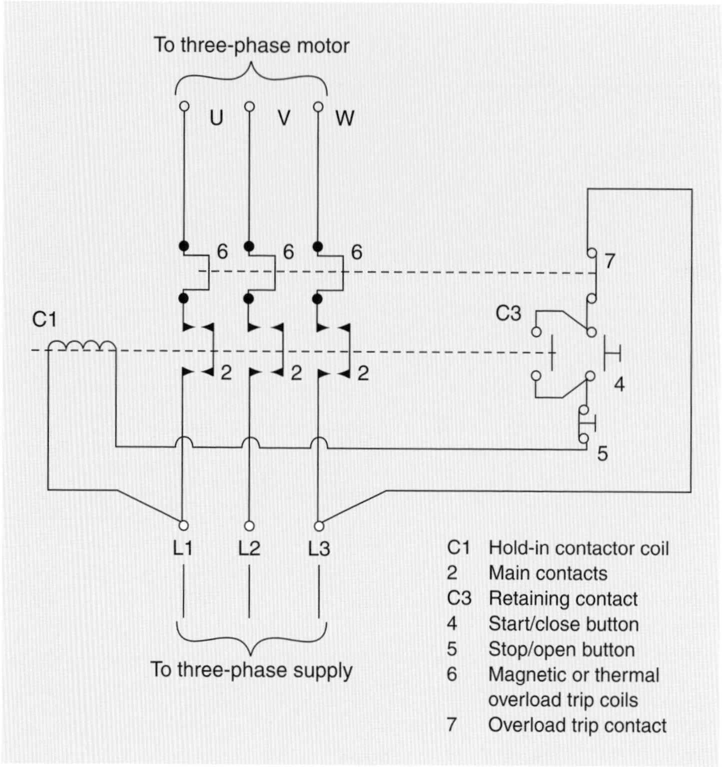

Figure 5.9 Schematic diagram – DOL motor starter.

Freehand working diagrams

Freehand working drawings or sketches are another important way in which we communicate our ideas. The drawings of the spring toggle bolt (Fig. 3.67) were done from freehand sketches. A freehand sketch may be done as an initial draft of an idea before a full working drawing is made. It is often much easier to produce a sketch of your ideas or intentions than to describe them or produce a list of instructions.

To convey the message or information clearly it is better to make your sketch large rather than too small. It should also contain all the dimensions necessary to indicate clearly the size of the finished object depicted by the sketch.

All drawings and communications must be aimed at satisfying the client's wishes for the project. It is the client who will pay the final bill which, in turn, pays your wages. The detailed arrangements of what must be done to meet the client's wishes are contained in the client's specification documents and all your company's efforts must be directed at meeting the whole specification, but no more.

On-site documentation

A lot of communications between and within larger organizations take place by completing standard forms or sending internal memos. Written messages have the advantage of being 'auditable'. An auditor can follow the paperwork trail to see, for example, who was responsible for ordering certain materials.

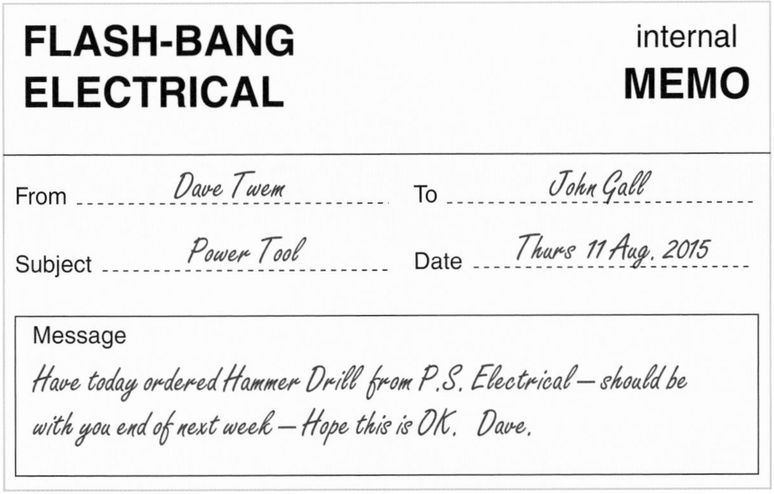
FLASH-BANG ELECTRICAL internal MEMO

From Dave Twem To John Gall

Subject Power Tool Date Thurs 11 Aug. 2015

Message

Have today ordered Hammer Drill from P.S. Electrical – should be with you end of next week – Hope this is OK. Dave.

Figure 5.10 Typical standard memo form.

When completing standard forms, follow the instructions given and ensure that your writing is legible. Do not leave blank spaces on the form, always specifying 'not applicable' or 'N/A' whenever necessary. Sign or give your name and the date as asked for on the form. Finally, read through the form again to make sure you have answered all the relevant sections correctly.

Internal memos are forms of written communication used within an organization; they are not normally used for communicating with customers or suppliers. Figure 5.10 shows the layout of a typical standard memo form used by Dave Twem to notify John Gall that he has ordered the hammer drill.

Letters provide a permanent record of communications between organizations and individuals. They may be handwritten for internal use but formal business letters give a better impression of the organization if they are typewritten. A letter should be written using simple, concise language, and the tone of the letter should always be polite even if it is one of complaint. Always include the date of the correspondence. The greeting on a formal letter should be 'Dear Sir/Madam' and conclude with 'Yours faithfully'. A less formal greeting would be 'Dear Mr Smith' and conclude 'Yours sincerely'. Your name and status should be typed below your signature.

Delivery notes

When materials are delivered to site, the person receiving the goods is required to sign the driver's **delivery note**. This record is used to confirm that goods have been delivered by the supplier, who will then send out an invoice requesting payment, usually at the end of the month.

The person receiving the goods must carefully check that all the items stated on the delivery note have been delivered in good condition. Any missing or damaged items must be clearly indicated on the delivery note before signing, because, by signing the delivery note the person is saying 'yes, these items were delivered to me as my company's representative on that date and in good condition and I am now responsible for these goods'. Suppliers will replace materials damaged in transit provided that they are notified within a set time period, usually three days. The person receiving the goods should try to quickly

Definition

A *delivery note* is used to confirm that goods have been delivered by the supplier, who will then send out an invoice requesting payment.

TIME SHEET

FLASH-BANG ELECTRICAL

Employee's name (Print) ....

Week ending ....

Day	Job number and/or address	Start time	Finish time	Total hours	Travel time	Expenses
Monday						
Tuesday						
Wednesday						
Thursday						
Friday						
Saturday						
Sunday						

Employee's signature .... Date ....

Figure 5.11 Typical time sheet.

determine their condition. Has the packaging been damaged, does the container 'sound' like it might contain broken items? It is best to check at the time of delivery if possible, or as soon as possible after delivery and within the notifiable period. Electrical goods delivered to site should be handled carefully and stored securely until they are installed. Copies of delivery notes are sent to Head Office so that payment can be made for the goods received.

Time sheets

A **time sheet** is a standard form completed by each employee to inform the employer of the actual time spent working on a particular contract or site. This helps the employer to bill the hours of work to an individual job. It is usually a weekly document and includes the number of hours worked, the name of the job and any travelling expenses claimed. Office personnel require time sheets such as that shown in Fig 5.11 so that wages can be made up.

Definition

A *time sheet* is a standard form completed by each employee to inform the employer of the actual time spent working on a particular contract or site.

Job sheets

A **job sheet** or job card such as that shown in Fig. 5.12 carries information about a job which needs to be done, usually a small job. It gives the name and address of the customer, contact telephone numbers, often a job reference number and a brief description of the work to be carried out. A typical job sheet work description might be:

Job 1 Upstairs lights not working.

Job 2 Funny fishy smell from kettle socket in kitchen.

An electrician might typically have a 'jobbing day' where he picks up a number of job sheets from the office and carries out the work specified.

Job 1, for example, might be the result of a blown fuse which is easily rectified, but the electrician must search a little further for the fault which caused the fuse to blow in the first place. The actual fault might, for example, be a decayed

Definition

A *job sheet* or job card carries information about a job which needs to be done, usually a small job.

JOB SHEET

Job Number ----------------

FLASH-BANG ELECTRICAL

Customer name --

Address of job --

Contact telephone No. --

Work to be carried out --

Any special instructions/conditions/materials used

Figure 5.12 Typical job sheet.

flex on a pendant drop which has become shorted out, blowing the fuse. The pendant drop would be re-flexed or replaced, along with any others in poor condition. The installation would then be tested for correct operation and the customer given an account of what has been done to correct the fault. General information and assurances about the condition of the installation as a whole might be requested and given before setting off to job 2.

The kettle socket outlet at job 2 is probably getting warm and, therefore, giving off that 'fishy' bakelite smell because loose connections are causing the bakelite socket to burn locally. A visual inspection would confirm the diagnosis. A typical solution would be to replace the socket and repair any damage to the conductors inside the socket box. Check the kettle plug top for damage and loose connections. Make sure all connections are tight before reassuring the customer that all is well; then, off to the next job or back to the office.

The time spent on each job and the materials used are sometimes recorded on the job sheet, but alternatively a daywork sheet can be used. This will depend upon what is normal practice for the particular electrical company. This information can then be used to 'bill' the customer for work carried out.

Daywork sheets or variation order

Definition

Daywork is one way of recording variations to a contract.

Daywork is one way of recording variations to a contract, that is, work done which is outside the scope of the original contract. If daywork is to be carried out, the site supervisor must first obtain a signature from the client's representative, for example, the architect, to authorize the extra work. A careful record must then be kept on the daywork sheets of all extra time and materials used so that the client can be billed for the extra work. A typical daywork sheet is shown in Fig. 5.13.

Reports

On large jobs, the foreman or supervisor is often required to keep a report of the relevant events which happen on the site – for example, how many people from your company are working on site each day, what goods were delivered, whether there were any breakages or accidents, and records of site meetings attended. Some firms have two separate documents: a site diary to record daily events and a weekly report which is a summary of the week's events extracted from the site diary. The site diary remains on-site and the weekly report is sent to Head Office to keep managers informed of the work's progress.

Telephone communications

Telephones and mobile phones today play one of the most important roles in enabling people to communicate with each other. You are never alone when you have a phone. If there is a problem, you can ring your supervisor or foreman for help. The advantage of a telephone message over a written message is its speed; the disadvantage is that no record is kept of an agreement made over the telephone. Therefore, business agreements made on the telephone are often followed up by written confirmation.

When *taking* a telephone call, remember that you cannot be seen and, therefore, gestures and facial expressions will not help to make you understood. Always

FLASH-BANG ELECTRICAL

VARIATION ORDER or DAYWORK SHEET

Client name ....

Job number/ref. ....

Date	Labour	Start time	Finish time	Total hours	Office use

Materials quantity	Description	Office use

Site supervisor or F.B. Electrical Representative responsible for carrying out work ....

Signature of person approving work and status e.g.

Client ☐ Architect ☐ Q.S. ☐ Main contractor ☐ Clerk of works ☐

Signature ....

Figure 5.13 Typical daywork sheet or variation order.

be polite and helpful when answering your company's telephone – you are your company's most important representative at that moment. Speak clearly and loud enough to be heard without shouting, sound cheerful and write down messages if asked. Always read back what you have written down to make sure that you are passing on what the caller intended.

Many companies now use standard telephone message pads such as that shown in Fig. 5.14 because they prompt people to collect all the relevant information. In this case, John Gall wants Dave Twem to pick up the Megger from Jim on Saturday and take it to the Bispham site on Monday. The person taking the call and relaying the message is Dave Low.

FLASH-BANG ELECTRICAL TELEPHONE MESSAGES

Date *Thurs 11 Aug. 2015* Time *09.30*

Message to *Dave Twem*

Message from (Name) *John Gall*

(Address) *Bispham Site*

Blackpool.

(Telephone No.) *(01253) 123456*

Message *Pick up Megger*

from Jim on Saturday and take to Bispham

site on Monday.

Thanks

Message taken by *Dave Lan*

Figure 5.14 Typical standard telephone message pad.

When *making* a telephone call, make sure you know what you want to say or ask. Make notes so that you have times, dates and any other relevant information ready before you make the call.

Company policies and employment rights

As a trainee in the electrical industry you will be employed by a member company and receive a weekly or monthly wage, which will be dependent upon your age and grade as agreed by the appropriate trade union, probably UNITE.

We have seen in the beginning of this book that there are many rules and regulations which your employer must comply with in order to make your workplace healthy and safe. There are also responsibilities that apply to you, as an employee (or worker) in the electrical industry, in order to assist your employer to obey the law.

As an employee you must:

- obey all lawful and reasonable requests;
- behave in a sensible and responsible way at work;
- work with care and reasonable skill.

Your employer must:

- take care for your safety;
- not ask you to do anything unlawful or unreasonable;

- pay agreed wages;
- not change your contract of employment without your agreement.

Most of the other things that can be expected of you are things like honesty, punctuality, reliability and hard work. Really, just common-sense things like politeness will help you to get on at work.

If you have problems relating to your employment rights you should talk them through with your supervisor or trade union representative at work.

Wages and tax

When you start work you will be paid either weekly or monthly. It is quite common to work a week in hand if you are paid weekly, which means that you will be paid for the first week's work at the end of the second week. When you leave that employment, if you have worked a week in hand, you will have a week's wage to come. Money that you have worked for belongs to you and cannot be kept by your employer if you leave without giving notice.

Every employee is entitled to a payslip along with their wages, which should show how much you have earned (gross), how much has been taken off for tax and national insurance and what your take-home pay (net) is.

If you are not given a payslip, ask for one, it is your legal right and you may be required to show payslips as proof of income. Always keep your payslips in a safe place.

We all pay tax on the money we earn (income tax). The government uses tax to pay for services such as health, education, defence, social security and pensions.

We are all allowed to earn a small amount of money tax free each year and this is called the personal allowance. The personal allowance for the tax year in which I am writing this book (2014/2015) is £10,000. So every pound that we earn above £10,000 is taxed. The tax year starts on 6 April each year and finishes on 5 April the following year. Your personal tax code enables the personal allowance to be spread out throughout the year and you pay tax on your wages on a system called PAYE, pay as you earn.

At the end of the tax year your employers will give you a form called a P60 which shows your tax code, how much you have earned and how much tax you have paid during a particular tax year. It is important to keep your P60 somewhere safe along with your payslips. If at some time you want to buy a house, a building society will want proof of your earnings, which these documents show.

When leaving a particular employment you must obtain from your employer a form P45. On starting new employment this form will be required by your new employer and will ensure that you do not initially pay too much tax.

Working hours

Employees cannot be forced to work more than 48 hours each week on average, and 40 hours for 16- to 18-year-old trainees. Trainees must also have 12 hours' uninterrupted rest from work each day. Older workers, required to work for more than six hours continuously, are entitled to a 20-minute rest break, to be taken within the six hours, and must have 11 hours' uninterrupted rest from work each day. If you think you are not getting the correct number of breaks, talk to your supervisor or trade union representative.

Sickness

If you are sick and unable to go to work, you should contact your employer or supervisor as soon as you can on the first day of illness. When you go back to work, if you have been sick for up to seven days, you will have to fill in a self-certification form. After seven days you will need a medical certificate from your doctor and you must send it to work as soon as you can. If you are sick for four days or more your employer must pay you statutory sick pay (SSP), which can be paid for up to 28 weeks. If you are still sick after 28 weeks you can claim incapacity benefit. To claim this you will need a form from your employer or social security office. If you have a sickness problem, talk to your supervisor or someone at work who you trust, or telephone the local social security office.

Accidents

It is the employer's duty to protect the health and safety and welfare of its employees, so if you do have an accident at work, however small, inform your supervisor, safety officer or first aid person. Make sure that the details are recorded in the accident/first aid book. Failure to do so may affect compensation if the accident proves to be more serious than you first thought.

Always be careful, use common sense and follow instructions. If in doubt, ask someone. A simple accident might prevent you from playing your favourite sport for a considerable period of time.

Holidays

Most employees are entitled to at least four weeks' paid holiday each year. Your entitlement to paid holidays builds up each month, so a month after you start work you are entitled to one-twelfth of the total holiday entitlement for the year. After two months it becomes two-twelfths and so on. Ask your supervisor or HR/Payroll contact to explain your holiday entitlement to you.

Problems at work

It is not unusual to find it hard to fit in when you start a new job. Give it a chance, give it time and things are likely to settle down. As a new person you might seem to get all the rotten jobs, but sometimes, being new, these are the only jobs that you can do for now.

In some companies there can be a culture of 'teasing', which may be OK if everyone is treated the same, but not so good if you are always the one being teased. If this happens, see if it stops after a while; if not, talk to someone about it. Don't give up your job without trying to get the problem sorted out.

If you feel that you are being discriminated against or harassed because of your race, sex or disability, then talk to your supervisor, trainer or someone you trust at work. There are laws about discrimination that are discussed in Chapter 3 of this book.

You can join a trade union when you are 16 years of age or over. Trade unions work towards fair deals for their members. If you join a trade union there will be subscriptions (subs) to pay. These are often reduced or suspended during the training period. As a member of a trade union you can get advice and support from them. If there is a problem of any kind at work, you can ask for the union's support. However, you cannot get this support unless you are a member.

Resignation/dismissal

Most employers like to have your resignation or 'Notice' in writing. Your contract of employment will tell you how much notice is expected. The minimum notice you should give is one week if you have been employed for one month or more by that employer. However, if your contract states a longer period, then that is what is expected.

If you have worked for one month or more, but less than two years, you are entitled to one week's notice. If you have worked for two years you are entitled to two weeks' notice and a further week's notice for every additional continuous year of employment (with the same employer) up to 12 weeks for 12 years' service.

If you are dismissed or 'sacked', you are entitled to the same periods of notice. However, if you do something very serious, like stealing or hitting someone, your employer can dismiss you without notice.

You can also be dismissed if you are often late or your behaviour is inappropriate to the type of work being done. You should have verbal or written warnings before you are dismissed.

If there are 20 or more employees at your place of employment then there should be a disciplinary procedure written down, which must be followed. If you do get a warning, then you might like to see this as a second chance to start again.

If you have been working for the same employer for one year or more, you can complain to an Employment Tribunal if you think you have been unfairly dismissed. If you haven't worked for the same employer for this length of time, then you should talk to your training officer or trade union.

I do not want to finish this section in a negative way, talking about problems at work, so let me finally say that each year over 8,000 young people are in apprenticeships in the electrical contracting industry and very few of them have problems. The small problems that may arise, because moving into full-time work is very different to school, can usually be resolved by your training officer or supervisor. Most of the trainees go on to qualify as craftsmen and enjoy a well-paid and fulfilling career in the electrical industry.

Limits of personal authority

Part 2 of the IET Regulations describes people as:

- **Ordinary persons** – that is, a person who is neither a skilled nor instructed person.
- **Instructed (electrically) persons** – that is, a person who is adequately advised or supervised by an electrically skilled person to enable him or her to perceive risks and avoid the hazards that electricity can create.
- **Skilled (electrically) person** – that is, a person who possesses relevent education and experience of electrical work to be able at all times to preceive risks and avoid the hazards which electricity can create.

When you begin your apprenticeship you will be an ordinary person, one who has no knowledge of the dangers of electricity. In just a few weeks or months you will become an instructed (electrically) person because you will be doing electrical work under the supervision of your supervisor. When you have completed your apprenticeship you will be a skilled (electrically) person, able to carry out electrical work safely.

When you start out, you will have little authority and knowledge and so it is reasonable that you are given little responsibility. As you progress through your apprenticeship you will be given more responsibility and authority as your knowledge of working safely and competently increases.

The Health and Safety Regulations make reference to workers having a *duty of care* for the health and safety of themselves and others in the workplace. The Electricity at Work Regulations identify one responsible person on-site as the *'duty holder'*. This recognizes the responsibility of an electrician to take on the control of electrical safety for the whole construction site.

Everyone has a duty of care but not everyone is a duty holder. The duty holder will be the electrician who takes on the responsibility for the electrical health and safety of everyone working on site. The chosen person will certainly be a skilled (electrically) person, probably someone with great experience and common sense, and whose other responsibilities might be to act as a supervisor, foreman or manager.

Handing over the installation to the client

Handing over the completed electrical installation to the client is a big day. The job is finished and the electrical contractor can now be paid in total for the work done. You are effectively saying to the client 'here are your electrical circuits, systems and equipment just as you ordered, let me show you how everything works'. You should then:

- demonstrate how everything works;
- hand over all test certificates and inspection schedules;
- hand over all manufacturers' instructions.

All these documents should be presented in a document file for the client to keep in a safe place for future reference.

Check your understanding

When you have completed the questions, check out the answers at the back of the book.

Note: more than one multiple-choice answer may be correct.

1 A standard form completed by every employee to inform the employer of the time spent working on a particular site is called a:
   a. job sheet
   b. time sheet
   c. delivery note
   d. daywork sheet.

2 A record which confirms that materials ordered have been delivered to site is called a:
   a. job sheet
   b. time sheet
   c. delivery note
   d. daywork sheet.

3 A standard form containing information about work to be done usually distributed by a manager to an electrician is called a:
   a. job sheet
   b. time sheet
   c. delivery note
   d. daywork sheet.

4 A standard form which records changes or extra work on a large project is called a:
   a. job sheet
   b. time sheet
   c. delivery note
   d. daywork sheet.

5 A scale drawing which shows the original intention for the position of electrical equipment is called a:
   a. wiring diagram
   b. detail assembly drawing
   c. site plan or layout drawing
   d. as-fitted drawing.

6 The scale drawing which shows the actual position of the electrical equipment upon completion of the contract is called a:
   a. wiring diagram
   b. detail assembly drawing
   c. site plan or layout drawing
   d. as-fitted drawing.

7 A scale drawing showing the position of equipment by graphical symbols is a description of a:
   a. block diagram
   b. site plan or layout diagram
   c. wiring diagram
   d. circuit diagram.

8 A diagram which shows the detailed connections between individual items of equipment is a description of a:
   a. block diagram
   b. site plan or layout diagram
   c. wiring diagram
   d. circuit diagram.

9 A diagram which shows most clearly how a circuit works, with all items represented by graphical symbols, is a description of a:
   a. block diagram
   b. site plan or layout diagram
   c. wiring diagram
   d. circuit diagram.

10 A site plan has a scale of 1:100. From the scale drawing you can see that a socket outlet must be positioned on a wall 40 cm from the corner of the room in which you are standing. How far from the corner of the room would you actually measure to the centre of the socket's fixing position taking the scale into account?
   a. 0.4 m
   b. 4.0 m
   c. 10.0 m
   d. 40.0 m.

11 Identify the statutory regulations from the following list:
   a. Health and Safety at Work Act 1974
   b. BSI Kite Safety Mark
   c. Electricity at Work Regulations 1989
   d. IET Regulations BS 7671: 2008.

12 Identify the non-statutory regulations from the following list which impact upon electrical systems:
   a. Health and Safety at Work Act 1974
   b. Electricity at Work Regulations 1989
   c. IET Regulations BS 7671: 2008
   d. BSI Kite Safety Mark.

13 The person who is the company representative responsible for electrical safety in the work environment is called the:
   a. Supervisor
   b. Duty of care
   c. Duty holder
   d. Competent person.

14 Everyone in the workplace has this responsibility for safety:
   a. First aid
   b. Slips, trips and falls
   c. Maintenance of tools and equipment
   d. Duty of care.

15 Employers of companies employing more than five people must:
   a. become a member of the NICEIC
   b. provide PPE if appropriate
   c. carry out a hazard risk assessment
   d. display a health and safety poster.

16 What method would you use to let the office know that the materials you were expecting have not yet arrived?

17 What method would you use to send a long list of materials required for the job you are on to the wholesalers for later delivery to the site? Use bullet points.

18 What are the advantages and disadvantages of having sources of technical information on:
   a. some form of electronic storage system such as a CD, DVD or USB memory stick, or
   b. hard copy such as a catalogue, drawings or *On Site Guide*?

   Would it make a difference if you were at the office or on a construction site?

19 State the advantages and disadvantages of:
   a. telephone messages
   b. written messages.

20 List at least two statutory regulations which have an impact upon all electrical activities and state your reasons why.

21 List at least one code of practice which has an impact upon all electrical activities and state your reason why.

22 State four requirements of the Construction (Health, Safety and Welfare) Regulations for construction sites.

23 Sketch and label six graphical symbols for equipment or accessories that you have used in your training so far.

24 How does the Data Protection Act protect our privacy?

25 How does a bar chart help with the organization of a work programme?

26 Why are good relationships important between yourself and the customer and other trades on-site when carrying out work activities?

27 Briefly state why time sheets, fully and accurately completed, are important to:

a. an employer
b. an employee.

28 State the reasons why you should always present the right image to a client, customer, or his or her representative.

29 Briefly describe what we mean by a schedule of work. Who would use a bar chart or schedule of work in your company and why?

Answers to Check your understanding questions

Chapter 1		
1. D	2. A	3. A, B
4. B, C	5. D	6. D
7. D	8. C	9. A, B
10. B, C, D	11. A, B	12. C, D
13. B, C, D	14. C, D	15. C
16. C	17. A, C	18. B, D
19. D	20. B	21. C
22. B, C, D	23. C	24. D
25. C	26. A	27. C
28. C	29. A, B, D	
30 to 48 Answers in text of Chapter 1.		

Chapter 2				
1. A	2. C	3. B	4. D	5. A
6. D	7. B	8. A	9. C	10. D
11. C	12. B	13. B	14. C	15. D
16. A, C	17. B, D	18. A, C	19. B, D	20. A, C
21. B, D	22. C	23. D	24. C	25. D
26. D	27. A	28. D	29. A	30. C
31. D	32. A	33. C	34. D	35. C
36. A	37. B	38. C	39. C	40. C
41. A, D	42. C	43. B	44. B	45. C
46 to 70 Answers in text of Chapter 2.				

Chapter 3				
1. A	2. D	3. C	4. B	5. B
6. B	7. C	8. A	9. D	10. D
11. C	12. C	13. B	14. B	15. C
16. D	17. A, C	18. A, B	19. B	20. A
21. C	22. B	23. B	24. C	25. B
26. C	27. D	28. D	29. C	30. D
31. C	32. D	33. A	34. B	35. C
36. D	37. C	38. D	39. B	40. C
41. A, D	42. A, D	43. D	44. C	45. A
46. B	47. C	48. D	49. B	50. C
51. D	52. B	53. C	54. D	55. C, D
56. C	57. D			
58 to 103 Answers in text of Chapter 3.				

Chapter 4				
1. B	2. C	3. C, D	4. B, C	5. C
6. D	7. B	8. D	9. B	10. D
11. A, D	12. C	13. C	14. C	15. A, B, D
16. A, B	17. A	18. D	19. C	20. C, D
21. B, D	22. A, B	23. C, D	24. C, D	
25 to 48 Answers in text of Chapter 4.				

Chapter 5				
1. B	2. C	3. A	4. D	5. C
6. D	7. B	8. C	9. D	10. B
11. A, C	12. C	13. C	14. D	15. B, C, D
16 to 29 Answers in text of Chapter 5.				

Preparing for assessment

City & Guilds will assess your knowledge and understanding of the topics covered in this electrical installation course in two ways:

1. Internally and externally set written assignments, internally marked, externally verified.
2. Online multiple-choice questions similar to the questions at the end of each chapter in this book. Each question will consist of a statement with four possible answers. **Only one answer will be correct in the on-line exam**.

Preparing for your online assessment examination

Before the exam:

- You should have been revising throughout the course – revision is not just something you start with a day or two before the exam;
- Don't cram all the subject areas into one revision session – narrow areas down into bite-sized chunks and tackle them one at a time;
- Don't try to revise or learn new things at the last minute;
- Do lots of sample papers before the exam;
- Look online for sample questions and answers;
- Make sure that you are used to the style, type and wording of the questions that you will have to answer in your exam;
- Make sure you know what documents are required and expected in the exam (e.g. the *On Site Guide* or Wiring Regulations);
- Go to bed early the night before your exams (no partying until they're all finished, not just one exam);
- If your exam is in the morning make sure you get up early enough to eat a decent breakfast, and don't eat just before exams, as this can make you sleepy;
- Drink water well before exams in order to ensure that you are fully hydrated, which in turn makes you more alert;
- Sip water during exams (if allowed);
- Go to the toilet before the exam starts – it's too late when you are in the room;
- Make sure you know where and when the exam is taking place, and how long it will take you to get there;
- Check that your method of transportation to the exam centre is going to work on the day of the exams (e.g. timetables for buses and trains, etc.);
- Arrive early at the exam centre, but not so early that you're hanging around outside for hours;
- Try to relax in the minutes leading up to the exam;
- Check all your personal stationery before going into the exam, ensuring that you've got everything you need (e.g. pens, a calculator, etc.).

Taking the online assessment:

- Listen carefully to the exam instructions that the invigilator gives you;
- You may not know the invigilator but don't be frightened to ask questions if you don't understand the instructions;
- Check the exam title details and codes to make sure that you're in the right exam;
- Make sure you know how many questions there are and the time limit of the exam. Use this information to allocate an appropriate amount of time for each question;
- Make sure you do the tutorial session before the real exam. This will ensure that you become familiar with the techniques required to successfully complete the online exams;
- Questions are not always clear, but should contain clues to the answer. Take your time reading each question when you get to it, and read questions more than once before answering them;
- Look at the four possible answers and read them more than once before picking an answer to run with; don't tick the first thing that you recognize in a list of answers – use a process of elimination to rule out wrong answers before settling on a final answer;
- If you are unsure about an answer put a flag on it and leave it until the end, when you can go back and spend more time on it;
- Answer the questions that you find easy first, which will leave you more time to spend on the questions that you find difficult;
- Check any answers that need a calculation at least twice;
- Never leave the exam with questions unanswered – guess if you have to, but use the process of elimination mentioned earlier first as this will give you more chance of getting it right;
- Don't try to be first out of the exam, take your time – there are no prizes for finishing first!
- Don't press the finish button until you are totally sure that you can't do any more. Once the finish button is pressed you cannot restart the test or go back to check your answers.

APPENDIX A
Abbreviations, symbols and codes

Abbreviations used in electronics for multiples and submultiples

T	Tera	10^{12}
G	Giga	10^{9}
M	Mega or Meg	10^{6}
k	Kilo	10^{3}
d	Deci	10^{-1}
c	Centi	10^{-2}
m	Milli	10^{-3}
μ	Micro	10^{-6}
n	Nano	10^{-9}
p	Pico	10^{-12}

Terms and symbols used in electronics

Term	Symbol
Approximately equal to	$\cong$
Proportional to	$\propto$
Infinity	∞
Sum of	$\sum$
Greater than	$>$
Less than	$<$
Much greater than	$\gg$
Much less than	$\ll$
Base of natural logarithms	e
Common logarithms of x	log x
Temperature	θ
Time constant	T
Efficiency	η
Per unit	p.u.

APPENDIX

B

Health and Safety Executive (HSE) publications and information

HSE Books, Information Leaflets and Guides may be obtained from
HSE Books, P.O. Box 1999, Sudbury, Suffolk CO10 6FS

HSE Infoline – Telephone: 0845 3450055 or write to
HSE Information Centre, Broad Lane, Sheffield S3 7HO

HSE home page on the World Wide Web
http:/www.hse.gov.uk

The Health and Safety Poster and other HSE publications are available from
www.hsebooks.com

Environmental Health Department of the Local Authority
Look in the local telephone directory under the name of the authority

HSE OFFICES

East & South East

Covers the counties of Bedfordshire, Berkshire, Buckinghamshire, Cambridgeshire, Essex, Hampshire, Hertfordshire, Isle of Wight, Norfolk, Suffolk, Oxfordshire, Kent, East & West Sussex and Surrey

Basingstoke

Priestley House
Priestley Road
Basingstoke
RG24 9NW
Fax: 01256 404100
Covers: Berkshire, Hampshire, Oxfordshire, Isle of Wight.

Bedford

Woodlands
Manton Lane
Manton Lane Industrial Estate
Bedford
Fax: 01234 220633
Covers: Hertfordshire, Cambridgeshire, Bedfordshire and Milton Keynes and Buckinghamshire.

Chelmsford

Redwing House
Hedgerows Business Park
Colchester Road
Springfield
Chelmsford
Essex
CM2 5PB
Fax: 01245 706222
Covers: Essex (except Barking, Havering, Redbridge and Waltham Forest these LAs are covered by London), Norfolk and Suffolk

Norwich

Rosebery Court
2nd Floor
St Andrew's Business Park
Norwich
Norfolk NR7 0HS
Covers: Norfolk, Suffolk, Essex (please also see Chelmsford)
Covers Topics: Agriculture, Food and Wood

Oxted

The Council Offices
Station Road East
Oxted
Surrey RH8 0BT
Covers: East & West Sussex, Surrey

Kent

International House
Dover Place
Ashford
Kent TN23 1HU
Fax: 01233 634827
Covers: Kent

London

Rose Court

2 Southwark Bridge
London
SE1 9HS
Fax: 020 7556 2201
Covers: London only

Westminster Office

Caxton House
Tothill Street
London
SW1H 9NA
Fax: 020 7227 3802

Midlands

Covers the counties of West Midlands, Leicestershire, Northamptonshire, Warwickshire, Derbyshire, Lincolnshire, Nottinghamshire, Hereford and Worcester

Birmingham

19 Ridgeway
9 Quinton Business Park
Quinton
Birmingham
B32 1AL
Fax: 0121 607 6349
Covers: West Midlands

Cheltenham

St James House
St James Square
Cheltenham
Gloucestershire
GL50 3PR
Fax: 01242 587 133
The Cheltenham Office covers Nuclear Directorate, and only queries associated with Nuclear Directorate should be sent to Cheltenham

Northampton

900 Pavilion Drive
Northampton Business Park
Northampton
NN4 7RG
Fax: 01604 738333
Covers: Leicestershire, Northampton, Rutland, Warwickshire

Nottingham

Kingsley Dunham Centre,
Nicker Hill,
Keyworth,
Nottingham
NG12 5GG
Fax: 0115 937 7500
Covers: Nottinghamshire, Derbyshire, Lincolnshire (most)
(North Lincolnshire covered by Sheffield office)

Stoke on Trent

Lyme Vale Court
Lyme Drive
Parklands Business Park
Newcastle Road
Trent Vale
Stoke on Trent
ST4 6NW
Fax: 01782 602400
Covers: Staffordshire and Shropshire

Worcester

See Birmingham office

North West

Covers the counties of Cheshire, Cumbria, Greater Manchester, Lancashire and Merseyside

Bootle Headquarters

Redgrave Court,
Merton Road,
Bootle,
Merseyside,
L20 7HS

Knowledge Centre

Knowledge Centre
Health and Safety Executive
(1G) Redgrave Court
Merton Road
Bootle
Merseyside, L20 7HS

Carlisle

2 Victoria Place
Carlisle
CA1 1ER
Fax: 01228 548482
Covers: Cumbria (except services, metals & minerals and fairgrounds)

Scotland

Covers all the Scottish unitary authorities and island councils

Edinburgh

Belford House
59 Belford Road
Edinburgh
EH4 3UE
Fax: 0131 247 2121
Covers: Borders, Lothian, Central Perth, Kinross, Fife and Dundee

Glasgow

1st Floor
Mercantile Chambers
53 Bothwell Street
Glasgow
G2 6TS
Fax: 0141 275 3100
Covers: West Scotland

Aberdeen

Field Operations Division
Lord Cullen House
Fraser Place
Aberdeen
AB25 3UB
Fax: 01224 252525
Covers: Angus, Aberdeenshire, Moray and Shetland

Aberdeen

Offshore Safety Division
Lord Cullen House
Fraser Place
Aberdeen
AB25 3UB
Fax: 01224 252555
Covers: Aberdeen

Inverness

Longman House
28 Longman Road
Longman Industrial Estate
Inverness
IV1 1SF
Fax: 01463 713459
Covers: Highlands and Orkney

Wales & South West

Covers Wales, Cornwall, Devon, Somerset, North Somerset, Bath and North East Somerset, Gloucestershire, South Gloucestershire, Bristol, Dorset, Swindon and Wiltshire

Cardiff

Government Buildings,
Phase 1,
Ty Glas,
Llanishen,
Cardiff
CF14 5SH
Fax: 029 2026 3120
Covers: Merthyr Tydfil, Rhondda Cynon Taff, Vale of Glamorgan, Bridgend, Neath Port Talbot, Powys, Blaenau Gwent, Caerphilly, Cardiff, Monmouthshire, Newport, Torfaen and part of Powys

Bristol

(This office will only receive visitors with a prior appointment)
2 Rivergate
Bristol
BS1 6EW
Covers: Bristol, Somerset, Bath and North East Somerset, North Somerset, Gloucestshire, South Gloucestershire, Dorset (public services only), Swindon and Wiltshire

Agriculture in North, Mid and East Devon

Construction in Devon (but not Cornwall or Plymouth City Council area – see Plymouth Office)

Plymouth

(This office will only receive visitors with a prior appointment)
North Quay House
Sutton Harbour
Plymouth
PL4 0RA
Fax: 01752 226024
Covers: Devon, Cornwall (all industries except public services and quarries)
Agriculture except in North, Mid and East Devon
Construction (all of Cornwall plus Plymouth City Council area, rest of Devon – see Bristol Office)

North Wales

Regent House,
Regent Street
Wrexham
LL11 1PR
Covers: Anglesey, Conwy, Denbighshire, Flintshire, Wrexham, Gwynedd and part of Powys

Dyfed

Tŷ Myrddin,
Old Station Road,
Carmarthen
Carmarthenshire. SA31 1LP
Fax: 01267 223267
Covers: Carmarthenshire, Pembrokeshire, Ceredigion and Swansea

Dorset

(This office will only receive visitors with a prior appointment)

14 New Fields
Stinsford Road
Nuffield Industrial Estate
Poole
Dorset BH17 0NF
Fax: 01202 667224
Covers: Dorset (except public services – covered by Bristol)

Yorkshire and North East

Covers the counties of Cleveland, Durham, North Yorkshire, Northumberland, West Yorkshire, Tyne & Wear, Humberside and South Yorkshire.

Leeds

The Lateral
8 City Walk
Leeds
LS11 9AT
Fax No: 0113 283 4382 (general enquiries)
Fax No: 0113 283 4296 (completed F10 forms)
Covers: West and North Yorkshire

Sheffield

Health and Safety Executive
Foundry House
3 Millsands
Riverside Exchange
Sheffield
S3 8NH
Fax: 0114 291 2379
Covers: East Yorkshire, North Lincolnshire, NE Lincolnshire, South Yorkshire, Kingston-upon-Hull

Newcastle

Alnwick House
Benton Park View
Newcastle-Upon-Tyne
NE98 1YX
Fax: 0191 202 6300
Covers: Northumberland, Tyne and Wear, Durham, Cleveland

York

HSE Chemical Regulation Directorate (CRD)
Mallard House
Kings Pool, 3 Peasholme Green
York
YO1 7PX
Fax: 01904 455733

APPENDIX C
Environmental organizations

The Department of Energy and Climate Change (DECC) for grants.

BS 7671:2008 Engineering Recommendations G 83/1 and G 59/1 Published by the Energy Network Association and the Department for Business, Enterprise and Regulatory Reform (BERR) for technical specifications.

Building Regulations England and Wales. The Department of Communities and Local Government at www.communities.gov.uk and for Scotland, The Scottish Building Standards Agency at www.scotland.gov.uk/bsd

The Labour government's 'feed in tariff' was introduced by the Climate Secretary Ed Miliband to encourage green electricity producers by paying a subsidy for every kWh of electricity generated by renewable energy fed back to the National Grid. The coalition government which came to power in 2010 are committed to supporting this policy.

For information on the feed in tariff scheme see the Office of the Gas and Electricity Markets (OFGEM) website at www.ofgem.gov.uk/fits.

Energy Saving Trust at 0207 222 0101 and www.energysavingtrust.org.uk for advice on grants and products.

Microgeneration product advice and their own certification scheme at 01752 823 600 and www.microgeneration.com.

The Carbon Trust at www.carbontrust.co.uk/energy offers free advice on loans to businesses which are upgrading to more energy-efficient equipment. The size of the loan will depend upon the CO2 savings.

Planning Guide for solar, PV and wind turbine installations can be found at www.planningportal.gov.uk.

Best practice guide for installing microgeneration systems can be found on the Electrical Safety Council website at www.esc.org.uk/bestpracticeguides.html.

Rainwater harvesting guidance and products can be found in abundance by Googling 'rainwater-harvesting'.

IET AMENDMENT 3 A summary of changes

D

- **Implementation** – Amendment 3 published in January 2015. The changes become law on 1 July 2015. First January 2016 is the date at which Regulation 421.1.200, domestic consumer units, comes into effect.
- **New numbering system** – The decimal point numbering system now represents 100 numbers for CENELEC European harmonized regulations and 200 numbers for UK only regulations. This system has been used for new regulations but will only be applied to existing regulations when they have been significantly rewritten.
- **Consumer units** – Recent fire statistics have shown that a large number of domestic fires involved plastic consumer units as a source of the fire. Consumer units are often located at the entrance or exit door or under stairs, raising the possibility that a fire starting as a result of faulty wiring could spread to coats or other household items and block the emergency escape routes. Regulation 421.1.200 now requires that consumer units be manufactured from non-combustible material, for example metal, or be enclosed in a non-combustible enclosure. Consumer units and switchgear must comply with BS EN 61439-3.
- **Wiring systems in escape routes** – 'An escape route is the route to follow for access to a safe area in the event of an emergency.' Regulation 521.200.1 now requires that wiring systems in escape routes be supported such that they will not be liable to premature collapse in the event of a fire. This will restrict the use of plastic clips and ties, and PVC trunking and conduit as the sole means of support in these areas. Cables in escape routes must be supported or restrained by metallic, fire resistant supports.
- **Bath or shower rooms** – Low voltage circuits passing through these special locations shall be 30 mA RCD protected, irrespective of whether they directly serve equipment or are just passing through. Regulation 701.411.3.3
- **Inspection testing and certification** – Chapter 6 and Appendix 6 now require the person carrying out this work to be an electrically skilled person competent in such work. The new schedule of inspections for new work has a more detailed schedule in Appendix 6. Condition reporting now requires the inspection of electrical equipment in roof spaces.
- **Protection of cables in walls** – section 522 has been rearranged and modified to make it clearer. Amendment 3 at 522.6.202 states that all installations buried at a depth of 50mm or less, which includes cables installed in partitions constructed of metal parts irrespective of depth, shall have 30 mA RCD protection, unless other methods detailed in 522.6.203 have been applied.
- **Compatibility** – Regulation 512.1.5 tells us that the installation designer must ensure that all the installed fixed equipment is designed and manufactured in accordance with the Electromagnetic Compatibility (EMC) Directive. Installations composed only of CE marked apparatus will conform to this directive.

- **Luminaires and lighting** – Section 559 of the Regulations now only covers the general requirements for lighting and luminaires in fixed installations. Outside lighting and installations and extra low voltage lighting installations have been moved to Part 7 of the Regulations within a new section 714 and 715. Changes to Section 559 include cables passing through luminaires which must be protected against the effects of heat and UV radiation. Where connections are not provided within the luminaire, a suitable means of connection must be made in accordance with 559.5.4. Display stand luminaires must now have protection against electric shock. This is provided by either SELV or PELV or with the additional protection of a 30 mA RCD. Additions have also been made under section 715 with regard to isolation, switching and control of ELV lighting installations.
- **IP rating of trunking installations** – regulation 521.10.1 now tells us that for a trunking system to meet the requirements of IP4X , the trunking and all accessories and bends must be rated IP4X. Where joints are manufactured on site the installer must confirm that the manufactured items meet at least IPXXD. The trunking cover must only be removable by means of a tool or deliberate action.

Glossary of terms

Acceleration *Acceleration is the rate of change in velocity with time.*

$$\text{Acceleration} = \frac{\text{Velocity}}{\text{Time}} = (\text{m/s}^2)$$

Accident *An accident may be defined as an uncontrolled event causing injury or damage to an individual or property.*

Alarm call points *Manually operated alarm call points should be provided in all parts of a building where people may be present, and should be located so that no one need walk for more than 30 m from any position within the premises in order to give an alarm.*

Alloy *An alloy is a mixture of two or more metals.*

Appointed person *An appointed person is someone who is nominated to take charge when someone is injured or becomes ill, including calling an ambulance if required. The appointed person will also look after the first aid equipment, including re-stocking the first aid box.*

Approved test instruments *The test instruments and test leads used by the electrician for testing an electrical installation must meet all the requirements of the relevant regulations. All testing must, therefore, be carried out using an 'approved' test instrument if the test results are to be valid. The test instrument must also carry a calibration certificate; otherwise the recorded results may be void.*

Basic protection *Basic protection is provided by the insulation of live parts in accordance with Section 416 of the IET Regulations.*

Bonding *The linking together of the exposed or extraneous metal parts of an electrical installation.*

Bonding conductor *A conductor providing protective bonding.*

Cable tray *Cable tray is a sheet-steel channel with multiple holes. The most common finish is hot-dipped galvanized but PVC-coated tray is also available. It is used extensively on large industrial and commercial installations for supporting MI and SWA cables which are laid on cable tray and secured with cable ties through the tray holes.*

Capacitive reactance *Capacitive reactance (X_C) is the opposition to an a.c. current in a capacitive circuit. It causes the current in the circuit to lead ahead of the voltage.*

Centrifugal force *Centrifugal force is the force acting away from the centre, the opposite to centripetal force.*

Centripetal force *Centripetal force is the force acting towards the centre when a mass attached to a string is rotated in a circular path.*

Circuit protective conductor (CPC) *A protective conductor connecting exposed conductive parts of equipment to the main earthing terminal.*

Cohesive or adhesive force *Cohesive or adhesive force is the force required to hold things together.*

Compact fluorescent lamps (CFLs) *CFLs are miniature fluorescent lamps designed to replace ordinary GLS lamps.*

Competent person *A competent person is anyone who has the necessary technical skills, training and expertise to safely carry out the particular activity.*

Compressive force *Compressive force is the force pushing things together.*

Conductor *A conductor is a material, usually a metal, in which the electrons are loosely bound to the central nucleus. These electrons can easily become 'free electrons' which allows heat and electricity to pass easily through the material.*

Conduit *A conduit is a tube, channel or pipe in which insulated conductors are contained.*

Corrosion *Corrosion is the destruction of a metal by chemical action.*

Delivery notes *A delivery note is used to confirm that goods have been delivered by the supplier, who will then send out an invoice requesting payment.*

Duty holder *This phrase recognizes the level of responsibility which electricians are expected to take on as a part of their job in order to control electrical safety in the work environment. Everyone has a duty of care, but not everyone is a duty holder. The person who exercises 'control over the whole systems, equipment and conductors' and is the electrical company's representative on-site is a duty holder.*

Earth *The conductive mass of the earth the electrical potential of which is taken as zero.*

Earthing *The act of connecting the exposed conductive parts of an installation to the main protective earthing terminal of the installation.*

Efficiency of any machine *The ratio of the output power to the input power is known as the* efficiency *of the machine. The symbol for efficiency is the Greek letter 'eta' (η). In general,*

$$\eta = \frac{\text{Power output}}{\text{Power input}}$$

Electric current *The drift of electrons within a conductor is known as an electric current, measured in amperes and given the symbol I.*

Electric shock *Electric shock occurs when a person becomes part of the electrical circuit.*

Electrical force *Electrical force is the force created by an electrical field.*

Electrical industry *The electrical industry is made up of a variety of individual companies, all providing a service within their own specialism to a customer, client or user.*

Emergency lighting *Emergency lighting is not required in private homes because the occupants are familiar with their surroundings, but in public buildings people are in unfamiliar surroundings. In an emergency people do not always act rationally, but well-illuminated and easily identified exit routes can help to reduce panic.*

Emergency switching *Emergency switching involves the rapid disconnection of the electrical supply by a single action to remove or prevent danger.*

Escape/standby lighting *Emergency lighting is provided for two reasons: to illuminate escape routes, called 'escape' lighting; and to enable a process or activity to continue after a normal lights failure, called 'standby' lighting.*

Expansion bolts *The most well-known expansion bolt is made by Rawlbolt and consists of a split iron shell held together at one end by a steel ferrule and a spring wire clip at the other end. Tightening the bolt draws up an expanding bolt inside the split iron shell, forcing the iron to expand and grip the masonry. Rawlbolts are for heavy-duty masonry fixings.*

Exposed conductive parts *The metalwork of an electrical appliance or the trunking and conduit of an electrical system which can be touched because they are not normally live, but which may become live under fault conditions.*

Extraneous conductive parts *The structural steelwork of a building and other service pipes such as gas, water, radiators and sinks.*

Faraday's law Faraday's law *states that when a conductor cuts or is cut by a magnetic field, an e.m.f. is induced in that conductor.*

Fault protection *Fault protection is provided by protective bonding and automatic disconnection of the supply (by a fuse or miniature circuit-breaker, MCB) in accordance with IET Regulations 411.3 to 6.*

Ferrous *A word used to describe all metals in which the main constituent is iron.*

Fire *Fire is a chemical reaction which will continue if fuel, oxygen and heat are present.*

Fire alarm circuits *Fire alarm circuits are wired as either normally open or normally closed. In a* normally open circuit, *the alarm call points are connected in parallel with each other so that when any alarm point is initiated the circuit is completed and the sounder gives a warning of fire. In a* normally closed circuit, *the alarm call points are connected in series to normally closed contacts. When the alarm is initiated, or if a break occurs in the wiring, the alarm is activated.*

First aid *First aid is the initial assistance or treatment given to a casualty for any injury or sudden illness before the arrival of an ambulance, doctor or other medically qualified person.*

First aider *A first aider is someone who has undergone a training course to administer first aid at work and holds a current first aid certificate.*

Flashpoint *The lowest temperature at which sufficient vapour is given off from a flammable substance to form an explosive gas–air mixture is called the flashpoint.*

Flexible conduit *Flexible conduit manufactured to BS 731-1: 1993 is made of interlinked metal spirals often covered with a PVC sleeving.*

Fluorescent lamp *A fluorescent lamp is a linear arc tube, internally coated with a fluorescent powder, containing a low-pressure mercury vapour discharge.*

Force *The presence of a force can only be detected by its effect on a body. A force may cause a stationary object to move or bring a moving body to rest.*

Friction force *Friction force is the force which resists or prevents the movement of two surfaces in contact.*

Functional switching *Functional switching involves the switching on or off, or varying the supply, of electrically operated equipment in normal service.*

Fuse *A fuse is the weakest link in the circuit. Under fault conditions it will melt when an over-current flows, protecting the circuit conductors from damage.*

Gravitational force *Gravitational force is the force acting towards the centre of the earth due to the effect of gravity.*

Hazard *A hazard is something with the 'potential' to cause harm, for example, chemicals, electricity or working above ground.*

Hazard risk assessment *Employers of more than five people must document the risks at work and the process is known as hazard risk assessment.*

Hazardous area *An area in which an explosive gas–air mixture is present is called a* hazardous area*, and any electrical apparatus or equipment within a hazardous area must be classified as flameproof to protect the safety of workers.*

Heating, magnetic or chemical *The three effects of an electric current: when an electric current flows in a circuit it can have one or more of the following three effects:* heating, magnetic *or* chemical.

Impedance *The total opposition to current flow in an a.c. circuit is called impedance and given the symbol Z.*

Inductive reactance *Inductive reactance (X_L) is the opposition to an a.c. current in an inductive circuit. It causes the current in the circuit to lag behind the applied voltage.*

Inertial force *Inertial force is the force required to get things moving, to change direction or stop.*

Inspection and testing techniques *The testing of an installation implies the use of instruments to obtain readings. However, a test is unlikely to identify a cracked socket outlet, a chipped or loose switch plate, a missing conduit-box lid or saddle, so it is also necessary to make a visual inspection of the installation. All existing installations should be periodically inspected and tested to ensure that they are safe and meet the IET Regulations (IET Regulations 610 to 634).*

Instructed (electrically) person *An instructed (electrically) person is a person adequately advised or supervised by electrically skilled persons to be able to perceive risks and avoid the hazards which electricity can create.*

Insulator *An insulator is a material, usually a non-metal, in which the electrons are very firmly bound to the nucleus and, therefore, will not allow heat or electricity to pass through it. Good insulating materials are PVC, rubber, glass and wood.*

Intrinsically safe circuit *An intrinsically safe circuit is one in which no spark or thermal effect is capable of causing ignition of a given explosive atmosphere.*

Intruder alarm systems *An intruder alarm system serves as a deterrent to a potential thief and often reduces home insurance premiums.*

Isolation *Isolation is defined as cutting off the electrical supply to a circuit or item of equipment in order to ensure the safety of those* *working on the equipment by making dead those parts which are live in normal service.*

Job sheets *A job sheet or job card carries information about a job which needs to be done, usually a small job.*

Lamp *A lamp is a device for converting electrical energy into light energy.*

Lever *A lever is any rigid body which pivots or rotates around a fixed axis or fulcrum. Load force × Distance from fulcrum = Effort force × Distance from fulcrum.*

Levers and turning force *A lever allows a heavy load to be lifted or moved by a small effort.*

Luminaire *A luminaire is equipment which supports an electric lamp and distributes or filters the light created by the lamp.*

Magnesium oxide *The conductors of mineral insulated metal sheathed (MICC) cables are insulated with compressed magnesium oxide.*

Magnetic field *The region of space through which the influence of a magnet can be detected is called the magnetic field of that magnet.*

Magnetic force *Magnetic force is the force created by a magnetic field.*

Magnetic hysteresis *Magnetic hysteresis loops describe the way in which different materials respond to being magnetized.*

Magnetic poles *The places on a magnetic material where the lines of flux are concentrated are called magnetic poles.*

Maintained emergency lighting *In a maintained system the emergency lamps are continuously lit using the normal supply when this is available, and change over to an alternative supply when the mains supply fails.*

Manual handling *Manual handling is lifting, transporting or supporting loads by hand or by bodily force.*

Mass *Mass is a measure of the amount of material in a substance, such as metal, plastic, wood, brick or tissue, which is collectively known as a body. The mass of a body remains constant and can easily be found by comparing it on a set of balance scales with a set of standard masses. The SI unit of mass is the kilogram (kg).*

Mechanics *Mechanics is the scientific study of 'machines', where a machine is defined as a device which transmits motion or force from one place to another.*

Metallic trunking *Metallic trunking is formed from mild steel sheet, coated with grey or silver enamel paint for internal use or a hot-dipped galvanized coating where damp conditions might be encountered.*

Mini-trunking *Mini-trunking is very small PVC trunking, ideal for surface wiring in domestic and commercial installations such as offices.*

Movement or heat detector *A movement or heat detector placed in a room will detect the presence of anyone entering or leaving that room.*

Mutual inductance *A mutual inductance of 1 henry exists between two coils when a uniformly varying current of 1 ampere per second in one coil produces an e.m.f. of 1 volt in the other coil.*

Non-ferrous *Metals which* do not *contain iron are called non-ferrous. They are non-magnetic and resist rusting. Copper, aluminium, tin, lead, zinc and brass are examples of non-ferrous metals.*

Non-maintained emergency lighting *In a non-maintained system the emergency lamps are only illuminated if the normal mains supply fails.*

Non-statutory regulations and codes of practice *Non-statutory regulations and codes of practice interpret the statutory regulations telling us how we can comply with the law.*

Ohm's law Ohm's law *says that the current passing through a conductor under constant temperature conditions is proportional to the potential difference across the conductor.*

Optical fibre cables *Optical fibre cables are communication cables made from optical-quality plastic, the same material from which spectacle lenses are manufactured. The energy is transferred down the cable as digital pulses of laser light, as against current flowing down a copper conductor in electrical installation terms.*

Ordinary person *An ordinary person is a person who is neither a skilled person nor an instructed person.*

Overload current *An overload current can be defined as a current which exceeds the rated value in an otherwise healthy circuit.*

Passive infra-red (PIR) detectors *PIR detector units allow a householder to switch on lighting units automatically whenever the area covered is approached by a moving body whose thermal radiation differs from the background.*

People *People may be described as an ordinary person, a skilled person, an instructed person or a competent person.*

Perimeter protection system *A perimeter protection system places alarm sensors on all external doors and windows so that an intruder can be detected as he or she attempts to gain access to the protected property.*

Person *A person can be described as ordinary, competent, instructed or skilled depending upon that person's skill or ability.*

Personal protective equipment (PPE) *PPE is defined as all equipment designed to be worn, or held, to protect against a risk to health and safety.*

Phasor *A phasor is a straight line, having definite length and direction, which represents to scale the magnitude and direction of a quantity such as a current, voltage or impedance.*

Plastic plugs *A plastic plug is made of a hollow plastic tube split up to half its length to allow for expansion. Each size of plastic plug is colour-coded to match a wood screw size.*

Polyvinylchloride (PVC) *PVC used for cable insulation is a thermoplastic polymer.*

Potential difference *The potential difference (p.d.) is the change in energy levels measured across the load terminals. This is also called the volt drop or terminal voltage, since e.m.f. and p.d. are both measured in volts.*

Power *Power is the rate of doing work.*

$$\text{Power} = \frac{\text{Work done}}{\text{Time taken}} \text{ (W)}$$

Power factor *Power factor (p.f.) is defined as the cosine of the phase angle between the current and voltage.*

Pressure or stress *Pressure or stress is a measure of the force per unit area.*

$$\text{Pressure or stress} = \frac{\text{Force}}{\text{Area}} \ (\text{N/m}^2)$$

Primary cell *A primary cell cannot be recharged. Once the active chemicals are exhausted, the cell must be discarded.*

Protective bonding *This is protective bonding for the purpose of safety.*

PVC/SWA cable installations *Steel wire armoured PVC insulated cables are now extensively used on industrial installations and often laid on cable tray.*

Reasonably practicable or absolute *If the requirement of the regulation is absolute, then that regulation must be met regardless of cost or any other consideration. If the regulation is to be met 'so far as is reasonably practicable', then risks, cost, time, trouble and difficulty can be considered.*

Relay *A relay is an electromagnetic switch operated by a solenoid.*

Resistance *In any circuit, resistance is defined as opposition to current flow.*

Resistivity *The resistivity (symbol ρ – the Greek letter 'rho') of a material is defined as the resistance of a sample of unit length and unit cross-section.*

Risk *A risk is the 'likelihood' of harm actually being done.*

Risk assessments *Risk assessments need to be* suitable *and* sufficient, *not perfect.*

Rubber *Rubber is a tough elastic substance made from the sap of tropical plants.*

Safety first – isolation *We must ensure the disconnection and separation of electrical equipment from every source of supply and that this disconnection and separation is secure.*

Secondary cells *A secondary cell has the advantage of being rechargeable. If the cell is connected to a suitable electrical supply, electrical energy is stored on the plates of the cell as chemical energy.*

Secure supplies *A UPS (uninterruptible power supply) is essentially a battery supply electronically modified to provide a clean and secure a.c. supply. The UPS is plugged into the mains supply and the computer systems are plugged into the UPS.*

Security lighting *Security lighting is the first line of defence in the fight against crime.*

Shearing force *Shearing force is the force which moves one face of a material over another.*

Shock protection *Protection from electric shock is provided by basic protection and fault protection.*

Short-circuit *A short-circuit is an over-current resulting from a fault of negligible impedance connected between conductors.*

SI units *SI units are based upon a small number of fundamental units from which all other units may be derived.*

Silicon rubber *Introducing organic compounds into synthetic rubber produces a good insulating material such as FP200 cables.*

Simple machines *A machine is an assembly of parts, some fixed, others movable, by which motion and force are transmitted. With the aid of a* *machine we are able to magnify the effort exerted at the input and lift or move large loads at the output.*

Single PVC insulated conductors *Single PVC insulated conductors are usually drawn into the installed conduit to complete the installation.*

Skilled (electrically) person *A skilled (electrically) person is a person with relevant education and sufficient experience to be able to perceive risks and to avoid the dangers which electricity can create.*

Skirting trunking *Skirting trunking is a trunking manufactured from PVC or steel in the shape of a skirting board which is frequently used in commercial buildings such as hospitals, laboratories and offices.*

Socket outlets *Socket outlets provide an easy and convenient method of connecting portable electrical appliances to a source of supply.*

Sounders *The positions and numbers of sounders should be such that the alarm can be distinctly heard above the background noise in every part of the premises.*

Space factor *The ratio of the space occupied by all the cables in a conduit or trunking to the whole space enclosed by the conduit or trunking is known as the* space factor.

Speed *Speed is concerned with distance travelled and time taken.*

Spring toggle bolts *A spring toggle bolt provides one method of fixing to hollow partition walls which are usually faced with plasterboard and a plaster skimming.*

Static electricity *Static electricity is a voltage charge which builds up to many thousands of volts between two surfaces when they rub together.*

Statutory regulations *Statutory regulations have been passed by Parliament and have, therefore, become laws.*

Step-down transformers *Step-down transformers are used to reduce the output voltage, often for safety reasons.*

Step-up transformers *Step-up transformers are used to increase the output voltage. The electricity generated in a power-station is stepped up for distribution on the National Grid network.*

Switching for mechanical maintenance requirements *The switching for mechanical maintenance requirements are similar to those for isolation except that the control switch must be capable of switching the full load current of the circuit or piece of equipment.*

Synthetic rubber *Synthetic rubber is manufactured, as opposed to being produced naturally.*

Tensile force *Tensile force is the force pulling things apart.*

Thermoplastic polymers *These may be repeatedly warmed and cooled without appreciable changes occurring in the properties of the material.*

Thermosetting polymers *Once heated and formed, products made from thermosetting polymers are fixed rigidly. Plug tops, socket outlets and switch plates are made from this material.*

Time sheets *A time sheet is a standard form completed by each employee to inform the employer of the actual time spent working on a particular contract or site.*

Transformer *A transformer is an electrical machine which is used to change the value of an alternating voltage.*

Trap protection *Trap protection places alarm sensors on internal doors and pressure pad switches under carpets on through routes between, for example, the main living area and the master bedroom.*

Trunking *A trunking is an enclosure provided for the protection of cables which is normally square or rectangular in cross-section, having one removable side. Trunking may be thought of as a more accessible conduit system.*

Velocity *In everyday conversation we often use the word* velocity *to mean the same as speed, and indeed the units are the same. However, for scientific purposes this is not acceptable since velocity is also concerned with direction.*

Visual inspection *The installation must be visually inspected before testing begins. The aim of the visual inspection is to confirm that all equipment and accessories are undamaged and comply with the relevant British and European Standards, and also that the installation has been securely and correctly erected.*

Weight *Weight is a measure of the force a body exerts on anything which supports it. Normally it exerts this force because it is being attracted towards the earth by the force of gravity.*

Work done *Work done is dependent upon the force applied times the distance moved in the direction of the force. Work done = Force × Distance moved in the direction of the force (J). The SI unit of work done is the newton metre or joule (symbol J).*

Index